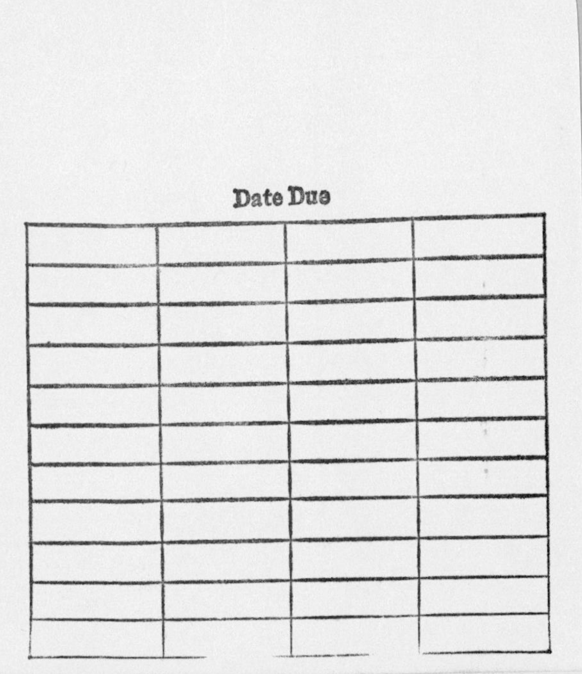
Date Due

NEW MEXICO
EASTERN
UNIVERSITY
E
N
M
U
LIBRARY

GEOMETRY AND CHRONOMETRY IN PHILOSOPHICAL PERSPECTIVE

GEOMETRY AND CHRONOMETRY

IN PHILOSOPHICAL PERSPECTIVE

BY

Adolf Grünbaum

UNIVERSITY OF MINNESOTA PRESS · MINNEAPOLIS

 Printed and bound in the United States of America at The Colonial Press, Inc., Clinton, Massachusetts

Library of Congress Catalog Card Number: 68-22363

PUBLISHED IN GREAT BRITAIN, INDIA, AND PAKISTAN BY THE OXFORD UNIVERSITY PRESS, LONDON, BOMBAY, AND KARACHI, AND IN CANADA BY THE COPP CLARK PUBLISHING CO. LIMITED, TORONTO

To my brother, Norbert Grünbaum, M.D., *and my sister, Suzanne Reines*

PREFACE

An essay of mine on the philosophy of geometry and chronometry appeared in 1962 as part of the third volume of the *Minnesota Studies in the Philosophy of Science* (*Scientific Explanation, Space, and Time*, edited by H. Feigl and G. Maxwell). This essay prompted Hilary Putnam to publish a lengthy critique in 1963 in the second volume of *The Delaware Seminar* (*Philosophy of Science*, edited by B. Baumrin). The kind proposal by the University of Minnesota Press to reissue my 1962 essay separately provided me with the welcome opportunity to publish a detailed reply to Putnam (Chapter III of the present volume) and an extension of the original essay (Chapter II), along with the 1962 text on which Putnam rested his critique (which appears here as Chapter I). Chapter II includes some corrections of the 1962 essay.

Apart from acknowledgments made within the text, I owe an immense debt to my physicist colleague Allen I. Janis for invaluable help with points and examples in Chapter III, §§2, 8, and 9. Furthermore, I had the benefit of discussions with Wilfrid Sellars of issues relating to Chapter II, §1, and with Wesley Salmon of matters pertaining to the interpretation of Putnam's views. As is evident from the length of my reply in Chapter III, I am indebted to Putnam for the great stimulus which his essay afforded me to clarify my views both to others and to myself, and to Robert S. Cohen and Wilfrid Sellars for encouragement to write this reply. To a lesser extent, this also applies to George Schlesinger with respect to Chapter

II. I am also obliged to Elizabeth (Mrs. John H.) McMunn for her excellent preparation of the typescript, to Jeanne Sinnen of the University of Minnesota Press for her imaginative and painstaking editing of the entire manuscript, and to Robert Herrick for his competent compilation of the index. Chapter II draws on material in my article "The Denial of Absolute Space and the Hypothesis of a Universal Nocturnal Expansion," published in the *Australasian Journal of Philosophy* (May 1967), and I owe thanks to the editors of that journal for permission to use the material here. Finally I wish to thank the National Science Foundation for the support of research and the following publishers for permission to quote from the works of their authors: University of California Press (Berkeley), publisher of Newton's *Principia*, edited by F. Cajori; Cambridge University Press, publisher of A. N. Whitehead's *The Concept of Nature*; Clarendon Press, publisher of E. A. Milne's *Modern Cosmology and the Christian Idea of God*; Dover Publications, Inc., publisher of A. D'Abro's *The Evolution of Scientific Thought from Newton to Einstein* and of D. M. Y. Sommerville's *The Elements of Non-Euclidean Geometry*; E. P. Dutton and Company and Methuen and Company, Ltd., publishers of Einstein's *Sidelights on Relativity*, translated by G. B. Jeffery and W. Perrett (in which Einstein's essay "Geometry and Experience" first appeared); and Interscience Publishers, which issued *The Delaware Seminar: Philosophy of Science*, in which Putnam's essay appeared.

A. G.

University of Pittsburgh
January 1968

CONTENTS

Chapter I. GEOMETRY, CHRONOMETRY, AND EMPIRICISM

Chapter III. REPLY TO HILARY PUTNAM'S "AN EXAMINATION OF GRÜNBAUM'S PHILOSOPHY OF GEOMETRY"

GEOMETRY AND CHRONOMETRY IN PHILOSOPHICAL PERSPECTIVE

I

GEOMETRY, CHRONOMETRY, AND EMPIRICISM

1. Introduction

The moment the mathematical discovery of the non-Euclidean geometries had deprived Euclideanism of its claim to *uniqueness*, the triumph of an empiricist account of physical geometry and chronometry seemed assured. Observational findings were presumed capable, at least in principle, of establishing the unique truth of a particular kind of metric geometry. Ironically, however, it soon became clear that the very mathematical discoveries which had heralded the demise of the classical rationalist and Kantian conceptions of geochronometry were a double-edged sword. Critics were quick to marshal these mathematical results against the renascent geometric empiricists who felt emboldened by their victory over the thesis that Euclidean geometry is certifiable a priori as the true description of physical space. The challenge came from several distinct versions of

NOTE: This chapter is reproduced without substantive change from Volume III of *Minnesota Studies in the Philosophy of Science: Scientific Explanation, Space, and Time*, edited by Herbert Feigl and Grover Maxwell (Minneapolis: University of Minnesota Press, 1962). Cross-references to Chapters II and III and references to my publications since 1962 have been inserted in asterisked footnotes.

I am indebted to Dr. Samuel Gulden of the department of mathematics, Lehigh University, for very helpful discussions of some of the issues treated in this chapter. I have also benefited from conversations with Professor Albert Wilansky of that department, with Professor E. Newman of the University of Pittsburgh, and with Professor Grover Maxwell and other fellow participants in the 1958 and 1959 conferences of the Minnesota Center for the Philosophy of Science. Grateful acknowledgment is made to the National Science Foundation of the United States for the support of research.

conventionalism whose espousal has issued in a proliferous and continuing philosophical debate.

In an endeavor to resolve the issues posed by the several variants of conventionalism, the present essay aims to answer the following question: In what sense and to what extent can the ascription of a particular metric geometry to physical space and the chronometry ingredient in physical theory be held to have an *empirical* warrant? To carry out this inquiry, we must ascertain whether and how empirical findings function restrictively so as to determine a *unique* geochronometry as the true account of the structure of physical space-time.

It will turn out that the status of the metrics of space and time is profoundly illuminated by (i) the distinction between *factual* and conventional ingredients of space-time theories and (ii) a precise awareness of the *warrant* for deeming *some* of the credentials of these theories to be conventional. Thus, we shall see that whatever the merits of the repudiation of the analytic-synthetic dichotomy and of the antithesis between theoretical and observation terms, it is grievously incorrect and obfuscating to deny as well the distinction between factual and conventional ingredients of sophisticated space-time theories in physics.

Among the writers who have held the empirical status of geochronometry to depend on whether the rigidity of rods and the isochronism of clocks are conventional, we find Riemann [75, pp. 274, 286], Clifford [14, pp. 49–50], Poincaré [61, 62, 63, 64, 65], Russell [81], Whitehead [97, 99, 101], Einstein [21, p. 161; 23, Section 1, pp. 38–40; 26, pp. 676–678], Carnap [8], and Reichenbach [72]. Their assessment of the epistemological status of *congruence* as pivotal has been rejected by Eddington [20, pp. 9–10]. On his view, the thesis that congruence (for line segments or time intervals) is conventional is true only in the trivial sense that "the meaning of every word in the language is conventional" [20, p. 9]: instead of being an insight into the status of spatial or temporal equality, the conventionality of congruence is a semantical platitude expressing our freedom to decree the referents of the *word* 'congruent,' a freedom which we can exercise in regard to any linguistic symbols whatever which have not already been preempted semantically. Thus, we are told that though the conventionality of congruence is merely an unenlightening triviality holding for the language of *any* field of inquiry whatever, it has been misleadingly inflated into a philosophical doctrine about the relation of spatiotemporal equality purporting to codify fundamental features endemic to the materials of geochronometry. Eddington's conclusion that only the use of the *word* 'congruent' but *not* the ascription of the congruence *relation* can be held to be a matter of convention has also been defended by a cognate argument which invokes the theory of models of uninterpreted formal calculi as follows: (i) physical geometry is a spatially interpreted abstract calculus, and this interpretation of a formal system was effected by semantical rules which are all equally conventional and among which the definition of the relation term 'congruent' (for line segments) does *not* occupy an epistemologically distinguished position, since we are just as free to give a *non*customary interpretation of the abstract sign 'point' as of the sign 'congruent'; (ii) this model theoretic conception makes it apparent that there can be no basis at all for an epistemological distinction *within* the system of *physical* geochronometry between factual statements, on the one hand, and supposedly conventional assertions of rigidity and isochronism on the other; (iii) the factual credentials of physical geometry or chronometry can no more be impugned by adducing the alleged conventionality of rigidity and isochronism than one could gainsay the factuality of genetics by incorrectly affirming the conventionality of the relation of uniting which obtains between two gametes when forming a zygote.

When defending the alternative metrizability of space and time and the resulting possibility of giving either a Euclidean or a non-Euclidean description of the same spatial facts, Poincaré had construed the conventionality of congruence as an epistemological discovery about the status of the relation of spatial or temporal equality. The proponent of the foregoing model theoretic argument therefore indicts Poincaré's defense of the feasibility of *choosing* the metric geometry as amiss, misleading, and unnecessary, deeming this choice to be automatically assured by the theory of models. And, by the same token, this critic maintains that there is just as little reason for our having posed the principal question of this essay as there would be for instituting a philosophical inquiry as to the sense in which genetics *as such* can be held to have an empirical warrant. Whereas the aforementioned group of critics has charged the conventionality of congruence with being only *trivially true*, such thinkers as Russell and Whitehead have strongly opposed that doctrine because they deemed it to be *importantly false*.

These strictures call for critical scrutiny, and their rebuttal is included among the polemical objectives of this chapter. Before turning to their refutation, which will be presented in §5 below, I shall (a) set forth the rationale of this chapter's principal concern; (b) articulate the meaning of the contention that rigidity and isochronism are conventional; (c) assess the respective merits of the several justifications which have been given for that thesis by its advocates; and (d) develop its import for (i) the epistemological status of geochronometry and of explanatory principles in dynamics, and (ii) alternative formulations of physical theory. §§6 and 7 will then come to grips with the articulation of the answer to the principal inquiry of this chapter.

2. The Criteria of Rigidity and Isochronism: The Epistemological Status of Spatial and Temporal Congruence

(i) *The Clash between Newton's and Riemann's Conceptions of Congruence and the Role of Conventions in Geochronometry.**

The metrical comparisons of separate spatial and temporal intervals required for geochronometry involve *rigid* rods and *isochronous* clocks. Is this involvement of a transported congruence standard to

* The ideas of subsection (i) are developed in considerably greater detail in §§2, 3, and 8 of Chapter III below.

which separate intervals can be referred a matter of the mere *ascertainment* of an otherwise intrinsic equality or inequality obtaining among these intervals? Or is reference to the congruence standard essential to the very *existence* of these relations? More specifically, we must ask the following questions:

1. What is the warrant for the claim that a solid rod remains *rigid* under transport in a spatial region *free* from inhomogeneous thermal, elastic, electromagnetic, and other "deforming" or "perturbational" influences? The geometrically pejorative characterization of thermal and other inhomogeneities in space as "deforming" or "perturbational" is due to the fact that they issue in a dependence of the coincidence behavior of transported solid rods on the latter's *chemical composition,* and *mutatis mutandis* in a like dependence of the rates of clocks.

2. What are the grounds for asserting that a clock which is *not perturbed* in the sense just specified is *isochronous?*

This pair of questions and their far-reaching philosophical ramifications will occupy us in §§2–5. It will first be in §7 that we shall deal with the further issues posed by the logic of making *corrections* to compensate for deformations and rate variations exhibited by rods and clocks respectively when employed geochronometrically under *perturbing* conditions.

In the *Principia,* Newton states his thesis of *the intrinsicality of the metric* in "container" space and the corresponding contention for absolute time as follows:

. . . the common people conceive those quantities [i.e., time, space, place, and motion] under no other notions but from the relation they bear to sensible objects. And thence arise certain prejudices, for the removing of which it will be convenient to distinguish them into absolute and relative, true and apparent, mathematical and common [54, p. 6]. . . . because the parts of space cannot be seen, or distinguished from one another by our senses, therefore in their stead we use sensible measures of them. For from the positions and distances of things from any body considered as immovable, we define all places; and then with respect to such places, we estimate all motions, considering bodies as transferred from some of those places into others. And so, instead of absolute places and motions, we use relative ones; and that without any inconvenience in common affairs; but in philosophical disquisitions, we ought to abstract from our senses, and consider things themselves, distinct from what are only sensible measures of them. For it may be that there is no body really at rest, to which the places and motions of others may be referred [54, p. 8]. . . . those . . . defile the purity of mathematical and philosophical truths, who confound real quantities with their relations and sensible measures [54, p. 11]. . . .

I. Absolute, true, and mathematical time, of itself, and from its own nature, flows equably[1] without relation to anything external and by another name is called duration: relative, apparent, and common time, is some sensible and external (whether accurate or unequable) measure of duration by the means of motion, which is commonly used instead of true time; such as an hour, a day, a month, a year.

II. Absolute space, in its own nature, without relation to anything external, remains always similar and immovable. Relative space is some movable dimension or measure of the absolute spaces; which our senses determine by its position to bodies; and which is commonly taken for immovable space; such is the dimension of a subterraneous, an aerial, or celestial space, determined by its position in respect of the earth. Absolute and relative space are the same in figure and magnitude; but they do not remain always numerically the same. For if the earth, for instance, moves, a space of our air, which relatively and in respect of the earth remains always the same, will at one time be one part of the absolute space into which the air passes; at another time it will be another part of the same, and so, absolutely understood, it will be continually changed [54, p. 6]. . . . Absolute time, in astronomy, is distinguished from relative, by the equation or correction of the apparent time. For the natural days are truly unequal, though they are commonly considered as equal, and used for a measure of time; astronomers correct this inequality that they may measure the celestial motions by a more accurate time. It may be, that there is no such thing as an equable motion, whereby time may be accurately measured. All motions may be accelerated

[1] It is Newton's conception of the attributes of "equable" (i.e., congruent) time intervals which will be subjected to critical examination and found untenable in this essay. But I likewise reject Newton's view that the concept of "flow" has relevance to the time of physics, as distinct from the time of psychology: see [32, Sec. 4].

and retarded, but the flowing of absolute time is not liable to any change. The duration or perseverance of the existence of things remain the same, whether the motions are swift or slow, or none at all: and therefore this duration ought to be distinguished from what are only sensible measures thereof; and from which we deduce it, by means of the astronomical equation [54, pp. 7–8].

Newton's fundamental contentions here are that (a) the *identity* of points in the physical container space in which bodies are located and of the instants of receptacle time at which physical events occur is autonomous and *not* derivative: physical things and events do *not* first define, by their own identity, the points and instants which constitute their loci or the loci of other things and events, and (b) receptacle space and time each has its own *intrinsic metric*, which exists quite independently of the existence of material rods and clocks in the universe, devices whose function is *at best* the purely epistemic one of enabling us to ascertain the intrinsic metrical relations in the receptacle space and time contingently containing them. Thus, for example, even when clocks, unlike the rotating earth, run "equably" or uniformly, these periodic devices merely *record* but do *not define* the temporal metric. And what Newton is therefore *rejecting* here is a *relational* theory of space and time which asserts that (a) bodies and events first *define* points and instants by conferring their identity upon them, thus enabling them to serve as the loci of other bodies and events, and (b) instead of having an intrinsic metric, physical space and time are metrically amorphous pending explicit or tacit appeal to the bodies which are first to define their respective metrics. To be sure, Newton would *also* reject quite emphatically any identification or isomorphism of absolute space and time, on the one hand, with the *psychological* space and time of conscious awareness whose respective metrics are given by unaided ocular congruence and by psychological estimates of duration, on the other. But one overlooks the essential point here, if one is led to suppose with F. S. C. Northrop (see, for example, [56, pp. 76–77]) that the relative, apparent, and common space and time which Newton contrasts with absolute, true, and mathematical space and time are the private visual space and subjective psychological time of immediate sensory experience. For Newton makes it unambiguously clear, as shown by the quoted passages, that his *relative* space and time are indeed that *public* space and time which is defined by the system of *relations* between *material* bodies and events, and *not* the egocentrically private space and time of phenomenal experience. The "sensible" measures discussed by Newton as constitutive of "relative" space and time are those furnished by the public bodies of the physicist, *not* by the unaided ocular congruence of one's eyes or by one's mood-dependent psychological estimates of duration. This interpretation of Newton is fully attested by the following specific assertions of his:

i. "Absolute and relative space are the same in figure and magnitude," a declaration which is incompatible with Northrop's interpretation of *relative* space as "the immediately sensed spatial extension of, and relation between, sensed data (which is a purely private space, varying with the degree of one's astigmatism or the clearness of one's vision)" [56, p. 76].

ii. As examples of merely "*relative*" times, Newton cites any "sensible and external (whether accurate or unequable [nonuniform]) measure of duration" such as "an hour, a day, a month, a year" [54, p. 6]. And he adds that the apparent time commonly used as a measure of time is based on natural days which are "truly unequal," true equality being allegedly achievable by astronomical corrections compensating for the nonuniformity of the earth's rotational motion caused by tidal friction, etc.[2] But Northrop erroneously takes Newton's relative time to be the "immediately sensed time" which "varies from person to person, and even for a single person passes very quickly under certain circumstances and drags under others" and asserts incorrectly that Newton identified with absolute time the public time "upon which the ordinary time of social usage is based."

[2] The logical status of the criterion of uniformity implicitly invoked here will be discussed at length in §4(i) below.

iii. Newton illustrates *relative* motion by reference to the kinematic relation between a body on a moving ship, the ship, and the earth, these relations being defined in the customary manner of physics *without* phenomenal space or time.

Northrop is entirely right in going on to say that Einstein's conceptual innovations in the theory of relativity cannot be construed, as they have been in certain untutored quarters, as the abandonment of the distinction between physically public and privately or egocentrically sensed space and time. But Northrop's misinterpretation of the Newtonian conception of "relative" space and time prevents him from pointing out that Einstein's philosophical thesis can indeed be characterized epigrammatically as the enthronement of the very relational conception of the space-time framework which Newton sought to proscribe by his use of the terms "*relative*," "apparent," and "common" as philosophically disparaging epithets!

Long before the theory of relativity was propounded, a relational conception of the metric of space and time diametrically opposite to Newton's was enunciated by Riemann in the following words:

Definite parts of a manifold, which are distinguished from one another by a mark or boundary are called quanta. Their quantitative comparison is effected by means of counting in the case of discrete magnitudes and by measurement in the case of continuous ones.[3] Measurement consists in bringing the magnitudes to be compared into coincidence; for measurement, one therefore needs a means which can be applied (transported) as a standard of magnitude. If it is lacking, then two magnitudes can be compared only if one is a [proper] part of the other and then only according to more or less, not with respect to how much. . . . in the case of a discrete manifold, the principle [criterion] of the metric relations is already implicit in [intrinsic to] the concept of this manifold, whereas in the case of a continuous manifold, it must be brought in from elsewhere [extrinsically]. Thus, either the reality underlying space must form a discrete manifold or the reason for the metric relations must be sought extrinsically in binding forces which act on the manifold. [75, pp. 274, 286.]

[3] Riemann apparently does not consider sets which are neither discrete nor continuous, but we shall consider the significance of that omission below.

Although we shall see in §2(iii) that Riemann was mistaken in supposing that the first part of this statement will bear critical scrutiny as a characterization of continuous manifolds *in general*, he does render here a fundamental feature of the continua of *physical space* and *time*, which are manifolds whose elements, taken singly, all have zero magnitude. And this basic feature of the spatiotemporal continua will presently be seen to invalidate decisively the Newtonian claim of the intrinsicality of the metric in empty space and time. When now proceeding to state the upshot of Riemann's declaration for the spatiotemporal congruence issue before us, we shall *not* need to be concerned with either of the following two facets of his thesis: (1) the inadequacies arising from Riemann's treatment of discrete and continuous types of order as *jointly exhaustive* and (2) the prophetic character of his suggestion that the "reason for the metric relations must be sought extrinsically in the binding forces which act on the manifold" as a precursor of Einstein's original quest to implement Mach's Principle in the general theory of relativity [25; 36, pp. 526–527].

Construing Riemann's statement as applying not only to lengths but also, *mutatis mutandis*, to areas and to volumes of higher dimensions, he gives the following *sufficient* condition for the intrinsic definability and nondefinability of a metric without claiming it to be necessary as well: in the case of a discretely ordered set, the "distance" between two elements can be defined *intrinsically* in a rather natural way by the cardinality of the smallest number of intervening elements.[4] On the other hand, upon confronting the extended continuous manifolds of physical space and time, we see that neither the cardinality of intervals nor any of their other topological properties provide a basis for an *intrinsically* defined metric. The first part of this conclusion was tellingly emphasized by Cantor's proof of the equicardinality of all positive intervals independently of their length. Thus, there is no *intrinsic* attribute of the space between the end

[4] The *basis* for the discrete ordering is not here at issue: it can be conventional, as in the case of the letters of the alphabet, or it may arise from special properties and relations characterizing the objects possessing the specified order.

points of a line-segment *AB*, or any relation between these two points themselves, in virtue of which the interval *AB* could be said to contain the same amount of space as the space between the termini of another interval *CD* not coinciding with *AB*. Corresponding remarks apply to the time continuum. Accordingly, the continuity we postulate for physical space and time furnishes a *sufficient* condition for their *intrinsic metrical amorphousness*.[5]

This intrinsic metric amorphousness is made further evident by reference to the axioms for spatial congruence [99, pp. 42–50]. These axioms preempt "congruent" (for intervals) to be a *spatial equality predicate* by assuring the reflexivity, symmetry, and transitivity of the congruence relation in the class of spatial intervals. But although having thus preempted the use of "congruent," the congruence axioms still allow an *infinitude* of *mutually exclusive* congruence classes of intervals, where it is to be understood that any *particular* congruence class is a *class of classes* of congruent intervals whose lengths are specified by a *particular* distance function $ds^2 = g_{ik}\, dx^i\, dx^k$. And we just saw that there are no intrinsic metric attributes of intervals

[5] Clearly, this does *not* preclude the existence of sufficient conditions *other than continuity* for the intrinsic metrical amorphousness of sets. But one *cannot* invoke densely ordered, *denumerable* sets of points (instants) in an endeavor to show that discontinuous sets of such elements may likewise lack an intrinsic metric: even without measure theory, ordinary analytic geometry allows the deduction that the length of a *denumerably* infinite point set is intrinsically zero. This result is evident from the fact that since each point (more accurately, each unit point set or degenerate subinterval) has length *zero*, we obtain zero as the *intrinsic* length of the densely ordered denumerable point set upon summing, in accord with the usual limit definition, the sequence of zero lengths obtainable by denumeration (cf. Grünbaum [33, pp. 297–298]). More generally, the measure of a denumerable point set is always zero (cf. Hobson [41, p. 166]) unless one succeeds in developing a very restrictive intuitionistic measure theory of some sort.

These considerations show incidentally that space intervals cannot be held to be merely denumerable aggregates. Hence in the context of our post-Cantorean meaning of "continuous," it is actually not as damaging to Riemann's statement as it might seem prima facie that he neglected the denumerable dense sets by incorrectly treating the discrete and continuous types of order as *jointly exhaustive*. Moreover, since the distinction between denumerable and super-denumerable dense sets was almost certainly unknown to Riemann, it is likely that by "continuous" he merely intended the property which we now call "dense." Evidence of such an earlier usage of "continuous" is found as late as 1914 (cf. Russell [80, p. 138]).

which could be invoked to single out *one* of these congruence classes as unique.

How then can we speak of the assumedly continuous physical space as having *a* metric or *mutatis mutandis* suppose that the physical time continuum has a unique metric? The answer can be none other than the following:[6] Only the choice of a particular *extrinsic* congruence standard can determine a unique congruence class, the *rigidity* or self-congruence of that standard under transport being *decreed by convention,* and similarly for the periodic devices which are held to be *isochronous* (uniform) clocks. Thus the role of the spatial or temporal congruence standard *cannot* be construed with Newton or Russell ([81]; cf. also §5(i) below) to be the mere ascertainment of an otherwise intrinsic equality obtaining between the intervals belonging to the congruence class defined by it. Unless one of two segments is a subset of the other, the congruence of two segments is a matter of convention, stipulation, or definition and not a factual matter concerning which empirical findings could show one to have been mistaken. And hence there can be no question at all of an *empirically* or factually determinate metric geometry or chronometry until *after* a physical stipulation of congruence.[7]

[6] The conclusion which is about to be stated will appear unfounded to those who follow A. N. Whitehead in rejecting the "bifurcation of nature," which is assumed in its premises. But in §5 below, the reader will find a detailed rebuttal of the Whiteheadian contention that *perceptual* space and time do have an intrinsic metric and that once the allegedly illegitimate distinction between physical and perceptual space (or time) has been jettisoned, an intrinsic metric can hence be meaningfully imputed to physical space and time. Cf. also A. Grünbaum, "Whitehead's Philosophy of Science," *Philosophical Review*, 71:218–229 (1962).

[7] A. d'Abro [16, p. 27] has offered an *unsound* illustration of the thesis that the metric in a continuum is conventional: he considers a stream of sounds of varying pitch and points out that a congruence criterion based on the successive auditory octaves of a given musical note would be at variance with the congruence defined by equal differences between the associated frequencies of vibration, since the frequency differences between successive octaves are *not* equal. But instead of constituting an example of the alternative metrizability of the *same* mathematically continuous manifold of elements, d'Abro's illustration involves the metrizations of *two different* manifolds only *one* of which is continuous in the mathematical sense. For the auditory contents sustaining the relation of being octaves of one another are elements of a merely *sensory* "continuum." Moreover,

In the case of geometry, the specification of the intervals which are stipulated to be congruent is given by the distance function $ds = \sqrt{g_{ik}\, dx^i\, dx^k}$, congruent intervals being those which are assigned equal lengths ds by this function. Whether the intervals defined by the coincidence behavior of the transported rod are those to which the distance function assigns *equal* lengths ds or not will depend on our selection of the functions g_{ik}. Thus, if the components of the metric tensor g_{ik} are suitably chosen in any given coordinate system, then the transported rod will have been stipulated to be congruent to itself everywhere independently of its position and orientation. On the other hand, by an appropriately different choice of the functions g_{ik}, the length ds of the transported rod will be made to *vary* with position or orientation instead of being constant. Once congruence has been defined via the distance function ds, the geodesics (straight lines)[8] associated with that choice of congruence are determined, since the family of geodesics is defined by the variational requirement $\delta \int ds = 0$, which takes the form of a differential equation whose solution is the equation of the family of geodesics. The geometry characterizing the relations of the geodesics in question is likewise determined by the distance function ds, because the Gaussian curvature K of every surface element at any point in space is fixed by the functions g_{ik} ingredient in the distance function ds.

There are therefore alternative metrizations of the same factual coincidence relations sustained by a transported rod, and *some* of these alternative definitions of congruence will give rise to different metric geometries than others. Accordingly, via an appropriate defi-

---

we shall see in subsection (iii) that while holding for the mathematical continua of physical space and time, whose elements (points and instants) are respectively alike both qualitatively and in magnitude, the thesis of the conventionality of the metric cannot be upheld for *all* kinds of mathematical continua, Riemann and d'Abro to the contrary notwithstanding.

[8] The geodesics are called "straight lines" when discussing their relations in the context of synthetic geometry. But this identification must *not* be taken to entail that *every* geodesic connection of two points is a line of shortest distance between them. For once we abandon the restriction to Euclidean geometry, being a geodesic connection is only a necessary and not also a sufficient condition for being the shortest distance [86, pp. 140–143].

nition of congruence we are free to choose as the description of a *given* body of spatial facts any metric geometry compatible with the existing topology. Moreover, we shall find later on (§3(iii)) that there are infinitely many incompatible definitions of congruence which will implement the choice of any one metric geometry, be it the Euclidean one or one of the non-Euclidean geometries.

An illustration will serve to give concrete meaning to this general formulation of the conventionality of spatial congruence. Consider a physical surface such as part or all of an infinite blackboard and suppose it to be equipped with a network of Cartesian coordinates. The customary metrization of such a surface is based on the congruence defined by the coincidence behavior of transported rods: line segments whose termini have coordinate differences dx and dy are assigned a length ds given by $ds = \sqrt{dx^2 + dy^2}$, and the geometry associated with this metrization of the surface is, of course, Euclidean. But we are also at liberty to employ a different metrization in part or all of this space. Thus, for example, we could equally legitimately metrize the portion *above* the x-axis by means of the new metric

$$ds = \sqrt{\frac{dx^2 + dy^2}{y^2}}.$$

This alternative metrization is incompatible with the customary one: for example, it makes the lengths $ds = dx/y$ of horizontal segments whose termini have the *same* coordinate differences dx depend on *where* they are along the y-axis. Consequently, the new metric would commit us to regard a segment for which $dx = 2$ at $y = 2$ as *congruent* to a segment for which $dx = 1$ at $y = 1$, although the customary metrization would regard the length ratio of these segments to be 2:1. But, of course, the new metric does *not* say that a transported solid rod will coincide successively with the intervals belonging to the congruence class defined by that metric; instead it allows for this *non*coincidence by making the length of the rod a suitably nonconstant function of its position. And this noncustomary congruence definition, which was suggested by Poincaré, confers a *hyperbolic* geometry on the half plane $y > 0$ of the customarily Euclidean plane: the associated geodesics in the half plane are a family of hyperbolically related lines whose infinitude is assured by the behavior of the new metric as $y \to 0$ and whose *Euclidean* status is that of semi*circular* arcs.[9] It is now clear that the hyperbolic metrization of the semiblackboard possesses not only mathematical but also epistemological credentials as good as those of the Euclidean one.

It might be objected that although *not* objectionable *epistemologi-*

[9] The reader can convince himself that the new metrization issues in a hyperbolic geometry by noting that now $g_{11} = 1/y^2$, $g_{12} = g_{21} = 0$, and $g_{22} = 1/y^2$ and then using these components of the metric tensor to obtain a *negative* value of the Gaussian curvature K via Gauss's formula. (For a statement of this formula, see, for example, F. Klein [44, p. 281].)

To determine what particular curves in the semiblackboard are the geodesics of our hyperbolic metric

$$ds = \frac{\sqrt{dx^2 + dy^2}}{y},$$

one must substitute this ds and carry out the variation in the equation $\delta \int ds = 0$, which is the defining condition for the family of geodesics. The desired geodesics of our new metric must therefore be given by the equation

$$\delta \int \frac{\sqrt{1 + \left(\frac{dy}{dx}\right)^2}}{y} dx = 0.$$

It is shown in the calculus of variations (cf., for example [48, pp. 193–195]) that this variational equation requires the following differential equation—known as Euler's equation—to be satisfied, if we put

$$I \equiv \frac{\sqrt{1 + \left(\frac{dy}{dx}\right)^2}}{y}: \qquad \frac{\partial I}{\partial y} - \frac{d}{dx} \frac{\partial I}{\partial \left(\frac{dy}{dx}\right)} = 0.$$

Upon substituting our value of I, we obtain the differential equation of the family of geodesics:

$$\frac{d^2y}{dx^2} + \frac{1}{y}\left[1 + \left(\frac{dy}{dx}\right)^2\right] = 0.$$

The solution of this equation is of the form $(x - k)^2 + y^2 = R^2$, where k and R are constants of integration, and thus represents—*Euclideanly speaking*—a family of circles centered on and perpendicular to the x-axis, the upper semicircles being the geodesics of Poincaré's remetrized half plane above the x-axis.

cally, there is a pedantic artificiality and even perverse *complexity* in all congruence definitions which do not assign equal lengths *ds* to the intervals defined by the coincidence behavior of a solid rod. The grounds of this objection would be that (a) there are no convenient and familiar natural objects whose coincidence behavior under transport furnishes a physical realization of the bizarre, noncustomary congruences, and (b) after correcting for the chemically dependent distortional idiosyncrasies of various kinds of solids in inhomogeneous thermal, electric, and other fields, *all* transported solid bodies furnish the *same* physical intervals, and thus they realize only one of the infinitude of incompatible mathematical congruences. *Mutatis mutandis*, the same objection might be raised to any definition of temporal congruence which does not accord with the cycles of standard material clocks. The reply to this criticism is twofold:

1. The prima-facie plausibility of the demand for *simplicity* in the choice of the congruence definition gives way to second thoughts the moment it is realized that the desideratum of simplicity requires consideration not only of the congruence definition but also of the latter's bearing on the form of the associated system of geometry and physics. And our discussions in §§4 and 6 will show that a bizarre definition of congruence may well have to be countenanced as the price for the attainment of the over-all simplicity of the total theory. Specifically, we anticipate §4(ii) by just mentioning here that although Einstein merely alludes to the possibility of a noncustomary definition of *spatial* congruence in the general theory of relativity without actually availing himself of it [21, p. 161], he does indeed utilize in that theory what our putative objector deems a highly artificial definition of *temporal* congruence, since it is *not* given by the cycles of standard material clocks.

2. It is particularly instructive to note that the cosmology of E. A. Milne [51, p. 22] postulates the actual existence in nature of two metrically different kinds of clocks whose respective periods constitute physical realizations of *incompatible* mathematical congruences. Specifically, Milne's assumptions lead to the result that there

is a nonlinear relation

$$\tau = t_0 \log\left(\frac{t}{t_0}\right) + t_0$$

between the time τ defined by periodic astronomical processes and the time t defined by atomic ones, t_0 being an appropriately chosen arbitrary constant. The nonlinearity of the relation between these two kinds of time is of paramount importance here, because it assures that two intervals which are congruent in *one* of these two time scales will be *incongruent* in the other. Clearly, it would be utterly gratuitous to regard one of these two congruences as bizarre, since each of them is presumed to have a physical realization. And the choice between these scales is incontestably conventional, for it is made quite clear in Milne's theory that their associated different metric descriptions of the world are factually equivalent and hence equally true.

What would be the verdict of the Newtonian proponent of the intrinsicality of the metric on the examples of alternative metrizability which we gave both for space (Poincaré's hyperbolic metrization of the half plane) and also for time (general theory of relativity and Milne's cosmology)? He would first note correctly that once it is understood that the term "congruent," as applied to intervals, is to denote a reflexive, symmetrical, and transitive relation in this class of geometrical configurations, then the use of this term is restricted to designating a spatial equality relation. But then the Newtonian would proceed to claim *unjustifiably* that the spatial equality obtaining between congruent line segments of physical space (or between regions of surfaces and of 3-space respectively) consists in their each containing *the same intrinsic amount of space*. And having introduced this false premise, he would feel entitled to contend that (1) it is *never* legitimate to choose arbitrarily what specific intervals are to be regarded as congruent, and (2) as a corollary of this lack of choice, there is no room for selecting the lines which are to be regarded as straight and hence no choice among alternative geometric descriptions of actual physical space, since the geodesic requirement

$\delta \int ds = 0$ which must be satisfied by the straight lines is subject to the restriction that only the members of the unique class of *intrinsically equal* line segments may ever be assigned the same length ds. By the same token, the Newtonian asserts that only "truly" (intrinsically) equal time intervals may be regarded as congruent and therefore holds that there exists only *one* legitimate metrization of the time continuum, a conclusion which he then attempts to buttress further by adducing certain *causal* considerations from Newtonian dynamics which will be refuted in §4(i) below.

It is of the utmost importance to realize clearly that the thesis of the conventionality of congruence is, in the first instance, a claim concerning *structural properties of physical space and time;* only the semantical *corollary* of that thesis concerns the *language* of the geochronometric description of the physical world. Having failed to appreciate this fact, some philosophers were led to give a shallow caricature of the debate between the Newtonian, who affirms the factuality of congruence on the strength of the alleged intrinsicality of the metric, and his Riemannian conventionalistic critic. According to the burlesqued version of this controversy, the Riemannian is offering no more than a semantical truism, and the Newtonian assertion of metric absolutism can be dismissed as an evident absurdity on purely semantical grounds, since it is the denial of that mere truism. More specifically, the detractors suppose that their trivialization of the congruence issue can be vindicated by pointing out that we are, of course, free to decree the referents of the *unpreempted word* "congruent," because such freedom can be exercised with respect to *any* as yet semantically uncommitted term or string of symbols whatever. And, in this way, they misconstrue the conventionality of congruence as merely an inflated special case of a semantical banality holding for any and all linguistic signs or symbols, a banality which we shall call "trivial semantical conventionalism," or, in abbreviated form, "TSC."

No one, of course, will wish to deny that *qua uncommitted signs,* the terms "spatially congruent" and "temporally congruent" are fully on a par in regard to the *trivial* conventionality of the semantical rules governing their use with all linguistic symbols whatever. And thus a sensible person would hardly wish to contest that the unenlightening affirmation of the conventionality of the use of the *unpreempted word* "congruent" is indeed a subthesis of TSC. But it is a serious obfuscation to identify the Riemann-Poincaré doctrine that the *ascription* of the congruence or equality *relation* to space or time intervals is conventional with the platitude that the use of the *unpreempted word* "congruent" is conventional. And it is therefore totally incorrect to conclude that the Riemann-Poincaré tenet is merely a gratuitously emphasized special case of TSC. For what these mathematicians are advocating is *not* a doctrine about the semantical freedom we have in the use of the uncommitted sign "congruent." Instead, they are putting forward the initially nonsemantical claim that the continua of physical space and time each lack an intrinsic metric. And the metric amorphousness of these continua then serves to *explain* that even *after* the word "congruent" has been preempted semantically as a spatial or temporal *equality* predicate by the axioms of congruence, congruence remains ambiguous in the sense that these axioms still allow an infinitude of mutually exclusive congruence classes of intervals. Accordingly, fundamentally *non*semantical considerations are used to show that only a conventional choice of *one* of these congruence classes can provide a unique standard of length equality. In short, the conventionality of congruence is a claim *not* about the noise "congruent" but about the character of the conditions relevant to the obtaining of the relation denoted by the term "congruent." For alternative metrizability is *not* a matter of the freedom to use the semantically uncommitted noise "congruent" as we please; instead, it is a matter of the *nonuniqueness* of a relation term already *preempted* as the *physico-spatial* (or *temporal*) equality predicate. And *this nonuniqueness arises from the lack of an intrinsic metric in the continuous manifolds of physical space and time.*

The epistemological status of the Riemann-Poincaré conventionality of congruence is fully analogous to that of Einstein's conventionality of simultaneity. And if the reasoning used by critics of the former in an endeavor to establish the banality of the former *were* actually sound, then, as we shall now show, it would follow by a precisely similar argument that Einstein's enunciation of the conventionality of simultaneity [23, Section 1] was no more than a turgid statement of the platitude that the uncommitted word "simultaneous" (or "gleichzeitig") may be used as we please. In fact, in view of the complete epistemological affinity of the conventionality of congruence with the conventionality of simultaneity, which we are about to exhibit, it will be useful subsequently to combine these two theses under the name "geochronometric conventionalism" or, in abbreviated form, "GC."

We saw in the case of spatial and temporal congruence that congruence is conventional in a sense *other than* that prior to being preempted semantically, the *sign* "congruent" can be used to denote anything we please. *Mutatis mutandis,* we now wish to show that precisely the same holds for the conventionality of metrical simultaneity. Once we have furnished this demonstration as well, we shall have established that *neither* of the component claims of conventionality in our compound GC thesis is a subthesis of TSC.

We proceed in Einstein's manner in the special theory of relativity and first decree that the *noise* "topologically simultaneous" denote the relation of *not* being connectible by a physical causal (signal) chain, a relation which may obtain between two physical events. We now ask: Is this definition *unique* in the sense of assuring that one and only one event at a point Q will be topologically simultaneous with a given event occurring at a point P elsewhere in space? The answer to this question depends on *facts of nature,* namely on the range of the causal chains existing in the physical world. Thus, once the above definition is given, its *uniqueness* is not a matter subject to regulation by semantical convention. If now we assume with Einstein as a fact of nature that light *in vacuo* is the fastest causal chain, then this postulate entails the nonuniqueness of the definition of "topological simultaneity" given above and thereby also prevents *topological* simultaneity from being a transitive relation. On the other hand, if the facts of the physical world had the structure assumed by Newton, this *non*uniqueness would *not* arise. Accordingly, the structure of the facts of the world postulated by relativity prevents the above definition of "topological simultaneity" from also serving, as it stands, as a metrical synchronization rule for clocks at the spatially separated points P and Q. Upon coupling this result with the relativistic assumption that transported clocks do *not* define an absolute metrical simultaneity, we see that the facts of the world leave the equality relation of metrical simultaneity *indeterminate*, for they do not confer upon topological simultaneity the uniqueness which it would require to serve as the basis of metrical simultaneity as well. Therefore, the assertion of that indeterminateness and of *the corollary that metrical simultaneity is made determinate by convention* is *in no way* tantamount to the purely semantical assertion that the mere uncommitted *noise* "metrically simultaneous" must be given a physical interpretation before it can denote *and* that this interpretation is trivially a matter of convention.

Far from being a claim that a mere linguistic noise is still awaiting an assignment of semantical meaning, the assertion of the factual indeterminateness of metrical simultaneity concerns *facts of nature* which find expression in the residual nonuniqueness of the definition of "topological simultaneity" once the latter has already been given. And it is thus impossible to construe this residual nonuniqueness as being attributable to taciturnity or tight-lippedness on Einstein's part in telling us what he means by the noise "simultaneous." Here, then, we are confronted with a kind of logical *gap* needing to be filled by definition which is precisely analogous to the case of congruence, where the continuity of space and time issued in the residual nonuniqueness of the congruence axioms.

When I say that metrical simultaneity is not wholly factual but contains a conventional ingredient, what am I asserting? I am claiming none other than that the residual nonuniqueness or logical gap *cannot* be removed by an appeal to facts but only by a *conventional* choice of a *unique* pair of events at *P* and at *Q* as *metrically* simultaneous from within the class of pairs that are topologically simultaneous. And when I assert that it was a great philosophical (as well as physical) achievement for Einstein to have discovered the conventionality of metrical simultaneity, I am crediting Einstein *not* with the triviality of having decreed semantically the meaning of the *noise* "metrically simultaneous" (or "gleichzeitig") but with the recognition that, contrary to earlier belief, the facts of nature are such as to deny the required kind of semantical univocity to the already preempted term "simultaneous" ("gleichzeitig"). In short, Einstein's insight that metrical simultaneity is conventional is a contribution to the theory of time rather than to semantics, because *it concerns the character of the conditions relevant to the obtaining of the relation denoted by the term "metrically simultaneous."*

The *conventionality* of metrical simultaneity has just been formulated without any reference whatever to the relative motion of different Galilean frames and does *not* depend upon there being a relativity or nonconcordance of simultaneity as between different Galilean frames. On the contrary, it is the conventionality of simultaneity which provides the logical framework within which the *relativity* of simultaneity can first be understood: if *each* Galilean observer adopts the particular metrical synchronization rule adopted by Einstein in Section 1 of his fundamental paper [23]—a rule which corresponds to the value $\epsilon = \frac{1}{2}$ in the Reichenbach notation [72, p. 127]—then the relative motion of Galilean frames issues in their choosing as metrically simultaneous *different pairs of events* from within the class of topologically simultaneous events at *P* and *Q*, a result embodied in the familiar Minkowski diagram.

In discussing the definition of simultaneity [23, Section 1], Einstein italicized the words "*by definition*" in saying that the *equality* of the to-and-fro velocities of light between two points *A* and *B* is a matter of definition. Thus, he is asserting that metrical simultaneity is a matter of definition or convention. Do the detractors really expect anyone to believe that Einstein put these words in italics to convey to the public that the *noise* "simultaneous" can be used as we please? Presumably they would recoil from this conclusion. But how else could they solve the problem of making Einstein's avowedly conventionalist conception of metrical simultaneity compatible with their semantical trivialization of GC? H. Putnam, one of the advocates of the view that the conventionality of congruence is a subthesis of TSC, has sought to meet this difficulty along the following lines: in the case of the congruence of intervals, one would never run into trouble upon using the *customary* definition;[10] but in the case of simultaneity, actual contradictions would be encountered upon using the customary classical definition of metrical simultaneity, which is based on the transport of clocks and is vitiated by the dependence of the clock rates (readings) on the transport velocity. But Putnam's retort will not do. For the appeal to Einstein's recognition of the inconsistency of the classical definition of metrical simultaneity accounts only for his abandonment of the latter but does *not* illuminate—as does the thesis of the conventionality of simultaneity—the logical status of the *particular set* of definitions which Einstein put in its place. Thus, the Putnamian retort does *not* recognize that the logical status of Einstein's synchronization rules is not at all adequately rendered by saying that whereas the classical definition of metrical simultaneity was inconsistent, Einstein's rules have the virtue of consistency. For what needs to be elucidated is *the nature of the logical step* leading to Einstein's particular synchronization scheme within the wider framework of the *alternative consistent sets of rules for* metrical simultaneity any one of which is allowed by the nonuniqueness of topological simultaneity. Precisely this elucidation is given, as we have seen, by the thesis of the conventionality of metrical simultaneity.

We see therefore that the *philosophically illuminating* conventionality of an affirmation of the congruence of two intervals or of the

[10] We shall see in our discussion of time measurement on the rotating disk of the general theory of relativity in §4(ii), that there is one sense of "trouble" for which Putnam's statement would *not* hold.

metrical simultaneity of two physical events does *not* inhere in the arbitrariness of what linguistic *sentence* is used to express the proposition that a relation of equality obtains among intervals, or that a relation of metrical simultaneity obtains between two physical events. Instead, the important conventionality lies in the fact that even *after* we have specified what respective linguistic sentences will express these propositions, a convention is ingredient in each of the propositions expressed, i.e., in the very *obtaining* of a congruence relation among intervals or of a metrical simultaneity relation among events.

These considerations enable us to articulate the misunderstanding of the conventionality of congruence to which the proponent of its *model theoretic* trivialization (cf. §1) fell prey. It will be recalled that this critic argued somewhat as follows: "The theory of models of uninterpreted formal calculi shows that there can be no basis at all for an epistemological distinction *within* the system of physical geometry (or chronometry) between factual statements, on the one hand, and supposedly conventional statements of rigidity (or isochronism), on the other. For we are just as free to give a *non*customary spatial interpretation of, say, the abstract sign 'point' in the formal geometrical calculus as of the sign 'congruent,' and hence the physical interpretation of the relation term 'congruent' (for line segments) cannot occupy an epistemologically distinguished position among the semantical rules effecting the interpretation of the formal system, all of which are on a par in regard to conventionality." But this objection overlooks that (a) the obtaining of the spatial congruence relation provides scope for the role of convention because, independently of the particular formal geometrical calculus which is being interpreted, the term "congruent" functions as a spatial *equality predicate* in its *non*customary spatial interpretations no less than in its customary ones; (b) consequently, suitable alternative spatial interpretations of the term "congruent" and correlatively of "straight line" ("geodesic") show that it is always a live option (subject to the restrictions imposed by the existing topology) to give *either* a Euclidean or a *non*-Euclidean description of the same body of physico-geometrical facts; and (c) by contrast, the possibility of alternative spatial interpretations of such *other* primitives of rival geometrical calculi as "point" does *not* generally issue in this option. Our concern is to note that, even disregarding inductive imprecision, the empirical facts themselves do *not* uniquely dictate the truth of either Euclidean geometry or of one of its non-Euclidean rivals in virtue of the lack of an intrinsic metric. Hence in this context the different spatial interpretations of the term "congruent" (and hence of "straight line") in the respective geometrical calculi play a philosophically different role than do the interpretations of such other primitives of these calculi as "point," since the latter generally have the *same* spatial meaning in both the Euclidean and non-Euclidean descriptions. The preeminent status occupied by the interpretation of "congruent" in this context becomes apparent once we cease to look at physical geometry as a spatially interpreted system of abstract *synthetic* geometry and regard it instead as an interpreted system of abstract *differential* geometry of the Gauss-Riemann type: by choosing a particular distance function $ds = \sqrt{g_{ik}\, dx^i\, dx^k}$ for the line element, we specify not only what segments are congruent and what lines are straights (geodesics) but the entire geometry, since the metric tensor g_{ik} fully determines the Gaussian curvature K. To be sure, if one were discussing *not* the alternative between a Euclidean and non-Euclidean description of the same *spatial* facts but rather the set of *all* models (including *non*spatial ones) of a *given* calculus, say the Euclidean one, then indeed the physical interpretation of "congruent" and of "straight line" would not merit any more attention than that of other primitives like "point." [11]

We have argued that the continuity postulated for physical space and time issues in the metric amorphousness of these manifolds and

[11] We have been speaking of certain *uninterpreted* formal calculi as systems of synthetic or differential *geometry*. It must be understood, however, that *prior* to being given a *spatial* interpretation, these abstract deductive systems no more qualify as *geometries*, strictly speaking, than as systems of genetics or of anything else; they are called "geometries," it would seem, only "because the name seems good, on emotional and traditional grounds, to a sufficient number of competent people" [90, p. 17].

thus makes for the conventionality of congruence, much as the conventionality of nonlocal metrical simultaneity in special relativity is a consequence of the postulational *fact* that light is the fastest causal chain *in vacuo* and that clocks do not define an absolute metrical simultaneity under transport. But it might be objected that unlike the latter Einstein postulate, the postulational ascription of continuity (in the mathematical sense) to physical space and time instead of some discontinuous structure cannot be regarded, even in principle, as a *factual* assertion in the sense of being either true or false. For surely, this objection continues, there can be no *empirical* grounds for accepting a geometry postulating continuous intervals in preference to one which postulates discontinuous intervals consisting of, say, only the algebraic or only the rational points. The rejection of the latter kind of denumerable geometry in favor of a nondenumerable one affirming continuity therefore has no kind of factual warrant but is based solely on considerations of *arithmetic convenience* within the analytic part of geometry. And hence the *topology* is no less infested with features springing from conventional choice than is the metric. Hence the exponent of this criticism concludes that it is not only a misleading emphasis but outright incorrect for us to persevere in the course taken by Carnap [8] and Reichenbach [72] and to discern conventional elements in geometry only in the *metrization* of the topological substratum while deeming the latter to be *factual.*

This plea for a conventionalist conception of continuity is not convincing, however. Admittedly, the justification for regarding continuity as a broadly *inductive framework principle* of physical geometry cannot be found in the direct verdicts of measuring rods, which could hardly disclose the super-denumerability of the points on the line. And prima facie there is a certain measure of plausibility in the contention that the postulation of a super-denumerable infinity of irrational points in addition to a denumerable set of rational ones is dictated solely by the desire for such arithmetical convenience as having closure under operations like taking the square root. But, even disregarding the Zenonian difficulties which may vitiate denumerable geometries logically (cf. footnote 5 above), these considerations lose much of their force, it would seem, as soon as one applies the *acid test* of a convention to the conventionalist conception of continuity in physical geometry: the feasibility of one or more *alternate* formulations dispensing with the particular alleged convention and yet permitting the successful rendition of the same total body of experiential findings, such as in the case of the choice of a particular system of *units* of measurement. Upon applying this test, what do we find? Attempts to dispense with the continuum of classical mathematics (geometry and analysis) by providing adequate substitutes for the mathematics used by the total body of advanced modern physical theory have been programmatic rather than successful. And this *not* for want of effort on the part of their advocates. Thus, for example, the neointuitionistic endeavors to base mathematics on more restrictive foundations involve mutilations of mathematical physics whose range A. A. Fraenkel has characterized as follows: "intuitionistic restriction of the concept of continuum and of its handling in analysis and geometry, though carried out in quite different ways by various intuitionistic schools, always goes as far as to exclude vital parts of those two domains. (This is not altered by Brouwer's peculiar way of admitting the continuum *per se* as a 'medium of free growth.')" [12] The impressive difficulties encountered in these endeavors to provide a viable denumerable substitute for such topological components of the geometry as continuity thus insinuate the following suspicion: the empirical facts codified in terms of the classical mathematical apparatus in our most sophisticated and best confirmed physical theories support continuity in a *broadly inductive* sense as a *framework principle* to the exclusion of the prima-facie rivals of the continuum.[13] Pending the elaboration

[12] A. A. Fraenkel and Y. Bar-Hillel [29, p. 200]. Chapter IV of this work gives an admirably comprehensive and lucid survey of the respects in which neointuitionist restrictions involve truncations of the system of classical mathematics.

[13] It would be an error to believe that this conclusion requires serious qualification in the light of recent suggestions of space (or time) quantization. For as

of a successful alternative to the continuum, therefore, the charge that in a geometry the topological component of continuity is no less conventional than the metric itself seems to be unfounded.

(*ii*) *Physical Congruence, Testability, and Operationism.*

We have grounded the conventionality of spatial and temporal congruence on the continuity of the manifolds of space and time. And, in thus arguing that "true," absolute, or intrinsic rigidity and isochronism are *non*existing properties in these respective continua, we did *not* adduce any phenomenalist or homocentric-operationist criterion of factual meaning. For we did *not* say that the actual and potential failure of *human testing operations* to disclose "true rigidity" *constitutes* either its nonexistence or its meaninglessness. It will be well, therefore, to compare our Riemannian espousal of the conventionality of rigidity and isochronism with the reasoning of those who arrive at this conception of congruence by arguing from *nontestability* or from an operationist view of scientific concepts. Thus W. K. Clifford writes: "we have defined length or distance by means of a measure which can be carried about *without changing its length.* But how then is this property of the measure to be tested? . . . Is it possible . . . that lengths do really change by mere moving about, without our knowing it? Whoever likes to meditate seriously upon this question will find that it is wholly devoid of meaning" [14, pp. 49, 50].

We saw that within our Riemannian framework of ideas, length is relational rather than absolute in a twofold sense: (i) length obviously

H. Weyl has noted [95, p. 43]: "so far it [i.e., the atomistic theory of space] has always remained mere speculation and has never achieved sufficient contact with reality. How should one understand the metric relations in space on the basis of this idea? If a square is built up of miniature tiles, then there are as many tiles along the diagonal as there are along the side; thus the diagonal should be equal in length to the side." And Einstein has remarked (see *The Meaning of Relativity* (Princeton: Princeton University Press, 1955), pp. 165–166) that "From the quantum phenomena it appears to follow with certainty that a finite system of finite energy can be completely described by a finite set of numbers (quantum numbers). This does not seem to be in accordance with a continuum theory, and must lead to an attempt to find a purely algebraic theory for the description of reality. But nobody knows how to obtain the basis of such a theory."

depends numerically on the units used and is thus arbitrary to within a constant factor, and (ii) in virtue of the lack of an intrinsic metric, *sameness* or *change* of the length possessed by a body in different places and at different times *consists* in the sameness or change respectively of the *ratio* (relation) of that body to the conventional standard of congruence. Whether or not this ratio changes is quite independent of any human discovery of it: the number of times which a given body *B* will contain a certain unit rod is a property of *B* that is *not* first conferred on *B* by human operations. And thus the relational character of length derives, in the first instance, *not* from how we human beings discover lengths but from the failure of the continuum of physical space to possess an intrinsic metric, a failure obtaining quite independently of our measuring activities. In fact, it is this relational character of length which prescribes and regulates the kinds of human operations appropriate to its discovery. Since, to begin with, there exists no property of true rigidity to be disclosed by any human test, no test could possibly reveal its presence. Accordingly, the unascertainability of true rigidity *by us humans* is a *consequence* of its *nonexistence* in physical space and *evidence* for that nonexistence but *not* constitutive of it.

On the basis of this nonhomocentric relational conception of length, the utter vacuousness of the following assertion is evident at once: overnight *everything* has *expanded* (i.e., increased its length) but such that all length *ratios* remained unaltered. That such an alleged "expansion" will elude any and all human test is then obviously *explained* by its not having obtained: the *increase* in the ratios between all bodies and the congruence standard which would have constituted the expansion avowedly did *not* materialize.*

We see that the relational theory of length and hence the particular assertion of the vacuousness of a universal nocturnal expansion *do not depend* on a grounding of the meaning of the metrical concepts of length and duration on *human* testability or manipulations of

* A very detailed discussion of the physical status of the hypothesis of a universal nocturnal expansion is given in Chapter II, §1 below.

rods and clocks in the manner of Bridgman's homocentric operationism.[14] Moreover, there is a further sense in which the Riemannian recognition of the need for a specification of the congruence criterion does *not* entail an operationist conception of congruence and length: as we noted preliminarily at the beginning of §2 and will see in detail in §7, the definition of "congruence" on the basis of the coincidence behavior *common* to *all kinds* of transported solid rods provides a rule of correspondence (coordinative definition) *through the mediation of hypotheses and laws* that are *collateral* to the abstract geometry receiving a physical interpretation. For the physical laws used to compute the corrections for thermal and other substance-specific deformations of solid rods made of different kinds of materials *enter integrally* into the statement of the physical meaning of "congruent." Thus, in the case of "length" no less than in most other cases, operational definitions (in any distinctive sense of the term "operational") are a quite idealized and limiting species of correspondence rules. Further illustrations of this fact are given by Reichenbach, who cites the definitions of the unit of length on the basis of the wave length of cadmium light and also in terms of a certain fraction of the circumference of the earth and writes: "Which distance serves as a unit for actual measurements can ultimately be given only by reference to some actual distance. . . . We say with regard to the measuring rod . . . that only 'ultimately' the reference is to be conceived in this form, because we know that by means of the interposition of conceptual relations the reference may be rather remote" [72, p. 128]. An even stronger repudiation of the operationist account of the definition of "length" because of its failure to allow for the role of auxiliary theory is presented by K. R. Popper, who says: "As to the doctrine of operationalism—which demands that scientific terms, such as length . . . should be defined in terms of the appropriate experimental procedure—it can be shown quite easily that all so-called operational definitions will be

[14] For arguments supporting the conclusion that homocentric operationism is similarly dispensable and, in fact, unsuccessful in giving an account of the conceptual innovations of the special theory of relativity, see [35].

circular. . . . the circularity of the operational definition of length . . . may be seen from the following facts: (a) the 'operational' definition of *length* involves *temperature* corrections, and (b) the (usual) operational definition of *temperature* involves measurements of *length*" [66, p. 440].[15]

*(iii) The Inadequacies of the Nongeometrical Portion of Riemann's Theory of Manifolds.**

In pointing out earlier in §2 that the status of *spatial* and *temporal* congruence is decisively illuminated by Riemann's theory of continuous manifolds, I stated that this theory will *not* bear critical scrutiny as a characterization of continuous manifolds in *general.* To justify and clarify this indictment, we shall now see that continuity cannot be held with Riemann to furnish a sufficient condition for the intrinsic metric amorphousness of any manifold *independently of the character of its elements.* For, as Russell saw correctly [79, Sections 63 and 64], there are continuous manifolds, such as that of colors (in the physicist's sense of spectral frequencies) in which the individual elements differ qualitatively from one another and have inherent magnitude, thus allowing for metrical comparison of *the elements themselves.* By contrast, in the continuous manifolds of *space* and of *time*, neither points nor instants have any inherent magnitude allowing an individual metrical comparison between them, since all points are alike, and similarly for instants. Hence in these manifolds metrical comparisons can be effected only among the *intervals* between the elements, *not* among the homogeneous elements themselves. And the continuity of *these* manifolds then assures the nonintrinsicality of the metric for their intervals.

To exhibit further the bearing of the character of the elements of a continuous manifold on the feasibility of an intrinsic metric in it, I shall contrast the status of the metric in space and time, on the

[15] In §7, we shall see how the circularity besetting the operationist conception of length is circumvented within our framework when making allowance for thermal and other deformations in the statement of the definition of congruence.

* The ideas presented in subsection (iii) are further elaborated in §4.0 of Chapter III.

one hand, with its status in (a) the continuum of real numbers, arranged according to magnitude, and (b) the quasi-continuum of masses, mass being assumed to be a property of bodies in the Newtonian sense clarified by Mach's definition.[16]

The assignment of real numbers to points of physical space in the manner of the introduction of generalized curvilinear coordinates effects only a *coordinatization* but *not* a *metrization* of the manifold of physical space. No informative metrical comparison among individual points could be made by comparing the magnitudes of their real number coordinate names. On the other hand, within the continuous manifold formed by the real numbers themselves, when arranged according to magnitude, every real number is singly distinguished from and metrically comparable to every other by its inherent magnitude. And the measurement of mass can be seen to constitute a counterexample to Riemann's metrical philosophy from the following considerations.

In the Machian definition of Newtonian (gravitational and inertial) mass, the *mass ratio* of a particle B to a standard particle A is given by the magnitude ratio of the acceleration of A due to B to the acceleration of B due to A. Once the space-time metric and thereby the accelerations are fixed in the customary way, this ratio for any particular body B is *independent*, among other things, of how far apart B and A may be during their interaction. Accordingly, any affirmations of the mass equality (mass "congruence") or inequality of two bodies will hold independently of the extent of their spatial separation. Now, the set of medium-sized bodies form a *quasi*-continuum with respect to the dyadic relations "being more massive than," and "having the same mass," i.e., they form an array which is a continuum *except* for the fact that *several* bodies can occupy the same place in the array by sustaining the mass-"congruence" relation to each other. Without having such a relation of mass-equality *ab initio*, the set of bodies do not even form a quasi-continuum. We complete the metrization of this quasi-continuum by choosing a unit

[16] For a concise account of that definition, see [58, pp. 56–58].

of mass (e.g., 1 gram) and by availing ourselves of the numerical mass ratios obtained by experiment. There is no question here of the lack of an intrinsic metric in the sense of a choice of making the mass *difference* between a given pair of bodies equal to that of another pair or not. In the resulting continuum of real mass numbers, the elements themselves have inherent magnitude and can hence be compared individually, thus defining an intrinsic metric. Unlike the point elements of space, the elements of the set of bodies are *not* all alike mass-wise, and hence the metrization of the quasi-continuum which they form with respect to the relations of being more massive than and having the same mass can take the form of directly comparing the individual elements of that quasi-continuum rather than only intervals between them.

If one did wish for a spatial (or temporal) analogue of the metrization of masses, one should take as the set to be metrized *not* the continuum of points (or instants) but the quasi-continuum of all spatial (or temporal) *intervals*. To have used such intervals as the *elements* of the set to be metrized, we must have had a prior criterion of spatial congruence and of "being longer than" in order to arrange the intervals in a quasi-continuum which can then be metrized by the assignment of length numbers. This metrization would be the space or time analogue of the metrization of masses.

3. An Appraisal of R. Carnap's and H. Reichenbach's Philosophy of Geometry

(*i*) *The Status of Reichenbach's 'Universal Forces,'* * *and His 'Relativity of Geometry.'*

In *Der Raum* [8, p. 33], Carnap begins his discussion of *physical* space by inquiring whether and how a line in this space can be identified as straight. Arguing from *testability* and not, as we did in §2, from the continuity of that manifold, he answers this inquiry as follows: "It is impossible in principle to ascertain this, if one restricts oneself to the unambiguous deliverances of experience and does not

* A further discussion of the status of Reichenbach's universal forces is to be found in §9 of Chapter III.

introduce freely-chosen conventions in regard to objects of experience" [8, p. 33]. And he then points out that the most important convention relevant to whether certain physical lines are to be regarded as straights is the specification of the metric ("Mass-setzung"), which is conventional because it could "never be either confirmed or refuted by experience." Its statement takes the following form: "A particular body and two fixed points on it are chosen, and it is then agreed what length is to be assigned to the interval between these points under various conditions (of temperature, position, orientation, pressure, electrical charge, etc.). An example of the choice of a metric is the stipulation that the two marks on the Paris standard meter bar define an interval of $100. f(T; \phi, \lambda, h; \ldots)$ cm; . . . a unit must also be chosen, but that is not our concern here which is with the choice of the body itself and with the function $f(T, \ldots)$" [8, pp. 33–34].

Once a particular function f has been chosen, the coincidence behavior of the selected transported body permits the determination of the metric tensor g_{ik} appropriate to that choice, thereby yielding a *congruence* class of intervals and the associated geometry. Accordingly, Carnap's thesis is that the question as to the geometry of physical space is indeed an *empirical* one but subject to an important proviso: it becomes empirical only *after* a physical definition of congruence for line segments has been given *conventionally* by stipulating (to within a constant factor depending on the choice of unit) what length is to be assigned to a transported solid rod in different positions of space.

Like Carnap, Reichenbach invokes testability [72, p. 16] to defend this qualified empiricist conception of geometry and speaks of "the relativity of geometry" [72, Section 8] to emphasize the dependence of the geometry on the definition of congruence. Carnap had lucidly conveyed the conventionality of congruence by reference to our freedom to choose the function f in the metric. But Reichenbach couches this conception in *metaphorical* terms by speaking of "universal forces" [72, Sections 3, 6, 8] whose metrical "effects" on measuring rods are then said to be a matter of convention as follows: the customary definition of congruence in which the rod is held to be of equal length everywhere (*after* allowance for substance-specific thermal effects and the like) corresponds to equating the universal forces to zero; on the other hand, a *non*customary definition of congruence, according to which the length of the rod varies with position or orientation (*after* allowance for thermal effects, etc.), corresponds to assuming an appropriately specified *non*-vanishing universal force whose mathematical characterization will be explained below. Reichenbach did not anticipate that this metaphorical encumbrance of his formulation would mislead some people into making the ill-conceived charge that noncustomary definitions of congruence are based on *ad hoc* invocations of universal forces. Inasmuch as this charge has been leveled against the conventionality of congruence, it is essential that we now divest Reichenbach's statement of its misleading potentialities.

Reichenbach [72, Section 3] invites consideration of a large hemisphere made of glass which merges into a huge glass plane, as shown in cross section by the surface *G* in the accompanying diagram, which consists of a plane with a hump. Using solid rods, human beings on this surface would readily determine it to be a Euclidean plane with a central hemispherical hump. He then supposes an opaque plane *E* to be located below the surface *G* as shown in the diagram. Vertical light rays incident upon *G* will cast shadows of all objects on that glass surface onto *E*. As measured by *actual* solid rods, *G*-people will find *A′B′* and *B′C′* to be equal, while their projections *AB* and *BC* on the Euclidean plane *E* would be unequal. Reichenbach now wishes to prepare the reader for the recognition of the conventionality of congruence by having him deal with the following kind of question. Might it not be the case that (1) the *inequality* of *AB* and *BC* is only *apparent*, these intervals and other projections like them

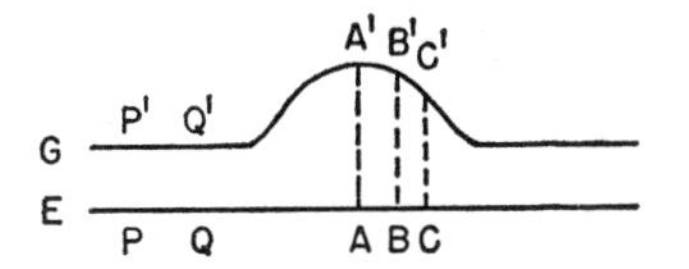

in the region R of E under the hemisphere being *really* equal, so that the *true* geometry of the surface E is *spherical* in R and Euclidean only outside it; (2) the equality of $A'B'$ and $B'C'$ is only *apparent*, the *true* geometry of surface G being plane Euclidean *throughout*, since in the apparently hemispherical region R' of G *real* equality obtains among those intervals which are the upward vertical projections of E-intervals in R that are equal in the *customary* sense of our daily life; and (3) on each of the two surfaces, transported measuring rods respectively fail to coincide with really equal intervals in R and R' respectively, because they do not remain truly congruent to themselves under transport, being deformed under the influence of undetectable forces which are "universal" in the sense that (a) they affect all materials alike, and (b) they penetrate all insulating walls?

On the basis of the conceptions presented in §2 above, *which involve no kind of reference to universal forces*, one can fulfill Reichenbach's desire to utilize this question as a basis for expounding the conventionality of congruence by presenting the following considerations. The legitimacy of making a distinction between the real (true) and the apparent geometry of a surface *turns on the existence of an intrinsic metric*. If there were an intrinsic metric, there would be a basis for making the distinction between real (true) and apparent equality of a rod under transport, and thereby between the true and the apparent geometry. But inasmuch as there is not, the question whether a given surface is really a Euclidean plane with a hemispherical hump or only apparently so must be *replaced* by the following question: *on a particular convention of congruence* as specified by a choice of one of Carnap's functions f, does the coincidence behavior of the transported rod on the surface in question yield the geometry under discussion or not? Thus the question as to the geometry of a surface is inherently ambiguous without the introduction of a congruence definition. And in view of the conventionality of spatial congruence, we are entitled to metrize G and E *either* in the customary way *or* in other ways so as to describe E as a Euclidean plane with a hemispherical hump R in the center and G as a Euclidean plane *throughout*. To assure the correctness of the latter *non*-customary descriptions, we need only *decree* the congruence of those respective intervals which our questioner called "really equal" as opposed to apparently equal in parts (1) and (2) of his question respectively. Accordingly, without the presupposition of an intrinsic metric there can be no question of an absolute or "real" deformation of *all* kinds of measuring rods alike under the action of universal forces, and, *mutatis mutandis*, the same statement applies to clocks. Since a rod undergoes no kind of objective physical change in the supposed "presence" of universal forces, that "presence" signifies no more than that we assign a different length to it in different positions or orientations by convention. Hence, just as the conversion of the length of a table from meters to feet does not involve the action of a force on the table as the "cause" of the change, so also reference to universal forces as "causes" of "changes" in the transported rod can have no literal but only *metaphorical* significance. Moreover, mention of universal forces is an entirely dispensable *façon de parler* in this context as is evident from the fact that the rule assigning lengths to the transported rod which vary with its position and orientation can be given by specifying Carnap's function f.

Reichenbach, on the other hand, chooses to formulate the conventionality of congruence by first distinguishing between what he calls "differential" and "universal" forces and then using "universal forces" metaphorically in his statement of the epistemological status of the metric. By "differential" forces [72, Section 3] he means thermal and other influences which we called "perturbational" (cf. §2(i)) and whose presence issues in the dependence of the coincidence behavior of transported rods on the latter's chemical composition. Since we conceive of physical geometry as the system of metric relations which are *independent* of chemical composition, we correct for the substance-specific deformations induced by differential forces [1; 72, p. 26; 76; 77, pp. 327–329]. Reichenbach defines "universal forces" as having the twin properties of affecting all materials in the same way and being all-permeating because there are no walls capable of providing insulation against them. There is precedent for a literal rather than metaphorical use of universal forces to give a congruence definition: in order to provide a *physical realization* of a noncustomary congruence definition which would metrize the interior of a sphere of radius R so as to be a model of an infinite 3-dimensional hyperbolic space, Poincaré [63, pp. 75–77] postulates that (a) each concentric sphere of radius $r < R$ is held at a constant absolute temperature $T \propto R^2 - r^2$, while the optical index of refraction is inversely proportional to $R^2 - r^2$, and (b) *contrary to actual fact*, all kinds of bodies within the sphere have the *same* coefficient of thermal expansion. It is essential to see that the expansions and contractions of these bodies under displacement have a *literal* meaning in this context, because they are *relative* to the actual displacement behavior of our normally Euclidean kinds of bodies and are linked to thermal sources.[17] Indeed, this fact is essential to Poincaré's description of his model.

But Reichenbach's *metaphorical* use of universal forces for giving the congruence definition and exhibiting the dependence of the geometry on that definition takes the following form: "Given a geometry G' to which the measuring instruments conform [*after* allowance for the effects of thermal and other "differential" influences], we can imagine a universal force F which affects the instruments in such a way that the actual geometry is an arbitrary geometry G, while the observed deviation from G is due to a universal deformation of the measuring instruments" [72, p. 33]. And he goes on to say that if g'_{ik} ($i = 1, 2, 3$; $k = 1, 2, 3$) are the empirically obtained metrical coefficients of the geometry G' and g_{ik} those of G, then the force tensor F is given mathematically by the tensor equation $g'_{ik} + F_{ik} = g_{ik}$, where the g'_{ik}, which had yielded the observed geometry G', are furnished experimentally by the

[17] A precisely analogous *literal* use of universal forces is also made by Reichenbach [72, pp. 11–12] to convey pictorially a physical realization of a congruence definition which would confer a spherical geometry on the region R of the surface E discussed above.

measuring rods[18] and where the F_{ik} are the "correction factors"[19] $g_{ik} - g'_{ik}$ which are added correctionally to the g'_{ik} so that the g_{ik} are obtained.[20] But since Reichenbach emphasizes that it is a matter of *convention* whether we equate F_{ik} to zero or not [72, pp. 16, 27–28, 33], this formulation is merely a metaphorical way of asserting that the following is a matter of convention: whether congruence is said to obtain among intervals having equal lengths ds given by the metric $ds^2 = g'_{ik}\, dx^i\, dx^k$—which entails G' as the geometric description of the observed coincidence relations—or among intervals having equal lengths ds given by the metric $ds^2 = g_{ik}\, dx^i\, dx^k$ which yields a different geometry G.[21] Clearly then, to equate the universal

[18] For details on this procedure, see for example [72, Secs. 39 and 40].

[19] The quotation marks are also in Reichenbach's text.

[20] We shall see in §3(iii) that Reichenbach was *mistaken* in asserting [72, pp. 33–34] that for a given surface or 3-space a particular metric geometry *determines* (1) a *unique* definition of congruence and, once a unit has been chosen, (2) a *unique* set of functions g_{ik} as the representations of the metric tensor in any particular coordinate system.

It will turn out that there are infinitely many *incompatible* congruence definitions and as many correspondingly different metric tensors which impart the *same* geometry to physical space. Hence, while a given metric tensor yields a unique geometry, a geometry G does *not* determine a metric tensor *uniquely* to within a constant factor depending on the choice of unit. And thus it is incorrect for Reichenbach to speak here of the components g_{ik} of a particular metric tensor as "those of G" [72, p. 33n] and to suppose that a unique F is specified by the requirement that a certain geometry G prevail in place of the observed geometry G'.

[21] It is to be clearly understood that the g_{ik} yield a congruence relation *incompatible* with the one furnished by the g'_{ik}, because in any *given* coordinate system they are *different* functions of the given coordinates and *not proportional* to one another. (A difference consisting in a mere proportionality could *not* involve a difference in the congruence classes but only in the unit of length used.) The incompatibility of the congruences furnished by the two sets of g's is a necessary though not a sufficient condition (see the preceding footnote) for the nonidentity of the associated geometries G and G'.

The difference between the two metric tensors corresponding to *incompatible* congruences must not be confounded with a mere difference in the *representations* in different coordinate systems of the *one* metric tensor corresponding to a *single* congruence criterion (for a given choice of a unit of length): the former is illustrated by the *incompatible* metrizations $ds^2 = dx^2 + dy^2$ and

$$ds^2 = \frac{dx^2 + dy^2}{y^2}$$

in which the g's are *different* functions of the *same* rectangular coordinates, whereas the latter is illustrated by using first rectangular and then polar coordi-

forces to zero is merely to choose the metric based on the tensor g'_{ik} which was obtained from measurements in which the rod was called congruent to itself everywhere. In other words, to stipulate $F_{ik} = 0$ is to choose the customary congruence standard based on the rigid body. On the other hand, apart from *one exception* to be stated presently, to stipulate that the components F_{ik} may not all be zero is to adopt a noncustomary metric given by a tensor g_{ik} corresponding to a specified *variation* of the length of the rod with position and orientation.

That there is one exception, which, incidentally, Reichenbach does not discuss, can be seen as follows: A given congruence determines the metric tensor up to a constant factor depending on the choice of unit and conversely. Hence two metric tensors will correspond to different congruences if and only if they differ other than by being proportional to one another. Thus, if g_{ik} and g'_{ik} are proportional by a factor different from 1, these two tensors furnish metrics differing only in the choice of unit and hence yield the same congruence. Yet in the case of such proportionality of g_{ik} and g'_{ik}, the F_{ik} *cannot* all be zero. For example, if we consider the line element $ds^2 = a^2\, d\phi^2 + a^2 \sin^2\phi\, d\theta^2$ on the surface of a sphere of radius $a = 1$ meter $= 100$ cm, the mere change of units from meters to centimeters would change the metric $ds^2 = d\phi^2 + \sin^2\phi\, d\theta^2$ into $ds^2 = 10000\, d\phi^2 + 10000 \sin^2\phi\, d\theta^2$. And if these metrics are identified with g'_{ik} and g_{ik} respectively, we obtain

$$\begin{aligned}
F_{11} &= g_{11} - g'_{11} = 10000 - 1 = 9999 \\
F_{12} &= F_{21} = g_{12} - g'_{12} = 0 \\
F_{22} &= g_{22} - g'_{22} = 10000 \sin^2\phi - \sin^2\phi = 9999 \sin^2\phi.
\end{aligned}$$

It is now apparent that the F_{ik}, being given by the difference between the g_{ik} and the g'_{ik}, will *not* all vanish *and* that also these two metric

nates to express the *same metric* as follows: $ds^2 = dx^2 + dy^2$ and $ds^2 = d\rho^2 + \rho^2\, d\theta^2$. In the latter case, we are not dealing with different *metrizations* of the space but only with different *coordinatizations* (*para*metrizations) of it, one or more of the g's being different functions of the coordinates in one coordinate system from what they are in another but so chosen as to yield an *invariant* ds.

tensors will yield the *same* congruence, if and only if these tensors are proportional by a factor different from 1. Therefore, a necessary and sufficient condition for obtaining *incompatible* congruences is that *both* there be at least *one* nonvanishing component F_{ik} *and* the metric tensors g_{ik} and g'_{ik} *not* be proportional to one another. The exception to our statement that a noncustomary congruence definition is assured by the failure of at least one component of the F_{ik} to vanish is therefore given by the case of the proportionality of the metric tensors.[22]*

Although Reichenbach's metaphorical use of "universal forces" issued in misleading and wholly unnecessary complexities which will be pointed out presently, he himself was in no way victimized by them. Writing in 1951 concerning the invocation of universal forces in this context, he declared: "The assumption of such forces means merely a change in the coordinative definition of congruence" [73, p. 133]. It is therefore quite puzzling that, in 1956, Carnap, who had lucidly expounded the same ideas *non*metaphorically as we saw, singled out Reichenbach's characterization and recommendation of the customary congruence definition in terms of equating the universal forces to zero as a praiseworthy part of Reichenbach's outstanding work *The Philosophy of Space and Time* [72]. In his Preface to the latter work, Carnap says: "Of the many fruitful ideas which Reichenbach contributed . . . I will mention only one, which seems to me of great interest for the methodology of physics but which has so far not found the attention it deserves. This is the principle of the elimination of universal forces. . . . Reichenbach proposes to accept as a general methodological principle that we choose that form of a theory among physically equivalent forms (or, in other

[22] There is a most simple illustration of the fact that if the metric tensors are *not* proportional while as few as *one* of the components F_{ik} is *non*vanishing, the congruence associated with the g_{ik} will be incompatible with that of the g'_{ik} and will hence be noncustomary. If we consider the metrics $ds^2 = dx^2 + dy^2$ and $ds^2 = 2dx^2 + dy^2$, then comparison of an interval for which $dx = 1$ and $dy = 0$ with one for which $dx = 0$ and $dy = 1$ will yield congruence on the first of these metrics but not on the second.

* See also Chapter II, footnote 6.

words, that definition of 'rigid body' or 'measuring standard') with respect to which all universal forces disappear" [9, p. vii].

The misleading potentialities of including metaphorical "universal forces" in the statement of the congruence definition manifest themselves in three ways as follows: (1) The formulation of noncustomary congruence definitions in terms of deformations by universal forces has inspired the erroneous charge that such congruences are *ad hoc* because they involve the *ad hoc* postulation of nonvanishing universal forces. (2) In Reichenbach's statement of the congruence definition to be employed to explore the spatial geometry in a *gravitational* field, universal forces enter in both a *literal* and a *metaphorical* sense. The conflation of these two senses issues in a seemingly contradictory formulation of the customary congruence definition. (3) Since the variability of the curvature of a space would manifest itself in alterations in the coincidence behavior of all kinds of solid bodies under displacement, Reichenbach speaks of bodies displaced in such a space as being subject to universal forces "destroying coincidences" [72, p. 27]. In a manner analogous to the gravitational case, the conflation of this literal sense with the metaphorical one renders the definition of rigidity for this context paradoxical. We shall now briefly discuss these three sources of confusion in turn.

1. If a congruence definition itself had factual content, so that alternative congruences would differ in factual content, then it would be significant to say of a congruence definition that it is *ad hoc* in the sense of being an *evidentially unwarranted* claim concerning facts. But inasmuch as ascriptions of spatial congruence to noncoinciding intervals are not factual but conventional, neither the customary nor any of the noncustomary definitions of congruence can possibly be *ad hoc*. Hence the abandonment of the former in favor of the latter kind of definition can be no more *ad hoc* than the regraduation of a centigrade thermometer into a Fahrenheit one or than the change from Cartesian to polar coordinates. By formulating noncustomary congruence definitions in terms of the metaphor of universal forces,

Reichenbach made it possible for people to misconstrue his metaphorical sense as literal. And once this error had been committed, its victims tacitly regarded the customary congruence definition as factually true and felt justified in dismissing other congruences as *ad hoc* on the grounds that they involved the *ad hoc* assumption of (literally conceived) universal forces. Thus we find that Ernest Nagel, for example, overlooks that the invocation of universal forces to preserve Euclidean geometry can be no more *ad hoc* than a change in the *units* of length (or of temperature) whereby one assures a particular numerical value for the length of a given object in a particular situation. For after granting that, if necessary, Euclidean geometry can be retained by an appeal to universal forces, Nagel writes: "Nevertheless, universal forces have the curious feature that their presence can be recognized *only* on the basis of geometrical considerations. The assumption of such forces thus has the appearance of an *ad hoc* hypothesis, adopted solely for the sake of salvaging Euclid" [55, p. 264].

2. Regarding the geometry in a gravitational field, Reichenbach says the following: "We have learned . . . about the difference between *universal* and *differential* forces. These concepts have a bearing upon this problem because we find that gravitation is a universal force. It does indeed affect all bodies in the same manner. This is the physical significance of the equality of gravitational and inertial mass" [72, p. 256]. It is entirely correct, of course, that a uniform gravitational field (which has not been transformed away in a given space-time coordinate system) is a universal force in the *literal* sense *with respect to a large class of effects* such as the free fall of bodies. But there are other effects, such as the bending of elastic beams, with respect to which gravity is clearly a *differential* force in Reichenbach's sense: a wooden bookshelf will sag more under gravity than a steel one. And this shows, incidentally, that Reichenbach's classification of forces into universal and differential is not mutually exclusive. Of course, just as in the case of any other force having differential effects on measuring rods, allowance is made for dif-

ferential effects of gravitational origin in laying down the congruence definition.

The issue is therefore twofold: (1) Does the fact that gravitation is a universal force in the *literal* sense indicated above have a bearing on the spatial geometry, and (2) in the presence of a gravitational field is the logic of the spatial congruence definition any different in regard to the role of *metaphorical* universal forces from what it is in the absence of a gravitational field? Within the particular context of the general theory of relativity—hereafter denoted by "GTR"—there is indeed a *literal* sense in which the gravitational field of the sun, for example, is causally relevant geometrically as a universal force. And the literal sense in which the coincidence behavior of the transported customary rigid body is objectively different in the vicinity of the sun, for example, from what it is in the absence of a gravitational field can be expressed in two ways as follows: (i) *relatively to the congruence defined by the customary rigid body*, the spatial geometry in the gravitational field is *not* Euclidean—contrary to pre-GTR physics—but is Euclidean in the absence of a gravitational field; (ii) the geometry in the gravitational field is Euclidean if and only if the customary congruence definition is supplanted by one in which the length of the rod varies *suitably* with its position or orientation,[23] whereas it is Euclidean relatively to the customary congruence definition for a vanishing gravitational field.

It will be noted, however, that formulation (i) makes no mention at all of any deformation of the rod by universal forces as that body is transported from place to place in a given gravitational field. Nor need there be any metaphorical reference to universal forces in the statement of the customary congruence definition ingredient in formulation (i). For that statement can be given as follows: in the presence no less than in the absence of a gravitational field, congruence is conventional, and hence we are free to adopt the customary congruence in a gravitational field as a basis for determining the

[23] For the gravitational field of the sun, the function specifying a noncustomary congruence definition issuing in a Euclidean geometry is given in [8, p. 58].

spatial geometry. By encumbering his statement of a congruence definition with metaphorical use of "universal forces," Reichenbach enables the unwary to infer incorrectly that a rod subject to the universal force of gravitation in the specified *literal* sense *cannot* consistently be regarded as *free* from deforming universal forces in the *metaphorical* sense and hence cannot serve as the congruence standard. This conflation of the literal and metaphorical senses of "universal force" thus issues in the mistaken belief that in the GTR the customary spatial congruence definition cannot be adopted consistently for the gravitational field. And those who were led to this misconception by Reichenbach's metaphor will therefore deem self-contradictory the following consistent assertion by him: "We do not speak of a change produced by the gravitational field in the measuring instruments, but regard the measuring instruments as 'free from deforming forces' in spite of the gravitational effects" [72, p. 256]. Moreover, those victimized by the metaphorical part of Reichenbach's language will be driven to reject as inconsistent Einstein's characterization of the geometry in the gravitational field in the GTR as given in formulation (i) above. And they will insist erroneously that formulations (i) and (ii) are *not* equally acceptable alternatives on the grounds that formulation (ii) is uniquely correct

The confounding of the literal and metaphorical senses of "universal force" by reference to gravity is present, for example, in Ernest Nagel's treatment of universal forces with resulting potentialities of confusion. Thus, he incorrectly cites the force of gravitation in its role of being *literally* a "universal force" as a *species* of what are only *metaphorically* "universal forces." Specifically, in speaking of the assumption of universal forces whose "presence can be recognized *only* on the basis of geometrical considerations" because they are assumed "solely for the sake of salvaging Euclid"—i.e., "universal forces" in the *metaphorical* sense—he says: "'Universal force' is not to be counted as a 'meaningless' phrase, for it is evident that a procedure is indicated for ascertaining whether such forces are present or not. Indeed, gravitation in the Newtonian theory of mechanics is just such a universal force; it acts alike on all bodies and cannot be screened" [55, p. 264, n. 19]. But this misleading formulation suggests the incorrect conclusion that a rod subject to a gravitational field cannot be held to be "free" from "universal forces" in performing its metrical function.

3. In a manner analogous to the gravitational case just discussed, we can assert the following: since congruence is conventional, we are at liberty to use the customary definition of it without regard to whether the geometry obtained by measurements expressed in terms of that definition is one of *variable* curvature or not. And thus we see that without the intrusion of a metaphorical use of "universal forces," the statement of the congruence definition need not take any cognizance of whether the resulting geometry will be one of constant curvature or not. A geometry of constant curvature or so-called congruence geometry is characterized by the fact that the so-called axiom of free mobility holds in it: for example, on the surface of a sphere, a triangle having certain angles and sides in a given place can be moved about without any change in their magnitudes relatively to the customary standards of congruence for intervals and angles. On the other hand, on the surface of an egg, the failure of the axiom of free mobility to hold can be easily seen from the following indicator of the variability of the curvature of that 2-space: a circle and its diameter made of any kind of wire are so constructed that one end P of the diameter is attached to the circle while the other end S is free *not* to coincide with the opposite point Q on the circle though coinciding with it in a given initial position on the surface of the egg. Since the ratio of the diameter and the circumference of a circle *varies* in a space of variable curvature such as that of the egg surface, S will no longer coincide with Q if the circular wire and its attachment PS are moved about on the egg so as to preserve contact with the egg surface everywhere. The indicator thus exhibits an objective destruction of the coincidence of S and Q which is wholly independent of the indicator's chemical composition (under uniform conditions of temperature, etc.). One may therefore speak *literally*

here, as Reichenbach does [72, Section 6], of universal forces acting on the indicator by destroying coincidences. And since the customary congruence definition is entirely permissible as a basis for geometries of variable curvature, there is, of course, no inconsistency in giving that congruence definition by equating universal forces to *zero* in the *metaphorical* sense, even though the destruction of coincidences attests to the quite *literal* presence of causally efficacious universal forces. But Reichenbach invokes universal forces *literally* without any warning of an impending metaphorical reference to them in the congruence definition. And the reader is therefore both startled and puzzled by the seeming paradox in Reichenbach's declaration [72, p. 27] that "Forces *destroying coincidences* must also be set equal to zero, if they satisfy the properties of the universal forces mentioned on p. 13; only then is the problem of geometry uniquely determined." Again the risk of confusion can be eliminated by dispensing with the metaphor in the congruence definition.

While regarding Reichenbach's *The Philosophy of Space and Time* as still the most penetrating single book on its subject, the analysis we have given compels us to dissent from Nagel's judgment that, in that book, Reichenbach employs "The distinction between 'universal' and 'differential' forces . . . with great clarifying effect" [55, p. 264, n. 18].

Having divested Reichenbach's statements about universal forces of their misleading potentialities, we shall hereafter safely be able to discuss other issues raised by statements of his which are couched in terms of universal forces.

The first of these is posed by the following assertion by him: "We obtain a statement about physical reality only if in addition to the geometry G of the space its universal field of force F is specified. Only the combination $G + F$ is a testable statement" [72, p. 33]. In order to be able to appraise this assertion, consider a surface on which some set of generalized curvilinear (or "Gaussian") coordinates has been introduced and onto which a metric $ds^2 = g_{ik}\, dx^i\, dx^k$ is then put quite arbitrarily by a capricious choice of a set of functions g_{ik} of the given coordinates.[24] Assume now that the latter specification of the geometry G is *not* coupled with any information regarding F. Is it then correct to say that since this metrization provides no information at all about the coincidence behavior of a rod under transport on the surface, it conveys no factual information whatever about the surface or physical reality? That such an inference is mistaken can be seen from the following: depending upon whether the Gaussian curvature K associated with the stipulated g_{ik} is positive (spherical geometry), zero (Euclidean geometry), or negative (hyperbolic geometry), it is an objective fact about the surface that through a point outside a given geodesic of the chosen metric, there will be respectively 0, 1, or infinitely many other such geodesics which will not *intersect* the given geodesic. Whether or not certain lines on a surface intersect is, however, merely a *topological* fact concerning it. And hence we can say that although an arbitrary metrization of a space without a specification of F is not altogether devoid of factual content pertaining to that space, such a metrization can yield no objective facts concerning the space not already included in the latter's topology.

We can therefore conclude the following: if the description of a space (surface) is to contain empirical information concerning the coincidence behavior of transported rods in that space and if a metric $ds^2 = g_{ik}\, dx^i\, dx^k$ (and thereby a geometry G) is chosen whose congruences do *not* accord with those defined by the application of the transported rod, then indeed Reichenbach's assertion holds. Specifically, the chosen metric tensor g_{ik} and its associated geometry G must then be coupled with a specification of the different metric

[24] The coordinatization of a space, whose purpose it is merely to number points so as to convey their topological neighborhood relations of betweenness, does not, *as such*, presuppose (entail) a metric. However, the statement of a *rule* assuring that different people will independently effect the *same* coordinatization of a given space may require reference to the use of a rod. But even in the case of coordinates such as rectangular (Cartesian) coordinates, whose assignment is carried out by the use of a rigid rod, it is quite possible to ignore the manner in which the coordinatization was effected and to regard the coordinates purely topologically, so that a metric very *different* from $ds^2 = dx^2 + dy^2$ can then still be introduced quite consistently.

tensor g'_{ik} that would have been found experimentally, if the rod had actually been chosen as the congruence standard. But Reichenbach's provision of that specification via the universal force F is quite unnecessarily roundabout. For F is defined by $F_{ik} = g_{ik} - g'_{ik}$ and cannot be known without already knowing *both* metric tensors. Thus, there is a loss of clarity in pretending that the metric tensor g'_{ik}, which codifies the empirical information concerning the rod, is first obtained from the identity $g'_{ik} = g_{ik} - F_{ik}$.

(*ii*) *Reichenbach's Theory of Equivalent Descriptions.*

A metric geometry in conjunction with a specified congruence definition given by a statement about the F_{ik} describes the coincidence behavior of a transported rod. Since *the same facts* of coincidence can be described in a linguistically alternative way by a different geometry when coupled with a suitably different congruence definition, Reichenbach speaks of different geometric descriptions which have the same factual content as "logically equivalent" [69, pp 374–375] or simply as "equivalent" [73, pp. 133ff]. More generally, since not only spatial and temporal congruence but also metrical simultaneity are conventional in the sense of §2(i) above, geochronometries based on different definitions of congruence and/or simultaneity are equivalent. Among the equivalent descriptions of space, Reichenbach calls the one employing the customary definition of congruence the "normal system" and the particular geometry appropriate to it the "natural geometry" [73, p. 134]. The choice of a particular member from *within* a class of equivalent descriptions is a matter of convention and hence decided on the basis of convenience. But, as Reichenbach points out correctly, the decision as to which class among nonequivalent classes of equivalent descriptions is true is not at all conventional but is a matter of *empirical fact*. Thus, a Euclidean description and a certain non-Euclidean one *cannot* both be the *natural* geometry of a given space simultaneously: if a normal system is employed, they cannot both be true but will be the respective "normal" *representatives* of two *non*equivalent classes of equivalent descriptions. On the other hand, all of the members of a particular class of equivalent descriptions obviously have the same truth value.

Reichenbach's characterization of the empirical status of various geochronometries in terms of equivalence classes of descriptions is seen to be correct in the light of our preceding analysis, and it is summarized here because of its usefulness in the further pursuit of our inquiry.[25]

(iii) An Error in the Carnap-Reichenbach Account of the Definition of Congruence: The Nonuniqueness of Any Definition of Congruence Furnished by Stipulating a Particular Metric Geometry.

Upon undertaking the mathematical implementation of a given choice of a metric geometry in the context of the topology of the space we must ask: does the topology in conjunction with the desired metric geometry determine a *unique* definition of congruence and a *unique* metric tensor, the latter's uniqueness being understood as uniqueness to within a constant factor depending on the choice of unit but *not* as unique representability of the metric tensor (cf. footnote 21) by a particular set of functions g_{ik}? As we shall see, the answer to this question has an immediate bearing on the validity of a number of important assertions made by Carnap and Reichenbach in their writings on the philosophy of geometry. Preparatory to appraising these claims of theirs, we shall show that given the topology, the desired geometry does *not* uniquely specify a metric tensor and thus, in a particular coordinate system, *fails* to determine a unique set of functions g_{ik} to within a constant factor depending on the choice of unit. Our demonstration will take the form of show-

[25] Our endorsement of Reichenbach's theory of equivalent descriptions in the context of geochronometry should *not* be construed as an espousal of his application of it to the theory of the states of macro-objects like trees during times when no human being observes them [70, pp. 17–20]. Since a critical discussion of Reichenbach's account of unobserved macro-objects would carry us too far afield in this essay, suffice it to say the following here. In my judgment, Reichenbach's application of the theory of equivalent descriptions to unobserved objects lacks clarity on precisely those points on which its relevance turns. But as I interpret the doctrine, I regard it as fundamentally incorrect because its ontology is that of Berkeley's *esse est percipi*.

ing that besides the customary definition of congruence, which assigns the same length to the measuring rod everywhere and thereby confers a Euclidean geometry on an ordinary table top, there are infinitely many *other* definitions of congruence having the following property: they likewise yield a Euclidean geometry for that surface but are incompatible with the customary one by making the length of a rod depend on its orientation and/or position.

Thus, consider our horizontal table top equipped with a network of Cartesian coordinates x and y, but now metrize this surface by means of the *non*standard metric $ds^2 = \sec^2\theta\, dx^2 + dy^2$, where $\sec^2\theta$ is a constant greater than 1. Unlike the standard metric, this metric assigns to an interval whose coordinates differ by dx *not* the length dx but the *greater* length $\sec\theta\, dx$. Although this metric thereby makes the length of a given rod dependent on its orientation, we shall show that the *infinitely* many different *non*standard congruences generated by values of $\sec\theta$ which exceed 1 each impart a *Euclidean* geometry to the table top no less than does the standard congruence given by $ds^2 = dx^2 + dy^2$. Accordingly, our demonstration will show that the requirement of Euclideanism does *not* uniquely determine a congruence class of intervals but allows an *infinitude* of incompatible congruences. We shall therefore have established that there are infinitely many ways in which a measuring rod could squirm under transport on the table top as compared to its familiar *de facto* behavior while still yielding a Euclidean geometry for that surface.

To carry out the required demonstration, we first note the preliminary fact that the geometry yielded by a particular metrization is clearly independent of the particular coordinates in which that metrization is expressed. And hence if we expressed the standard metric $ds^2 = dx^2 + dy^2$ in terms of the primed coordinates x' and y' given by the transformations $x = x'\sec\theta$ and $y = y'$, obtaining $ds^2 = \sec^2\theta\, dx'^2 + dy'^2$, we would obtain a *Euclidean* geometry as before, since the latter equation would merely express the original standard metric in terms of the primed coordinates. Thus, when the same *invariant* ds of the standard metric is expressed in terms of both primed and unprimed coordinates, the metric coefficients g'_{ik} given by $\sec^2 \theta$, 0 and 1 yield a Euclidean geometry no less than do the unprimed coefficients 1, 0 and 1.

This elementary ancillary conclusion now enables us to see that the following *non*standard metrization (or remetrization) of the surface in terms of the *original*, unprimed rectangular coordinates must likewise give rise to a Euclidean geometry: $ds^2 = \sec^2 \theta \, dx^2 + dy^2$. For the value of the Gaussian curvature and hence the prevailing geometry depends *not* on the particular coordinates (primed or unprimed) to which the metric coefficients g_{ik} pertain but only on the *functional form* of the g_{ik} [42, p. 281], which is the same here as in the case of the g'_{ik} above.

More generally, therefore, the geometry resulting from the *standard* metrization is *also* furnished by the following kind of *non*standard metrization of a space of points, expressed in terms of the same (unprimed) coordinates as the standard one: the *non*standard metrization has unprimed metric coefficients g_{ik} which have the same *functional form* (to within an arbitrary constant arising from the choice of unit of length) as those primed coefficients g'_{ik} which are obtained by expressing the *standard* metric in some set or other of *primed* coordinates via a suitable coordinate transformation. In view of the large variety of allowable coordinate transformations, it follows at once that the class of *non*standard metrizations yielding a Euclidean geometry for a table top is far wider than the already infinite class given by $ds^2 = \sec^2 \theta \, dx^2 + dy^2$, where $\sec^2 \theta > 1$. Thus, for example, there is *identity of functional form* between the *standard* metric in *polar* coordinates, which is given by $ds^2 = d\rho^2 + \rho^2 \, d\theta^2$, and the *non*standard metric in *Cartesian* coordinates given by $ds^2 = dx^2 + x^2 \, dy^2$, since x plays the same role *formally* as ρ, and similarly for y and θ. Consequently, the latter *non*standard metric issues in a *Euclidean* geometry just as the former standard one does.

It is clear that the multiplicity of metrizations which we have proven for Euclidean geometry obtains as well for each of the non-Euclidean geometries. The failure of a geometry of two or more dimensions to determine a congruence definition uniquely does *not*, however, have a counterpart in the *one*-dimensional time continuum: the demand that Newton's laws hold in their customary metrical form determines a *unique* definition of temporal congruence. And hence it is feasible to rely on the law of translational or rotational inertia to define a time metric or "uniform time."

On the basis of this result, we can now show that a number of claims made by Reichenbach and Carnap respectively are false.

1. In 1951, Reichenbach wrote: "If we change the coordinative definition of congruence, a different geometry will result. This fact is called the *relativity of geometry*" [73, p. 132]. That this statement is false is evident from the fact that if, in our example of the table top, we change our congruence definition from $ds^2 = dx^2 + dy^2$ to any one of the infinitely many definitions incompatible with it that are given by $ds^2 = \sec^2 \theta \, dx^2 + dy^2$, precisely the same Euclidean geometry results. Thus, contrary to Reichenbach, the introduction of a nonvanishing universal force corresponding to an alternative congruence does *not* guarantee a change in the geometry. Instead, the correct formulation of the relativity of geometry is that in the form of the *ds* function the congruence definition uniquely determines the geometry, though *not* conversely, and that any one of the congruence definitions issuing in a geometry G' can always be replaced by infinitely many *suitably* different congruences yielding a specified different geometry G. In view of the unique fixation of the geometry by the congruence definition in the context of the facts of coincidence, the repudiation of a given geometry in favor of a different one does indeed require a change in the definition of congruence. And the new congruence definition which is expected to furnish the new required geometry can do so in one of the following two ways: (i) by determining a system of geodesics *different* from the one yielded by the original congruence definition, or (ii) if the geodesics determined by the new congruence definition are the *same* as those associated with the original definition, then the *angle* congruences must be different, i.e., the new congruence definition will have to

require a different congruence class of *angles*. (For the specification of the magnitudes assigned to angles by the components g_{ik} of the metric tensor, see [27, pp. 37–38].)

That (ii) constitutes a genuine possibility for obtaining a different geometry will be evident from the following example in which two *incompatible* definitions of congruence $ds_1^2 = g_{ik}\, dx^i\, dx^k$, and $ds_2^2 = g'_{ik}\, dx^i\, dx^k$, yield the *same* system of geodesics via the equations $\delta \int ds_1 = 0$ and $\delta \int ds_2 = 0$ and yet determine *different* geometries (Gaussian curvatures) because they require incompatible congruence classes of *angles* appropriate to these respective geometries. A horizontal surface which is a Euclidean plane on the *customary* metrization can alternatively be metrized to have the geometry of a hemisphere by projection from the *center* of a sphere through the lower half of the sphere whose south pole is resting on that plane. Upon calling congruent on the horizontal surface segments and angles which are the projections of equal segments and angles respectively on the lower hemisphere, the great circle arcs of the hemisphere map into the Euclidean straight lines of the plane such that *every* straight of the Euclidean description is *also* a straight (geodesic) of the new hemispherical geometry conferred on the horizontal surface. But the *angles* which are regarded as congruent on the horizontal surface in the new metrization are *not* congruent in the original metrization yielding a Euclidean description.

It must be pointed out, however, that if a change in the congruence definition *preserves* the geodesics, then its issuance in a *different* congruence class of *angles* is only a necessary and *not* a sufficient condition for imparting to the surface a metric geometry *different* from the one yielded by the original congruence definition. This fact becomes evident by reference to our earlier case of the table top's being a model of Euclidean geometry *both* on the customary metric $ds^2 = dx^2 + dy^2$ *and* on the different metric $ds'^2 = \sec^2 \theta\, dx^2 + dy^2$: the geodesics as well as the geometries furnished by these incompatible metrics are the same, but the angles which are congruent in the new metric are generally *not* congruent in the original one. That these two metrics issue in *incompatible* congruence classes of *angles* though in the *same* geometry can be seen as follows: a Euclidean triangle which is equilateral on the new metric ds' will *not* be equilateral on the customary one ds, and hence the three angles of such a triangle will all be congruent to each other in the former metric but not in the latter.

It is clear now that an *arbitrary* change in the congruence definition for either line segments or angles or both cannot as such guarantee a different geometry.

2. Reichenbach explicitly asserts incorrectly that the geometry uniquely determines a congruence definition appropriate to it. Says he: "There is nothing wrong with a coordinative definition established on the requirement that a certain kind of geometry is to result from the measurements. . . . A coordinative definition can also be introduced by the prescription what the result of the measurements is to be. 'The comparison of length is to be performed in such a way that Euclidean geometry will be the result'—this stipulation is a possible form of a coordinative definition" [72, pp. 33–34]. And in reply to Hugo Dingler's contention [18, p. 50] that the rigid body is uniquely specified by the geometry and only by the latter, Reichenbach mistakenly agrees [68, p. 52; 74, p. 35] that the geometry is sufficient to define congruence and contests only Dingler's further claim that it is necessary.[26]

Carnap [8, p. 54] discusses the dependencies obtaining between (a) the metric geometry, which he symbolizes by "R" in this German publication; (b) the topology of the space and the facts concerning the coincidences of the rod in it, symbolized by "T" for "Tatbestand"; and (c) the metric M ("Mass-setzung"), which entails a congruence definition and is given by the function f (and by the choice of a unit), as will be recalled from the beginning of our §3.[27]

[26] In §7, we shall assess the merits of Reichenbach's denial that the geometry is necessary, which he rests on the grounds that rigidity can be defined by the elimination of differential forces.

[27] Although both Carnap's metric M and the distance function

$$ds = \sqrt{g_{ik}\,dx^i\,dx^k}$$

And he concludes that the functional relations between R, M, and T are such "that if two of them are given, the third specification is thereby uniquely given as well" [8, p. 54]. Accordingly, he writes: $R = \phi_1(M, T)$, $M = \phi_2(R, T)$, and $T = \phi_3(M, R)$.

While the first of these dependencies does hold, our example of imparting a Euclidean geometry to a table top by each of two incompatible congruence definitions shows that not only the second but also the third of Carnap's dependencies fails to hold. For the mere specification of M to the effect that the rod will be called congruent to itself everywhere and of R as Euclidean does not tell us whether the coincidence behavior T of the rod on the table top will be such as to coincide successively with those intervals that are equal according to the formula $ds^2 = dx^2 + dy^2$ or with the *different* intervals that are equal on the basis of one of the metrizations $ds^2 = \sec^2\theta\, dx^2 + dy^2$ (where $\sec^2\theta \neq 1$). In other words, the stated specifications of M and R do *not* tell us whether the rod behaves on the table top as we know it to behave in actuality or whether it squirms in any one of infinitely many different ways as compared to its actual behavior.

3. As a corollary of our proof of the *non*uniqueness of the congruence definition, we can show that the following statement by Reichenbach is false: "If we say: actually a geometry G applies but we measure a geometry G', we define at the same time a force F which causes the difference between G and G'" [72, p. 27]. Using our previous notation, we note first that instead of determining a metric tensor g_{ik} uniquely (up to an arbitrary constant), the geometry G determines an infinite class α of such tensors differing other than by being proportional to one another. But since $F_{ik} = g_{ik} - g'_{ik}$ (where the g'_{ik} are furnished by the rod prior to its being regarded as "deformed" by any universal forces), the failure of G to determine a tensor g_{ik} uniquely (up to an arbitrary constant) issues in there

can provide a congruence definition, they cannot be deduced from one another without information concerning the coincidence behavior of the rod in the space under consideration. For further details, see Chapter III, §§2.14, 9.0, and 9.1.

being as many *different* universal forces F_{ik} as there are different tensors g_{ik} in the class α determined by G. We see, therefore, that contrary to Reichenbach, there are infinitely many different ways in which the measuring rod can be held to be "deformed" while furnishing the *same* geometry G.

4. Some Chronometric Ramifications of the Conventionality of Congruence

(*i*) *Newtonian Mechanics.*

On the conception of time congruence as conventional, the preference for the customary definition of isochronism—a preference not felt by Einstein in the general theory of relativity, as we shall see in subsection (ii)—can derive only from considerations of convenience and elegance so long as the resulting form of the theory is *not* prescribed. Hence, the thesis that isochronism is conventional *precludes* a difference in factual import (content) or in *explanatory power* between two descriptions one of which employs the customary isochronism while the other is a "translation" (transcription) of it into a language employing a time congruence incompatible with the customary one.

As a test case for this thesis of *explanatory parity*, the following kind of counterargument has been suggested in outline. On the Riemannian analysis, congruence must be regarded as conventional in the time continuum of Newtonian dynamics no less than in the theory of relativity. We shall therefore wish to compare in regard to explanatory capability the two forms of Newtonian dynamics corresponding to two different time congruences as follows.

The first of these congruences is defined by the requirement that Newton's laws hold, as modified by the addition of very small corrective terms expressing the so-called relativistic motion of the perihelia. This time congruence will be called "Newtonian," and the time variable whose values represent Newtonian time after a particular unit has been chosen will be denoted by "t." The second time congruence is defined by the rotational motion of the earth. It does not matter for our purpose whether we couple the latter congruence with a unit given by the mean solar second, which is the 1/86400 part of the mean interval between two consecutive meridian passages of the fictitious mean sun, or with a different unit given by the sidereal day, which is the interval between successive meridian passages of a star. What matters is that both the mean solar second and the sidereal day are based on the periodicities of the earth's rotational motion. Assume now that one or another of these units has been chosen, and let T be the time variable associated with that metrization, which we shall call "diurnal time." The important point is that the time variables t and T are *non*linearly related and are associated with *incompatible* definitions of isochronism, because the speed of rotation of the earth *varies relatively to the Newtonian time metric* in several distinct ways [13, pp. 264–267]. Of these, the best known is the relative slowing down of the earth's rotation by the tidal friction between the water in the shallow seas of the earth and the land under it. Upon calculating the positions of the moon, for example, via the usual theory of celestial mechanics, which is based on the Newtonian time metric, the observed positions of the moon in the sky would be found to be *ahead* of the calculated ones *if* we were to identify the time defined by the earth's rotation with the Newtonian time of celestial mechanics. And the same is true of the positions of the planets of the solar system and of the moons of Jupiter in amounts all corresponding to a slowing down on the part of the earth.

Now consider the following argument for the lack of explanatory parity between the two forms of the dynamical theory respectively associated with the t and T scales of time: "*Dynamical* facts will discriminate in favor of the t scale as opposed to the T scale. It is granted that it is *kinematically* equivalent to say

> (a) the earth's rotational motion has slowed down relatively to the "clocks" constituted by various revolving planets and satellites of the solar system,

or

(b) the revolving celestial bodies speed up their periodic motions relatively to the earth's uniform rotation.

But these two statements are *not* on a par explanatorily in the context of the *dynamical* theory of the motions in the solar system. For whereas the slowing down of the earth's rotation in formulation (a) *can* be understood as the dynamical effect of nearby masses (the tidal waters and their friction), no similar dynamical *cause* can be supplied for the accelerations in formulation (b). And the latter fact shows that a theory incorporating formulation (a) has greater explanatory power or factual import than a theory containing (b)." In precisely this vein, D'Abro, though stressing on the one hand that apart from convenience and simplicity there is nothing to choose between different metrics [16, p. 53], on the other hand adduces the provision of *causal* understanding by the *t* scale as an argument in its favor and thus seems to construe such differences of simplicity as involving *nonequivalent* descriptions:

> If in mechanics and astronomy we had selected at random some arbitrary definition of time, if we had defined as congruent the intervals separating the rising and setting of the sun at all seasons of the year, say for the latitude of New York, our understanding of mechanical phenomena would have been beset with grave difficulties. As measured by these new temporal standards, free bodies would no longer move with constant speeds, but would be subjected to periodic accelerations *for which it would appear impossible to ascribe any definite cause*, and so on. As a result, the law of inertia would have to be abandoned, and with it the entire doctrine of classical mechanics, together with Newton's law. Thus a change in our understanding of congruence would entail far-reaching consequences.
>
> Again, in the case of the vibrating atom, had some arbitrary definition of time been accepted, we should have had to assume that the same atom presented the most capricious frequencies. *Once more it would have been difficult to ascribe satisfactory causes* to these seemingly haphazard fluctuations in frequency; and a simple understanding of the most fundamental optical phenomena would have been well-nigh impossible [16, p. 78, my italics].

To examine this argument, let us set the two formulations of dynamics corresponding to the t and T scales respectively before us mathematically in order to have a clearer statement of the issue.

The differences between the two kinds of temporal congruence with which we are concerned arise from the fact that the functional relationship $T = f(t)$ relating the two time scales is *nonlinear*, so that time intervals which are congruent on the one scale are generally incongruent on the other. It is clear that this function is monotone increasing, and thus we know that permanently

$$\frac{dT}{dt} \neq 0.$$

Moreover, in view of the nonlinearity of $T = f(t)$, we know that dT/dt is *not* constant. Since the function f has an inverse, it will be possible to translate any set of laws formulated on the basis of either of the two time scales into the corresponding other scale. In order to see what form the customary Newtonian force law assumes in diurnal time, we must express the acceleration ingredient in that law in terms of diurnal time. But in order to derive the transformation law for the accelerations, we first treat the velocities. By the chain rule for differentiation, we have, using '$\mathbf{r}$' to denote the position vector

$$(1) \qquad \frac{d\mathbf{r}}{dt} = \frac{d\mathbf{r}}{dT}\frac{dT}{dt}.$$

Suppose a body is *at rest* in the coordinate system in which $\mathbf{r}$ is measured, *when Newtonian time is employed*, then this body will also be held to be *at rest diurnally:* since we saw that the second term on the right-hand side of equation (1) cannot be zero, the left-hand side of (1) will vanish if and only if the first term on the right-hand side of (1) is zero. Though rest in a given frame in the t scale will correspond to rest in that frame in the T scale as well, equation (1) shows that the *constancy* of the nonvanishing Newtonian velocity $d\mathbf{r}/dt$ will *not* correspond to a constant diurnal velocity $d\mathbf{r}/dT$, since the derivative dT/dt changes with both Newtonian and diurnal time. Now, differentiation of equation (1) with respect to the Newtonian time t yields

$$(2) \qquad \frac{d^2\mathbf{r}}{dt^2} = \frac{d\mathbf{r}}{dT}\frac{d^2T}{dt^2} + \frac{dT}{dt}\frac{d}{dt}\left(\frac{d\mathbf{r}}{dT}\right).$$

But, applying the chain-rule to the second factor in the second term on the right-hand side of (2), we obtain

$$(2a) \qquad \frac{d}{dt}\left(\frac{d\mathbf{r}}{dT}\right) = \frac{d^2\mathbf{r}}{dT^2}\frac{dT}{dt}.$$

Hence (2) becomes

$$(3) \qquad \frac{d^2\mathbf{r}}{dt^2} = \frac{d\mathbf{r}}{dT}\frac{d^2T}{dt^2} + \frac{d^2\mathbf{r}}{dT^2}\left(\frac{dT}{dt}\right)^2.$$

Solving for the diurnal acceleration, and using equation (1) as well as the abbreviations

$$f'(t) \equiv \frac{dT}{dt} \text{ and } f''(t) \equiv \frac{d^2T}{dt^2},$$

we find

$$(4) \qquad \underbrace{\frac{d^2\mathbf{r}}{dT^2}}_{\text{diurnal acceleration}} = \frac{1}{[f'(t)]^2}\underbrace{\frac{d^2\mathbf{r}}{dt^2}}_{\text{Newtonian acceleration}} - \overbrace{\frac{f''(t)}{[f'(t)]^3}\underbrace{\frac{d\mathbf{r}}{dt}}_{\text{Newtonian velocity}}}^{\text{secular term}}$$

Several ancillary points should be noted briefly in regard to equation (4) before seeing what light it throws on the form assumed by causal explanation within the framework of a diurnal description When the Newtonian force on a body is *not* zero because the body is accelerating under the influence of masses, the diurnal acceleration will generally also not be zero, save in the unusual case when

$$(5) \qquad \frac{d^2\mathbf{r}}{dt^2} = \frac{f''(t)}{f'(t)}\frac{d\mathbf{r}}{dt}.$$

Thus the causal influence of masses, which gives rise to the Newtonian accelerations in the usual description, is seen in (4) to make a definite contribution to the diurnal acceleration as well. But the new feature of the diurnal description of the facts lies in the possession of a *secular acceleration* by all bodies not at rest, even when no masses are inducing Newtonian accelerations, so that the first term on the right-hand side of (4) vanishes. And this secular acceleration is numerically *not* the same for all bodies but depends on their velocities $d\mathbf{r}/dt$ in the given reference frame and thus also on the reference frame.

The character and existence of this secular acceleration calls for several kinds of comment.

Its dependence on the velocity and on the reference frame should neither occasion surprise nor be regarded as a difficulty of any sort. As to the velocity dependence of the secular acceleration, consider a simple numerical example which removes any surprise: if instead of calling two successive hours on Big Ben equal, we remetrized time so as to assign the measure ½ hour to the second of these intervals, then all bodies having uniform speeds in the usual time metric will double their speeds on the new scale after the first interval, and the *numerical* increase or *acceleration* in the speeds of initially faster bodies will be greater than that in the speeds of the initially slower bodies. Now as for the dependence of the secular acceleration on the reference frame, in the context of the physical facts asserted by the Newtonian theory *apart from* its metrical philosophy it is a mere prejudice to require that, to be admissible, a formulation of that theory must agree with the customary one in making the acceleration of a body at any given time be the same in all Galilean reference frames ("Galilean relativity"). For not a single bona fide physical fact of the Newtonian world is overlooked or contradicted by a kinematics not featuring this Galilean relativity. It is instructive to be aware in this connection that even in the *customary* rendition of the *kinematics of special relativity*, a constant acceleration in a frame S' would *not* generally correspond to a constant acceleration in a frame S, because the component accelerations in S depend not only on the accelerations in S' but also on the component velocities in that system which would be changing with the time.

But what are we to say, apart from the dependence on the velocity and reference system, about the very presence of this "dynamically unexplained" or causally baffling *secular acceleration?* To deal with this question, we first observe merely for comparison that in the *customary* formulation of Newtonian mechanics, constant *speeds* (as distinct from constant *velocities*) fall into *two* classes with respect to being attributable to the dynamical action of perturbing masses: constant *rectilinear* speeds are affirmed to prevail in the absence of any mass influences, while constant *curvilinear* (e.g., circular) speeds are related to the (centripetally) accelerating actions of masses. Now in regard to the presence of a secular acceleration in the diurnal description, it is fundamental to see the following: Whereas on the version of Newtonian mechanics employing the *customary* metrizations (of time and space) *all* accelerations whatsoever in Galilean frames are of *dynamical* origin by being attributable to the action of specific masses, *this feature of Newton's theory is made possible not only by the facts but also by the particular time metrization chosen to codify them.* As equation (4) shows upon equating $d^2\mathbf{r}/dt^2$ to zero, the dynamical character of *all* accelerations is not vouchsafed by any causal facts of the world with which every theory would have to come to terms. For the diurnal description encompasses the objective behavior of bodies (point events and coincidences) as a function of the presence or absence of other bodies no less than does the Newtonian one, thereby achieving full explanatory parity with the latter in all logical (as distinct from pragmatic!) respects.

Hence the provision of a dynamical basis for *all* accelerations should *not* be regarded as an *inflexible epistemological requirement* in the elaboration of a theory explaining mechanical phenomena. *Disregarding* the pragmatically decisive consideration of convenience, there can therefore be no valid explanatory objection to the diurnal description, in which accelerations fall into *two* classes by being the superpositions, in the sense of equation (4), of a dynamically grounded *and* a *kinematically* grounded term. And, most important, since there is no *slowing down* of the earth's rotation on the diurnal metric, there can be no question in that description of specifying a *cause* for such a nonexistent deceleration; instead, a frictional *cause* is now specified for the earth's diurnally *uniform* rotation *and* for the liberation of heat accompanying this kind of uniform motion. For in the T-scale description it is *uniform* rotation which requires a dynamical cause constituted by masses interacting (frictionally) with the uniformly rotating body, and it is now a law of nature or entailed by such a law that all diurnally uniform rotations issue in the dissipation of heat. Of course, the mathematical representation of the frictional interaction will not have the customary Newtonian form: to obtain the diurnal account of the frictional dynamics of the tides, one would need to apply transformations of the kind given in our equation (4) to the quantities appearing in the relevant Newtonian equations for this case, which can be found in [43, Chapter 8] and in [88].

But, it will be asked, what of the Newtonian conservation principles, if the T scale of time is adopted? It is readily demonstrable by reference to the simple case of the motion of a free particle that while the Newtonian kinetic energy will be constant in this case, its formal diurnal homologue (as opposed to its diurnal equivalent!) will *not* be constant. Let us denote the constant *Newtonian* velocity of the free particle by "v_t," the subscript "t" serving to represent the use of the t scale, and let "v_T" denote the diurnal velocity corresponding to v_t. Since we know from equation (1) above that $v_t = v_T\, dT/dt$, where v_t is constant but dT/dt is *not*, we see that the diurnal *homologue* $\frac{1}{2}mv_T^2$ of the Newtonian kinetic energy cannot be constant in this case, although the diurnal equivalent $\frac{1}{2}m(v_T\, dT/dt)^2$ of the constant Newtonian kinetic energy $\frac{1}{2}mv_t^2$ is necessarily constant. Just as in the case of the Newtonian equations of motion themselves, so also in the case of the Newtonian conservation principle of mechanical energy, the diurnal equivalent or transcription explains all the facts explained by the Newtonian original. Hence our critic can derive no support at all from the fact that the formal diurnal homologues of Newtonian conservation principles generally do not hold. And we see, incidentally, that the time invariance of a physical quantity and hence the appropriateness of singling it out from among others as a form of "energy," etc., will depend not only on the facts but also on the time metrization used to render them. It obviously will not do, therefore, to charge the diurnal description with inconsistency via the *petitio* of *grafting* onto it the requirement that it incorporate the homologues of Newtonian conservation principles which are incompatible with it: a case in point is the charge that the diurnal description violates the conservation of energy because in its metric the frictional generation of heat in the tidal case is *not* compensated by any reduction in the speed of the earth's rotation! Whether the diurnal time metrization permits the deduction of conservation principles of a *relatively simple type* involving diurnally based quantities is a rather involved mathematical question whose solution is *not* required to establish our thesis that, apart from pragmatic considerations, the diurnal description enjoys explanatory parity with the Newtonian one.

We have been disregarding pragmatic considerations in assessing the explanatory capabilities of two descriptions associated with different time metrizations as to parity. But it would be an error to infer that in pointing to the equivalence of such descriptions in regard to factual content, we are committed to the view that there is no criterion for choosing between them and hence no reason for preferring any one of them to the others. Factual adequacy (truth) is, of course, the cardinal *necessary* condition for the acceptance of a scientific theory, but it is hardly a *sufficient* condition for accepting any one particular formulation of it which satisfies this necessary condition. As well say that a person pointing out that equivalent descriptions can be given in the decimal (metric) and English system of units cannot give telling reasons for preferring the former! Indeed, after first commenting on the *factual basis* of the *existence* of the Newtonian time congruence, we shall see that there are weighty *pragmatic* reasons for preferring that metrization of the time continuum. And these reasons will turn out to be entirely consonant with our twin contention that alternative metrizability allows linguistically different, equivalent descriptions *and* that geochronometric conventionalism is *not* a subthesis of trivial semantical conventionalism.

The *factual basis* of the Newtonian time metrization will be appreciated by reference to the following two considerations: (i) As we shall prove presently, it is a highly fortunate empirical fact, and *not* an a priori truth, that there *exists* a time metrization *at all* in which *all* accelerations with respect to inertial systems are of dynamic origin, as claimed by the Newtonian theory, and (ii) it is a further empirical fact that the time metrization having this remarkable property (i.e., "ephemeris time") is furnished physically by the earth's annual revolution around the sun (*not* by its diurnal rotation) albeit *not* in any observationally simple way, since due account must be taken computationally of the irregularities produced by the gravitational influences of the other planets.[28] That the existence of a time metrization in which *all* accelerations with respect to inertial systems are of *dynamical* origin cannot be guaranteed a priori is demonstrable as follows.

Suppose that, *contrary to actual fact*, it *were* the case that a *free* body did accelerate when its motion is described in the metric of ephemeris time t, it thus being assumed that there are accelerations in the customary time metric which are *not* dynamical in origin. More particularly, let us now posit that, contrary to actual fact, a *free* particle were to execute *one-dimensional* simple harmonic motion of the form $r = \cos \omega t$, where r is the distance from the origin. In that hypothetical eventuality, the acceleration of a *free* particle in the t scale would have the time-dependent value

$$\frac{d^2r}{dt^2} = -\omega^2 \cos \omega t.$$

And our problem is to determine whether there would then exist some other time metrization $T = f(t)$ possessing the Newtonian property that our *free* particle has a *zero* acceleration. We shall now find that the answer is definitely negative: under the hypothetical

[28] For details on the so-called ephemeris time given by this metric, see [10, 11, 12, 13].

empirical conditions which we have posited, there would indeed be no admissible, single-valued time metrization T at all in which all accelerations with respect to inertial systems would be of dynamical origin.

For let us now regard "T" in equation (4) of this subsection as the time variable associated with the sought-after metrization $T = f(t)$ in which the acceleration d^2r/dT^2 of our free particle would be *zero* We recall that equation (5) of this subsection was obtained from equation (4) by equating the T-scale acceleration dr^2/dT^2 to zero. Hence *if* our sought-after metrization exists at all, it would have to be the solution $T = f(t)$ of the scalar form of equation (5) as applied to our one-dimensional motion. That equation is

$$(6) \qquad \frac{d^2r}{dt^2} = \frac{f''(t)}{f'(t)} \frac{dr}{dt}.$$

Putting $v \equiv dr/dt$ and noting that

$$\frac{d}{dt} \log f'(t) = \frac{f''(t)}{f'(t)}, \text{ and } \frac{d}{dt} \log v = \frac{1}{v} \frac{dv}{dt},$$

equation (6) becomes

$$\frac{d}{dt} \log v = \frac{d}{dt} \log f'(t).$$

Integrating, and using $\log c$ as the constant of integration, we obtain

$$\log v = \log cf'(t),$$

or

$$v = cf'(t),$$

which is

$$\frac{dr}{dt} = c \frac{dT}{dt}.$$

Integration yields

$$(7) \qquad r = cT + d,$$

where d is a constant of integration. But, by our earlier hypothesis, $r = \cos \omega t$. Hence (7) becomes

$$(8) \qquad T = \frac{1}{c} \cos \omega t - \frac{d}{c}.$$

It is evident that the solution $T = f(t)$ given by equation (8) is *not* a *one-to-one* function: the *same* time T in the sought-after metrization would correspond to all those *different* times on the t scale at which the oscillating particle would return to the *same place* $r = \cos \omega t$ in the course of its periodic motion. And by thus violating the basic topological requirement that the function $T = f(t)$ be one-to-one, the T scale which does have the sought-after Newtonian property under our *hypothetical* empirical conditions is physically a quite inadmissible and hence unavailable metrization.

It follows that there is no a priori assurance of the existence of at least one time metrization possessing the Newtonian property that the acceleration of a free particle in inertial systems is zero. So much for the *factual basis* of the Newtonian time metrization.

Now, inasmuch as the employment of the time metrization based on the earth's annual revolution issues in Newton's relatively *simple* laws, there are powerful reasons of *mathematical tractability* and *convenience* for greatly preferring the time metrization in which *all* accelerations are of dynamical origin. In fact, the various refinements which astronomers have introduced in their physical standards for temporal congruence have been dictated by the demand for a definition of temporal congruence (or of a so-called invariable time standard) for which Newton's laws will hold in the solar system, including the relatively simple conservation laws interconnecting diverse kinds of phenomena (mechanical, thermal, etc.). And thus, as Feigl and Maxwell have aptly put it, one of the important criteria of descriptive simplicity which greatly restrict the range of "reasonable" conventions is seen to be the *scope* which a convention will allow for mathematically tractable laws.

(*ii*) *The General Theory of Relativity.**

In the special theory of relativity, only the customary time metri-

* The particular example of time congruence dealt with in subsection (ii) is discussed in greater detail in §§2.3, 2.4, and 8.2 of Chapter III. Other aspects of the status of congruence in the general theory of relativity are treated in the remainder of §2 of Chapter III.

zation is employed in the following sense: At any given point A in a Galilean frame, the length of a time interval between two events at the point A is given by the *difference* between the time coordinates of the two events as furnished by the readings of a standard clock at A whose periods are defined to be congruent. This is, of course, the precise analogue of the customary definition of spatial congruence which calls the rod congruent to itself everywhere, when at relative rest, after allowance for substance-specific perturbations. On the other hand, as we shall now see, there are contexts in which the *general* theory of relativity (GTR) utilizes a criterion of *temporal* congruence which is an analogue of a *non*customary kind of spatial congruence in the following sense: the length of a time interval separating two events at a clock depends not only on the *difference* between the time coordinates which the clock assigns to these events but also on the *spatial* location of the clock (though not on the time itself at which the interval begins or ends).

A case in point from the GTR involves a *rotating disk* to which we apply those principles which GTR takes over from the special theory of relativity. Let a set of standard material clocks be distributed at various points on such a disk. The infinitesimal application of the special relativity clock retardation then tells us the following: a clock at the *center* O of the disk will maintain the rate of a contiguous clock located in an inertial system I with respect to which the disk has angular velocity ω, *but* the same does *not* hold for clocks located at other points A of the disk which are at positive distances r from O. Such A clocks have various linear velocities ωr relatively to I in virtue of their common angular velocity ω. Accordingly, all A clocks (whatever their chemical constitution) will have readings lagging behind the corresponding readings of the respective I-system clocks adjacent to them by a *factor* of

$$\sqrt{1 - \frac{r^2\omega^2}{c^2}},$$

where c is the velocity of light. What would be the consequences o using the *customary* time metrization everywhere on the rotating disk and letting the duration (length) of a time interval elapsing at a given point A be given by the *difference* between the time coordinates of the termini of that interval as furnished by the readings of the standard clock at A? The adoption of the customary time metric would saddle us with a *most complicated* description of the propagation of light in the rotating system having the following undesirable features: (i) time would enter the description of nature explicitly in the sense that the one-way velocity of light would depend on the time, since the lagging rate of the clock at A issues in a temporal change in the magnitude of the one-way transit time of a light ray for journeys between O and A, and (ii) the number of light waves emitted at A during a unit of time on the A clock is *greater* than the number of waves arriving at the center O in one unit of time on the O clock [53, pp. 225–226]. To avoid the undesirably complicated laws entailed by the use of the simple customary definition of time congruence, the GTR jettisoned the latter. In its stead, it adopted the following more complicated, noncustomary congruence definition for the sake of the simplicity of the resulting laws: at any point A on the disk the length (duration) of a time interval is given *not* by the *difference* between the A clock coordinates of its termini but by the *product* of this increment *and* the rate factor

$$\frac{1}{\sqrt{1-\frac{r^2\omega^2}{c^2}}}$$

which depends on the spatial coordinate r of the point A. This rate factor serves to assign a *greater* duration to time intervals than would be obtained from the customary procedure of letting the length of time be given by the increment in the clock readings. In view of the *dependence* of the metric *on the spatial position* r, via the rate factor entering into it, we are confronted here with a *non*customary time metrization fully as consonant with the temporal *order* of the events at A as is the customary metric.

A similarly nonstandard time metric is used by Einstein in his GTR paper of 1911 [24, Section 3] in treating the effect of gravitation on the propagation of light. Analysis shows that the very same complexities in the description of light propagation which are encountered on the rotating disk arise here as well, if the standard time metric is used. These complexities are eliminated here in quite analogous fashion by the use of a noncustomary time metric. Thus, if we are concerned with light emitted on the sun and reaching the earth and if "$-\Phi$" represents the negative difference in gravitational potential between the sun and the earth, then we proceed as follows: prior to being brought from the earth to the sun, a clock is set to have a rate *faster* than that of an adjoining terrestrial clock by a factor of

$$\frac{1}{1 - \frac{\Phi}{c^2}}$$

(to a first approximation), where

$$\frac{\Phi}{c^2} < 1.$$

(*iii*) *The Cosmology of E. A. Milne.*

E. A. Milne, whose two logarithmically related t and τ scales of time were mentioned in §2(i), has attempted to erect the usual space-time structure of special relativity on the basis of a light signal kinematics of particle observers purportedly dispensing with the use of rigid solids and isochronous material clocks [51, 52]. In his *Modern Cosmology and the Christian Idea of God* [52, Chapter III], Milne begins his discussion of time and space by *incorrectly* charging Einstein with failure to realize that the concept of a rigid body as a body whose rest length is invariant under transport contains a conventional ingredient just as much as does the concept of metrical simultaneity at a distance.[29] Milne then proposes to improve upon a

[29] That Einstein was abundantly aware of this point is evident from his definition of the "practically rigid body" in [22, p. 9].

rigid body criterion of spatial congruence by proceeding in the manner of radar ranging and using instead the round-trip times required by light to traverse the corresponding closed paths, these times *not* being measured by material clocks but, in outline, as follows.[30] Each particle is equipped with a device for ordering the genidentical events belonging to it temporally in a linear Cantorean continuum. Such a device is called a "clock," and the single observer at the particle using such a local clock is called a "particle observer." If now A and B are two particle observers and light signals are sent from one to the other, then the time of arrival $\bar{t}'$ at B can be expressed as a function $\bar{t}' = f(t)$ of the time t of emission at A, and likewise the time of arrival t' at A is a function $t' = F(\bar{t})$ of the time $\bar{t}$ of emission at B. Particle observers equipped with clocks as defined are said to be "*equivalent*," if the so-called signal functions f and F are the same, and the clocks of equivalent particle observers are said to be *congruent*. It can be shown that if A and B are not equivalent, then B's clock can be regraduated by a transformation of the form $\bar{t}' = \psi(\bar{t})$ so as to render them equivalent [52, pp. 39–41]. The congruence of the clocks at A and B does not, of course, assure their synchronism. Milne now uses Einstein's definition of simultaneity [52, p. 42]: the time t_2 assigned by A to the arrival of a light ray at B which is emitted at time t_1 at A and returns to A at time t_3 after instantaneous reflection at B is defined to be $t_2 = \frac{1}{2}(t_1 + t_3)$. And he defines the distance r_2 of B, *by A's clock, upon the arrival of the light from A at B* to be given by the relation $r_2 = \frac{1}{2}c(t_3 - t_1)$, where c is an *arbitrarily* chosen constant [52, p. 42]. Since

$$\frac{r_2}{t_2 - t_1} = \frac{r_2}{t_3 - t_2} = c,$$

the constant c represents the velocity of the light signal in terms of the conventions adopted by A for measuring distance and time at a remote point B. Milne gives the following statement of his epistemological objections to Einstein's use of rigid rods and of his claim

[30] For more detailed summaries of Milne's light-signal kinematics see [92, pp. 309–310; 59, pp. 78–85].

that his light-signal kinematics provides a philosophically satisfactory alternative to it:

> . . . the concept of the transport of a rigid body or rigid length measure is itself an indefinable concept. In terms of one given standard metre, we cannot say what we mean by asking that a given 'rigid' length measure shall remain 'unaltered in length' when we move it from one place to another; for we have no standard of length at the new place. Again, we should have to specify standards of 'rest' everywhere, for it is not clear without consideration that the 'length' will be the same, even at the same place, for different velocities. The fact is that to say of a body or measuring-rod that it is 'rigid' is no definition whatever; it specifies no 'operational' procedure for testing whether a given length-measure after transport or after change of velocity is the same as it was before [52, p. 35]. . . .
>
> It is part of the debt we owe to Einstein to recognize that only 'operational' definitions are of any significance in science . . . Einstein carried out his own procedure completely when he analysed the previously undefined concept of simultaneity, replacing it by tests using the measurements which have actually to be employed to recognize whether two distant events are or are not simultaneous. But he abandoned his own procedure when he retained the indefinable concept of the length of a 'rigid' body, i.e. a length unaltered under transport. The two indefinable concepts of the transportable rigid body and of simultaneity are on exactly the same footing; they are fog-centres, inhibiting further vision, until analysed and shown to be equivalent to conventions [52, p. 35]. . . .
>
> It will be one of our major tasks to elucidate the type of graduation employed for graduating our ordinary clocks; that is to say, to inquire what is meant by, and if possible to isolate what is usually understood by, 'uniform time'. In other words, we wish to inquire which of the arbitrarily many ways in which the markings of our abstract clock may be graduated can be identified with the 'uniform time' of physics [52, p. 37]. . . .
>
> The question now arises: is it possible to arrange that the mode of graduation of observer B's clock corresponds to the mode of graduation of A's clock in such a way that a meaning can be attached to saying that B's clock is a copy of A's clock? If so, we shall say that B's clock has been made congruent with A's [52, p. 39]. . . .
>
> It will have been noticed that we have succeeded in making B's

clock a copy of A's without bringing B into permanent coincidence with A. We have made a copy of an arbitrary clock *at a distance*. This is something we cannot do with metre-scales or other length-measures. The problem of copying a clock is in principle simpler than the problem of copying a unit of length. We shall see in due course that with the construction of a copy of a clock at a distance we have solved the problem of comparing lengths [52, p. 41]. . . .

The important point is that epoch and distance (which we shall call coordinates) are purely conventional constructs, and have meaning only in relation to a particular form of clock graduation . . . But it is to be pointed out that when the mode of clock graduation reduces to that of ordinary clocks in physical laboratories, our coordinate conventions provide measures of epoch and distance which coincide with those based on the standard metre [52, pp. 42–43]. . . .

The reason why it is more fundamental to use clocks alone rather than both clocks and scales or than scales alone is that the concept of the clock is more elementary than the concept of the scale. The concept of the clock is connected with the concept of 'two times at the same place', whilst the concept of the scale is connected with the concept of 'two places at the same time'. But the concept of 'two places at the same time' involves a convention of simultaneity, namely, simultaneous events at the two places, but the concept of 'two times at the same place' involves no convention; it involves only the existence of an ego [52, p. 46]. . . .

Length is just as much a conventional matter as an epoch at a distance. Thus the metre-scale is not such a fundamental instrument as the clock. In the first place its length for *any* observer, as measured by the radar method, depends on the clock used by the observer; in the second place, different observers assign different lengths to it even if their clocks are congruent, owing to the fact that the test of simultaneity is a conventional one. The clock, on the other hand, once graduated, gives epochs at itself which are independent of convention.

Once we have set up a clock, arbitrarily graduated, distances for the observer using this clock become definite. If a rod, moved from one position of rest relative to this observer to another position of rest relative to the same observer, possesses in the two positions the same length, as measured by this observer using his own clock, as graduated, then the rod is said to have undergone a rigid-body-displacement by this clock. In this way we see that once we have fixed on a clock, a rigid-body-displacement becomes definable. But until we have provided a clock, there is no way of saying what we mean by a rigid body under displacement [52, pp. 47–48].

Now, if Milne is to make good his criticism of Einstein by erecting the space-time structure of special relativity on alternative epistemological foundations, he must provide us with inertial systems by means of the resources of his light-signal kinematics as well as with the measures of length and time on which the kinematics of special relativity is predicated. This means that he must be able to characterize inertial systems *within* the confines of his epistemological program as some kind of dense assemblage of equivalent particle observers filling space such that each particle observer is at rest relative to and synchronous with every other. We have already seen that he was wholly in error in charging Einstein with lack of awareness of the conventionality of spatial congruence as defined by the rigid rod. But that, much more fundamentally, he is mistaken in believing to have erected the kinematics of special relativity on an epistemologically more satisfactory base than Einstein did will now be made clear by reference to the following result pointed out by L. L. Whyte [102]: Using only light signals and temporal succession without either a solid rigid rod or an isochronous material clock, it is *not* possible to construct ordinary measures of length and time. For "a physicist using only light signals cannot discriminate inertial systems from these subjected to arbitrary 4-*D* similarity transformations.[31] The system of 'resting' mass points which can be so identified may be arbitrarily expanding and/or contracting relatively to a rod, and these superfluous transformations can only be eliminated by using a rod or a clock" [102, p. 161].

The significance of the result stated by Whyte is twofold: (i) If Milne dispenses with material clocks and bases his chronometry only on the congruences yielded by his light-signal clocks, then he cannot obtain inertial systems without a rigid rod in the following

[31] For a brief account of similarity transformations, and a further articulation of Whyte's point here, cf. [72, pp. 172–173].

sense. The rigid rod is *not* needed for the definition of *spatial congruence* within the system but *is* required to assure that the distance between one particular pair of points connected by it *at one time* t_0 is the same as *at some later time* t_1. In other words, the rod is rigid *at a given place* by remaining congruent to itself (by convention) as time goes on. And in this way the rod assures the time constancy of the distance between the two given points connected by it. This reliance on the rigid rod thus involves the use of the definition of simultaneity. Hence, if Milne were right in charging that the use of a rigid rod is beset by philosophic difficulties, then he indeed would be incurring these liabilities no less than Einstein does. On the other hand, suppose that (ii) Milne does use a material clock to define the time metric at a space point and thereby to particularize his clock graduations to the kind required for the elimination of the unwanted reference systems described by L. L. Whyte. This procedure is a far cry from his purely topological clock which "involves only the existence of an *ego*" [52, p. 46] in contradistinction to the rigid scale's involvement of a definition of simultaneity. And, in that case, his measurement of the equality of space intervals by means of the equality of the corresponding round-trip time intervals involves the following conventions: (a) the tacit use of a definition of simultaneity of noncoinciding events. For although a *round*-trip time on a given clock does *not*, of course, itself require such a simultaneity criterion, the measurement of a spatial distance in an inertial system by means of this time does: the distance yielded by the round-trip time on a clock at A is the distance r_2 between A and B *at the time* t_2 on the A clock *when the light pulse from* A *arrives at* B on its round-trip ABA, (b) successive equal differences in the readings of a given local clock are stipulated to be measures of equal time intervals and thereby of equal space intervals, and (c) equal differences on separated clocks of identical constitution are decreed to be measures of equal time intervals and thereby of equal space intervals.

To what extent then, if any, does Milne have a case against Einstein? It would appear from our analysis that the only justifiable criticism is not at all epistemological but concerns an innocuous point of axiomatic economy: once you grant Milne a material clock, he does not require the rigid rod at all, whereas Einstein utilizes the spatial congruence definition based on the rigid rod *in addition* to all of the conventions needed by Milne. Thus, Milne's kinematics, as supplemented by the use of a material clock, is constructed on a slightly narrower base of conventions than is Einstein's.[32]

It will be recalled that if measurements of spatial and temporal extension are to be made by means of solid rods and material clocks, allowance must be made computationally for thermal and other perturbations of these bodies so that they can define rigidity and isochronism. Calling attention to this fact and believing Milne's light-signal kinematics to be essentially successful, L. Page [59, pp. 78–79] deemed Milne's construction more adequate than Einstein's, writing: "the original formulation of the relativity theory was based on undefined concepts of space and time intervals which could not be identified unambiguously with actual observations. Recently Milne has shown how to supply the desired criterion [of rigidity and isochronism] by erecting the space-time structure on the foundations of a constant light-signal velocity." It is apparent in the light of our appraisal of Milne's kinematics that Page's claim is vitiated by Milne's need for a rigid rod or material clock as specified.

It should be noted, however, as Professor A. G. Walker has pointed out to me, that if Milne's construction is interpreted as applying *not* to special relativity kinematics but to his cosmological world model, then our criticisms are no longer pertinent. In terms of his logarithmically related τ and t scales of time, it turns out that upon measuring distances by the specified chronometric convention, the galaxies are at relative rest in τ-scale kinematics and in uniform relative motion in the t scale. Each of these time scales is unique up to a trivial change of units, and their associated descriptions of

[32] The preceding critique of Milne supplants my earlier brief critique [36, pp. 531–533] in which Milne's arguments were *misinterpreted* as indicative of lack of appreciation on his part of the conventionality of *temporal* congruence.

the cosmological world are equivalent in Reichenbach's sense. In this *cosmological* context, the problem of eliminating the superfluous transformations mentioned by Whyte therefore does not arise.

5. Critique of Some Major Objections to the Conventionality of Spatiotemporal Congruence

(*i*) *The Russell-Poincaré Controversy.*

During the years 1897–1900, B. Russell and H. Poincaré had a controversy which was initiated by Poincaré's review [62] of Russell's *Foundations of Geometry* of 1897, and pursued in Russell's reply [81] and Poincaré's rejoinder [65]. Russell criticized Poincaré's conventionalist conception of congruence and invoked the existence of an intrinsic metric as follows:[33]

> It seems to be believed that since measurement [i.e., comparison by means of the congruence standard] is necessary to *discover* equality or inequality, these cannot exist without measurement. Now the proper conclusion is exactly the opposite. Whatever one can discover by means of an operation must exist independently of that operation: America existed before Christopher Columbus, and two quantities of the same kind must *be* equal *or* unequal before being measured. Any method of measurement [i.e., any congruence definition] is good or bad according as it yields a result which is true or false. Mr. Poincaré, on the other hand, holds that measurement creates equality and inequality [i.e., that there is no intrinsic metric]. It follows [then] . . . that there is nothing left to measure and that equality and inequality are terms devoid of meaning [81, pp. 687–688].

We have argued that the Newtonian position espoused by Russell is untenable. But our critique of the model-theoretic trivialization of the conventionality of congruence shows that we must reject as inadequate the following kind of criticism of Russell's position, which he would have regarded as a *petitio principii:* "Russell's claim is an absurdity, because it is the denial of the truism that we are at liberty to give whatever physical interpretations we like to such

[33] This argument is implicitly endorsed by Helmholtz [91, p. 15].

abstract signs as 'congruent line segments' and 'straight line' and then to inquire whether the system of objects and relations thus arbitrarily named is a model of one or another abstract geometric axiom system. Hence, these linguistic considerations suffice to show that there can be no question, as Russell would have it, whether two non-coinciding segments are truly equal or not and whether measurement is being carried out with a standard yielding results that are true in that sense. Accordingly, awareness of the model-theoretic conception of geometry would have shown Russell that alternative metrizability of spatial and temporal continua should never have been either startling or a matter for dispute. And, by the same token, Poincaré could have spared himself a polemic against Russell in which he spoke misleadingly of the conventionality of congruence as a philosophical doctrine pertaining to the structure of space."

Since this model-theoretic argument fails to come to grips with Russell's root assumption of an intrinsic metric, he would have dismissed it as a *petitio* by raising exactly the same objections that the Newtonian would adduce (cf. §2(i)) against the alternative metrizability of space and time. And Russell might have gone on to point out that the model theoretician cannot evade the spatial equality issue by (i) noting that there are axiomatizations of each of the various geometries dispensing with the abstract relation term "congruent" (for line segments), and (ii) claiming that there can then be no problem as to what physical interpretations of that relation term are permissible. For a metric geometry makes metrical comparisons of equality and inequality, however covertly or circuitously these may be rendered by its language. It is quite immaterial, therefore, whether the relation of spatial equality between line segments is designated by the term "congruent" or by some other term or terms. Thus, for example, Tarski's axioms for elementary Euclidean geometry [87] do not employ the term "congruent" for this purpose, using instead a quaternary predicate denoting the equidistance relation between four points. Also, in Sophus Lie's group-theoretical treatment of metric geometries, the congruences are specified by groups of point transformations [3, pp. 153–154]. But just as Russell invoked his conception of an intrinsic metric to restrict the permissible spatial interpretations of "congruent line segments," so also he would have maintained that it is never arbitrary what quartets of physical points may be regarded as the denotata of Tarski's quaternary equidistance predicate. And he would have imposed corresponding restrictions on Lie's transformations, since the displacements defined by these groups of transformations have the logical character of spatial congruences. These considerations show that it will *not* suffice in this context simply to take the model-theoretic conception of geometry for granted and thereby to dismiss the Russell-Helmholtz claim peremptorily in favor of alternative metrizability. Rather what is needed is a *refutation* of the Russell-Helmholtz *root assumption* of an *intrinsic metric:* to exhibit the untenability of that assumption as we have endeavored to do in §2 is to provide the *justification* of the model-theoretic affirmation that a given set of physico-spatial facts may be held to be as much a realization of a Euclidean calculus as of a non-Euclidean one yielding the same topology.

The refutation presented in §2 requires supplementation, however, to invalidate A. N. Whitehead's *perceptualistic* version of Russell's argument. We therefore now turn to an examination of Whitehead's philosophy of congruence.

(ii) A. N. Whitehead's Unsuccessful Attempt to Ground an Intrinsic Metric of Physical Space and Time on the Deliverances of Sense.

Commenting on the Russell-Poincaré controversy [97, pp. 121–124], Whitehead maintains the following: (i) Poincaré's argument on behalf of alternative metrizability is unanswerable only if the philosophy of physical geometry and chronometry is part of an epistemological framework resting on an illegitimate bifurcation of nature, (ii) consonant with the rejection of bifurcation, we must ground our metric account of the space and time of nature not on the relations between material bodies and events as fundamental entities but on the more ultimate metric deliverances of sense perception, and (iii) perceptual time and space exhibit an intrinsic metric. Specifically, Whitehead proposes to point out "the factor in nature which issues in the preeminence of one congruence relation over the indefinite herd of other such relations" [97, p. 124] and writes:

> The reason for this result is that nature is no longer confined within space at an instant. Space and time are now interconnected; and this peculiar factor of time which is so immediately distinguished among the deliverances of our sense-awareness, relates itself to one particular congruence relation in space [97, p. 124]. . . . Congruence depends on motion, and thereby is generated the connexion between spatial congruence and temporal congruence [97, p. 126].

Whitehead's argument is thus seen to turn on his ability to show that *temporal* congruence cannot be regarded as conventional in physics. He believes to have justified this crucial claim by the following reasoning in which he refers to the conventionalist conception as "the prevalent view" and to his opposing thesis as "the new theory":

> The new theory provides a definition of the congruence of periods of time. The prevalent view provides no such definition. Its position is that if we take such time-measurements so that certain familiar velocities which seem to us to be uniform are uniform, then the laws of motion are true. Now in the first place no change could appear either as uniform or nonuniform without involving a definite determination of the congruence for time-periods. So in appealing to familiar phenomena it allows that there is some factor in nature which we can intellectually construct as a congruence theory. It does not however say anything about it except that the laws of motion are then true. Suppose that with some expositors we cut out the reference to familiar velocities such as the rate of rotation of the earth. We are then driven to admit that there is no meaning in temporal congruence except that certain assumptions make the laws of motion true. Such a statement is historically false. King Alfred the Great was ignorant of the laws of motion, but knew very well what he meant by the measurement of time, and achieved his purpose by means of burning candles. Also no one in past ages justified the use of sand in hour glasses by saying that some centuries later interesting

laws of motion would be discovered which would give a meaning to the statement that the sand was emptied from the bulbs in equal times. Uniformity in change is directly perceived, and it follows that mankind perceives in nature factors from which a theory of temporal congruence can be formed. The prevalent theory entirely fails to produce such factors [97, p. 137]. . . . On the orthodox theory the position of the equations of motion is most ambiguous. The space to which they refer is completely undetermined and so is the measurement of the lapse of time. Science is simply setting out on a fishing expedition to see whether it cannot find some procedure which it can call the measurement of space and some procedure which it can call the measurement of time, and something which it can call a system of forces, and something which it can call masses, so that these formulae may be satisfied. The only reason—on this theory—why anyone should want to satisfy these formulae is a sentimental regard for Galileo, Newton, Euler and Lagrange. The theory, so far from founding science on a sound observational basis, forces everything to conform to a mere mathematical preference for certain simple formulae.

I do not for a moment believe that this is a true account of the real status of the Laws of Motion. These equations want some slight adjustment for the new formulae of relativity. But with these adjustments, imperceptible in ordinary use, the laws deal with fundamental physical quantities which we know very well and wish to correlate.

The measurement of time was known to all civilised nations long before the laws were thought of. It is this time as thus measured that the laws are concerned with. Also they deal with the space of our daily life. When we approach to an accuracy of measurement beyond that of observation, adjustment is allowable. But within the limits of observation we know what we mean when we speak of measurements of space and measurements of time and uniformity of change. It is for science to give an intellectual account of what is so evident in sense-awareness. It is to me thoroughly incredible that the ultimate fact beyond which there is no deeper explanation is that mankind has really been swayed by an unconscious desire to satisfy the mathematical formulae which we call the Laws of Motion, formulae completely unknown till the seventeenth century of our epoch [97, pp. 139–140].

After commenting that purely mathematically, an infinitude of ncompatible spatial congruence classes of intervals satisfy the congruence axioms, Whitehead says:

This breakdown of the uniqueness of congruence for space . . . is to be contrasted with the fact that mankind does in truth agree on a congruence system for space and a congruence system for time which are founded on the direct evidence of its senses. We ask, why this pathetic trust in the yard measure and the clock? The truth is that we have observed something which the classical theory does not explain.

It is important to understand exactly where the difficulty lies. It is often wrongly conceived as depending on the inexactness of all measurements in regard to very small quantities. According to our methods of observation we may be correct to a hundredth, or a thousandth, or a millionth of an inch. But there is always a margin left over within which we cannot measure. However, this character of inexactness is *not* the difficulty in question.

Let us suppose that our measurements can be ideally exact; it will be still the case that if one man uses one qualifying [i.e., congruence] class γ and the other man uses another qualifying [i.e., congruence] class δ, and if they both admit the standard yard kept in the exchequer chambers to be their unit of measurement, they will disagree as to what other distances [at other] places should be judged to be equal to that standard distance in the exchequer chambers. Nor need their disagreement be of a negligible character [99, pp. 49–50]. . . .

When we say that two stretches match in respect to length, what do we mean? Furthermore we have got to include time. When two lapses of time match in respect to duration, what do we mean?

We have seen that measurement presupposes matching, so it is of no use to hope to explain matching by measurement [99, pp. 50–51]. . . .

Our physical space therefore must already have a structure and the matching must refer to some qualifying class of quantities inherent in this structure [99, p. 51].

. . . there will be a class of qualities γ one and only one of which attaches to any stretch on a straight line or on a point, such that matching in respect to this quality is what we mean by congruence.

The thesis that I have been maintaining is that measurement presupposes a perception of matching in quality. Accordingly in examining the meaning of any particular kind of measurement we have to ask, What is the quality that matches? [99, p. 57].

. . . a yard measure is merely a device for making evident the spatial congruence of the [extended] events in which it is implicated [99, p. 58].

Let us now examine the several strands in Whitehead's argument in turn. We shall begin by inquiring whether his historical observation that the human race possessed a time metric prior to the enunciation of Newton's laws during the seventeenth century can serve to invalidate Poincaré's contentions [64] that (1) time congruence in physics is conventional, (2) the definition of temporal congruence used in refined physical theory is given by Newton's laws, and (3) we have no direct intuition of the temporal congruence of non-adjacent time intervals, the belief in the existence of such an intuition resting on an illusion.

To see how unavailing Whitehead's historical argument is, consider first the spatial analogue of his reasoning. We saw in §3(iii) that although the demand that Newton's laws be true does uniquely define temporal congruence in the one-dimensional time continuum, it is *not* the case that the requirement of the applicability of Euclidean geometry to a table top similarly yields a unique definition of spatial congruence for that two-dimensional space. For the sake of constructing a *spatial* analogue to Whitehead's historical argument, however, let us assume that, *contrary to fact*, it were the case that the requirement of the Euclideanism of the table top did uniquely determine the *customary* definition of perfect rigidity. And now suppose that a philosopher were to say that the latter definition of spatial congruence, like all others, is conventional. What then would be the force of the following kind of Whiteheadian assertion: "Well before Hilbert rigorized Euclidean geometry and even much before Euclid less perfectly codified the geometrical relations between the bodies in our environment, men used not only their own limbs but also diverse kinds of solid bodies to certify spatial equality"? Ignoring now refinements required to allow for substance-specific distortions, it is clear that, under the assumed hypothetical conditions, we would be confronted with *logically independent* definitions of spatial equality[34] issuing in the *same* congruence class of intervals. The concordance of these definitions would indeed be an impressive empirical fact, but it could not possibly refute the claim that the one congruence defined alike by all of them is conventional.

Precisely analogous considerations serve to invalidate Whitehead's historical argument regarding time congruence, if we discount Milne's hypothesis of the incompatibility of the congruences defined by "atomic" and "astronomical" clocks (cf. §4(iii)) and consider the agreement obtained after allowance for substance-specific idiosyncrasies between the congruences defined by a class of devices located in vanishing or stationary gravitational fields. A candle always made of the same material, of the same size, and having a wick of the same material and size burns very nearly the same number of inches each hour. Hence as early as during the reign of King Alfred (872–900), burning candles were used as rough timekeepers by placing notches or marks at such a distance apart that a certain number of spaces would burn each hour [50, pp. 53–54]. Ignoring the relatively small variations of the rate of flow of water with the height of the water column in a vessel, the water clock or clepsydra served the ancient Chinese, Byzantines, Greeks, and Romans [67], as did the sand clock, keeping very roughly the same time as burning candles. Again, an essentially frictionless pendulum oscillating with constant amplitude at a point of given latitude on the earth defines the same time metric as do "natural clocks," i.e., quasi-closed periodic systems [5]. And, ignoring various refinements, similarly for the rotation of the earth, the oscillations of crystals, the successive round trips of light over a fixed distance in an inertial system, and the time based on the natural periods of vibrating atoms or "atomic clocks" [93; 42; 2; 46].

[34] As noted in fn. 26, we shall show in §7 in what sense the criterion of rigidity based on the solid body is logically independent of Euclidean geometry when cognizance *is* taken of substance-specific distortions.

Thus, unless Milne is right, we find a striking concordance between the time congruence defined by Newton's laws and the temporal equality furnished by several kinds of definitions logically independent of that Newtonian one. This agreement obtains as a matter of *empirical fact* (cf. §4(i)) for which the GTR has sought to provide an explanation through its conception of the metrical field, just as it has endeavored to account for the corresponding concordance in the coincidence behavior of various kinds of solid rods [16, pp. 78–79]. No one, of course, would wish to deny that of all the definitions of temporal congruence which yield the same time metric as the Newtonian laws, some were used by man well before these laws could be invoked to provide such a definition. Moreover, Whitehead might well have pointed out that it was only because it was possible to measure time in one or another of these pre-Newtonian ways that the discovery and statement of Newton's laws became possible. But what is the bearing of these genetic considerations and of the (presumed) fact that the same time congruence is furnished alike by each of the aforementioned logically independent definitions on the issue before us? It seems quite clear that they cannot serve as grounds for impugning the thesis that the equality obtaining among the time intervals belonging to the one congruence class in question is conventional in the Riemann-Poincaré sense articulated in this essay: this particular equality is no less conventional in virtue of being defined by a plethora of physical processes in addition to Newton's laws than if it were defined merely by one of these processes alone or by Newton's laws alone.

Can this conclusion be invalidated by adducing such agreement as does obtain under appropriate conditions between the metric of psychological time and the physical criterion of time congruence under discussion? We shall now see that the answer is decidedly in the negative.

Prior attention to the *source* of such concordance as does exist between the psychological and physical time metrics will serve our endeavor to determine whether the metric deliverances of psychological time furnish any support for Whitehead's espousal of an intrinsic metric of physical time.[35] *

It is well known that in the presence of strong emotional factors such as anxiety, exhilaration, and boredom, the psychological time metric exhibits great variability as compared to the Newtonian one of physics. But there is much evidence that when such factors are not present, physiological processes which are geared to the periodicities defining physical time congruence impress a metric upon man's psychological time and issue in rhythmic behavior on the part of a vast variety of animals. There are two main theories at present about the source of the concordance between the metrics of physical and psychobiological time. The older of these maintains that men and animals are equipped with an internal "biological clock" *not* dependent for its successful operation on the conscious or unconscious reception of sensory cues from outside the organism. Instead the success of the biological clock is held to depend only on the occurrence of metabolic processes whose rate is *steady in the metric of physical clock time* [30, 39, 40]. As applied to humans, this hypothesis was supported by experiments of the following kind People were asked to tap on an electric switch at a rate which they judged to be a fixed number of times per second. It was found over a relatively small range of body temperatures that the temperature coefficient of counting was much the same as the one characteristic

[35] It will be noted that Whitehead does *not* rest his claim of the intrinsicality of the temporal metric on his thesis of the *atomicity* of becoming. We therefore need not deal here with the following of his contentions: (i) becoming or the transiency of "now" is a feature of the time of physics, the bifurcation of nature being philosophically illegitimate, and (ii) there is no continuity of becoming but only becoming of continuity [101, p. 53]. But the reader is referred to F. S. C. Northrop's rebuttal to Whitehead's attack on bifurcation [57], to my demonstration of the irrelevance of becoming to physical (as distinct from psychological) time [32, Sec. 4], to my critique [37] of Whitehead's use of the "Dichotomy" paradox of Zeno of Elea to prove that time intervals are only potential and not actual continua, and to my essay "Whitehead's Philosophy of Science," *Philosophical Review*, 71:218–229 (1962), for a defense of bifurcation.

* My most recent treatment of the issues raised in footnote 35 is given in Chapter I of *Modern Science and Zeno's Paradoxes* (Middletown: Wesleyan University Press, 1967).

of chemical reactions: a twofold or threefold increase in rate for a 10°C rise in temperature [17]. The defenders of the conception that the biological clock is purely internal further adduce observations of the behavior of bees: both outdoors on the surface of the earth and at the bottom of a mine, bees learned to visit at the correct time each day a table on which a dish of syrup was placed for a short time at a fixed hour. Since the bees were found to be hungry for sugar all day long, neither the assumption that they experience periodic hunger, nor the appearance of the sun nor yet the periodicities of the cosmic ray intensity can explain the bees' success in time keeping. But dosing them with substances like thyroid extract and quinine, which affect the rate of chemical reactions in the body, was found to interfere with their ability to appear at the correct time.

More recently, however, doubt has been cast on the adequacy of the hypothesis of the purely internal clock. A series of experiments with fiddler crabs and other cold-blooded animals [6, 7] showed that these organisms hold rather precisely to a 24-hour coloration cycle (lightening-darkening rhythm) regardless of whether the temperature at which they are kept is 26 degrees, 16 degrees, or 6 degrees centigrade, although at temperatures near freezing, the color clock changes. It was therefore argued that if the rhythmic timing mechanism were indeed a biochemical one wholly inside the organism, then one would expect the rhythm to speed up with increasing temperature and to slow down with decreasing temperature. And the exponents of this interpretation maintain that since the period of the fiddler crab's rhythm remained 24 hours through a wide range of temperature, the animals must possess a means of measuring time which is independent of temperature. This, they contend, is "a phenomenon quite inexplicable by any currently known mechanism of physiology, or, in view of the long period-lengths, even of chemical reaction kinetics" [7, p. 159]. The extraordinary further immunity of certain rhythms of animals and plants to many powerful drugs and poisons which are known to slow down living processes greatly is cited as additional evidence to show that organisms have daily, lunar, and annual rhythms impressed upon them by external physical agencies, thus having access to outside information concerning the corresponding physical periodicities. The authors of this theory admit, however, that the daily and lunar-tidal rhythms of the animals studied do not depend upon any now known kind of external cues of the associated astronomical and geophysical cycles [7, pp. 153, 166]. And it is postulated [7, p. 168] that these physical cues are being received because living things are able to respond to additional kinds of stimuli at energy levels so low as to have been previously held to be utterly irrelevant to animal behavior. The assumption of such sensitivity of animals is thought to hold out hope for an explanation of animal navigation.

We have dwelled on the two current rival theories regarding the source of the ability of man (and of animals) to make successful estimates of duration introspectively in order to show that, on either theory, the metric of psychological time is tied causally to those physical cycles which serve to define time congruence in physics. Hence when we make the judgment that two intervals of physical time which are equal in the metric of standard clocks also *appear* congruent in the psychometry of mere sense awareness, this justifies only the following innocuous conclusion in regard to physical time: the two intervals in question are congruent by the physical criterion which had furnished the psychometric standard of temporal equality both genetically and epistemologically. How then can the metric deliverances of psychological time possibly show that the time of physics possesses an intrinsic metric, if, as we saw, no such conclusion was demonstrable on the basis of the cycles of physical clocks?

As for spatial congruence, what are we to say of Whitehead's argument [100, p. 56] that just as it is an objective datum of experience that two phenomenal color patches have the same color, i.e., are "color-congruent," so also we *see* that a given rod has the same length in different positions, thus making the latter congruence as objective a relation as the former? As Whitehead puts it: "It is at once evident that all these tests [of congruence by means of steel yard measures, etc., are] dependent on a direct intuition of permanence" [101, p. 501]. He would argue, for example, that in the accompanying diagram the horizontal segment *AC* could not be stipulated to be congruent to the vertical segment *AB*. For the deliverances of our visual intuition unequivocally show *AC* to be shorter than *AB* and *AB* to be congruent to *AD*, a fact also attested by the finding that a solid rod coinciding with *AB* to begin with and then rotated into the horizontal position would extend over *AC* and coincide with *AD*.

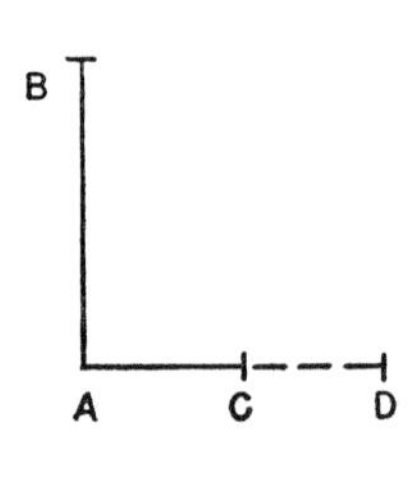

On this my first comment is to ask: What is the significance for the status of the metric of *physical* as distinct from *visual* space of these observational deliverances? And I answer that their significance is entirely consonant with the conventionalist view of *physical* congruence expounded above. The criterion for ocular congruence in our visual field was presumably furnished both genetically and epistemologically by ocular adaptation to the behavior of transported solids. For when pressed as to what it is about two congruent-*looking* intervals that enables them to sustain the relation of spatial equality, our answer will inevitably have to be this: the fact of their capacity to coincide successively with a transported solid rod. Hence when we make the judgment that two intervals of physical space with which transported solid rods coincide in succession also *look* congruent even when compared frontally purely by inspection, what this proves in regard to *physical space* is only that these intervals are congruent on the basis of the criterion of congruence which had furnished the basis for the ocular congruence to begin with, a criterion given by solid rods. But the visual deliverance of congruence does *not* constitute an ocular test of the "true" rigidity of solids under transport in the sense of establishing the factuality of the congruence defined by this class of bodies. Thus, it is a *fact* that in the diagram *AD* extends over (includes) *AC*, thus being longer. And it will be recalled that Riemann's views on the status of measurement in a spatial continuum require that every definition of "congruent" be consistent with this kind of inclusional fact. How then can visual data possibly interdict our calling AC congruent to AB and then allowing for the *de facto* coincidence of the rotated rod with AB and AD by assigning to the rod in the horizontal position a length which is suitably *greater* than the one assigned to it in the vertical orientation?

It will be recalled (cf. §5(i)) that Russell had unsuccessfully sought to counter Poincaré's position by answering the question "What is it that is measured?" on the basis of the affirmation of an intrinsic metric. Whitehead believes himself to have supplied the missing link in Russell's answer by having adduced the deliverances of *visual* space and of psychological time. It remains for us to consider briefly the further reasons put forward by Whitehead in his endeavor to show the following: transported rods and the successive periods of clocks can be respectively held to be *truly* unaltered or congruent to themselves, thereby rendering testimony of an intrinsic metric, because an intuitively apprehended matching relation obtains between the visual and psychotemporal counterparts of the respective space and time intervals in question.

Invoking visual congruence, Whitehead claims [97, p. 121] that an immediate perceptual judgment tells us that whereas an elastic thread does *not* remain unaltered under transport, a yard measure does. And from this he draws three conclusions [97, p. 121]: (i) "immediate judgments of congruence are presupposed in measurement," (ii) "the process of measurement is merely a procedure to extend the recognition of congruence to cases where these immediate judgments are not available," and (iii) "we cannot define congruence by measurement." [36] The valid core of assertions (i) and (iii) is that measurement presupposes (requires) a congruence criterion in terms of which its results are formulated. But this does not, of course, suffice to show that the congruence thus presupposed is nonconventional. Neither can the latter conclusion be established by White-

[36] Cf. also [100, pp. 54–55].

head's contention that we apprehend a *matching* relation among those intervals which are congruent according to the customary standards of rigidity (or isochronism). For the matching among intervals which are congruent relatively to a rigid rod is only with respect to such metrically nonintrinsic properties as the coincidence of each of them with that transported rod, or as the round-trip times required by light to traverse them in both directions in an inertial system. To this, Whitehead retorts with the declaration "There is a modern doctrine that 'congruence' *means* the possibility of coincidence. . . . although 'coincidence' is used as a *test* of congruence, it is not the *meaning* of congruence" [101, p. 501]. The issue raised by Whitehead here is, of course, one of *uniqueness* and *intrinsicness* of *equality* among intervals. We are therefore *not* concerned with the *separate* point that no one physical criterion can exhaustively specify "*the* meaning" of the open cluster concept of congruence as applied to any particular congruence class of intervals. Accordingly, we ignore here refinements that would allow for the open cluster character of physical congruence. And we point out that if there were a meaning in the ascription of an intrinsic metric to space, then it would be quite correct to regard coincidence as only a test of congruence in Whitehead's sense of ascertaining the existence of intrinsic equality. For in that case one would be able to speak of two separated intervals as matching spatially in the sense of containing the same intrinsic amount of space. But it is the existence of an intrinsic metric which is first at issue. And the position taken by Whitehead on that issue cannot be justified *without begging the question* by simply *asserting* that coincidence is only the test but not the meaning of congruence. Hence, his argument from visual congruence having failed, as we saw, Whitehead has not succeeded in refuting the conception that (1) the matching lies wholly in the objective coincidences of each of two (or more) intervals with the same transported rod and (2) the self-congruence of the rod under transport is a matter of convention.

It is significant, however, that there are passages in Whitehead where he comes close to the admission that the preeminent role of certain classes of physical objects as our standards of rigidity and isochronism is *not* tantamount to their making evident the *intrinsic* equality of certain spatial and temporal intervals. Thus speaking of the space-time continuum, he says: "This extensive continuum is one relational complex in which all potential objectifications find their niche. It underlies the whole world, past, present, and future. Considered in its full generality, apart from the additional conditions proper only to the cosmic epoch of electrons, protons, molecules, and star-systems, the properties of this continuum are very few and do not include the relationships of metrical geometry" [101, p. 103]. And he goes on to note that there are competing systems of measurement giving rise to alternative families of straight lines and corresponding by alternative systems of metrical geometry of which no one system is more fundamental than any other [101, p. 149]. It is in our present cosmic epoch of electrons, protons, molecules, and star systems that "more special defining characteristics obtain" and that "the ambiguity as to the relative importance of competing definitions of congruence" is resolved in favor of "one congruence definition" [101, p. 149]. Thus Whitehead maintains that among competing congruence definitions, "That definition which enters importantly into the internal constitutions of the dominating . . . entities is the important definition for the cosmic epoch in question" [101, p. 506]. This important concession thus very much narrows the gap between Whitehead's view and the Riemann-Poincaré conception defended in this essay, viz., that once a congruence definition has been given conventionally by means of the customary rigid body (or otherwise), then, assuming the usual physical interpretation of the remainder of the geometrical vocabulary, the question which metric geometry is true of physical space is one of objective physical fact. That the gap between the two views is narrowed by Whitehead's concession here becomes clear upon reading the following statement by him in the light of that concession. Speaking of Sophus Lie's treatment of congruence classes and their associated metric geometries in terms of groups of transformations between points (cf. §5(i)), Whitehead cites Poincaré and says:

The above results, in respect to congruence and metrical geometry, considered in relation to existent space, have led to the doctrine that it is intrinsically unmeaning to ask which system of metrical geometry is true of the physical world. Any one of these systems can be applied, and in an indefinite number of ways. The only question before us is one of convenience in respect to simplicity of statement of the physical laws. This point of view seems to neglect the consideration that science is to be relevant to the definite perceiving minds of men; and that (neglecting the ambiguity introduced by the invariable slight inexactness of observation which is not relevant to this special doctrine) we have, in fact, presented to our senses a definite set of transformations forming a congruence-group, resulting in a set of measure relations which are in no respect arbitrary. Accordingly our scientific laws are to be stated relevantly to that particular congruence-group. Thus the investigation of the type (elliptic, hyperbolic or parabolic) of this special congruence-group is a perfectly definite problem, to be decided by experiment [98, p. 265].

(iii) A. S. Eddington's Specious Trivialization of the Riemann-Poincaré Conception of Congruence and the Elaboration of Eddington's Thesis by H. Putnam and P. K. Feyerabend.

Though Whitehead's argument that observed matching relations attest the existence of an intrinsic metric is faulty, as we saw, that argument can no more be dismissed on the basis of the model-theoretic trivialization of the congruence issue than Russell's argument against Poincaré (cf. §5(i)). In fact, one should have supposed that those who maintain with Eddington that GC is a subthesis of TSC (cf. §2(i)) would have suspected that their critique had missed the point.* For what should have given them pause is that Russell, Whitehead, and, for that matter, Poincaré were clearly aware of the place of geometry in the theory of models of abstract calculi and yet carried on their philosophical polemic regarding the status of con-

* For an important clarification of the sense in which Eddington's views *differ* from what I have called "the model-theoretic trivialization of the congruence issue," see §3.0 of Chapter III below.

gruence. According to the triviality thesis, the stake in their controversy was no more than the pathetic one that Russell and Whitehead were advocating the customary linguistic usage of the term "congruent" (for line segments) while Poincaré was maintaining that we need not be bound by the customary usage but are at liberty to introduce bizarre ones as well. Thus, commenting on Poincaré's statement that we can always avail ourselves of alternative metrizability to give a Euclidean interpretation of any results of stellar parallax measurements ([63, p. 81]; cf. §6 below), Eddington writes:

Poincaré's brilliant exposition is a great help in understanding the problem now confronting us. He brings out the interdependence between geometrical laws and physical laws, which we have to bear in mind continually.[37] We can add on to one set of laws that which we subtract from the other set. I admit that space is conventional—for that matter, the meaning of every word in the language is conventional. Moreover, we have actually arrived at the parting of the ways imagined by Poincaré, though the crucial experiment is not precisely the one he mentions. But I deliberately adopt the alternative, which, he takes for granted, everyone would consider less advantageous. I call the space thus chosen *physical space*, and its geometry *natural geometry*, thus admitting that other conventional meanings of space and geometry are possible. If it were only a question of the meaning of space—a rather vague term—these other possibilities might have some advantages. But the meaning assigned to length and distance has to go along with the meaning assigned to space. Now these are quantities which the physicist has been accustomed to measure with great accuracy; and they enter fundamentally into the whole of our experimental knowledge of the world. . . . Are we to be robbed of the terms in which we are accustomed to describe that knowledge? [20, pp. 9–10.]

We see that Eddington objects to Poincaré's willingness to guarantee the retention of Euclidean geometry by resorting to an alternative metrization: in the context of general relativity, the retention of Euclideanism would indeed require a congruence definition different from the customary one. Regarding the possibility of a remetriza-

[37] This interdependence will be analyzed in §6 below.

tional retention of Euclidean geometry as merely illustrative of being able to avail oneself of the conventionality of all language, Eddington would rule out such a procedure on the grounds that the customary definition of spatial congruence which would be supplanted by it retains its usefulness.

Earlier in this essay (cf. §2(i)), we presented a critique of Eddington's claim that GC is a subthesis of TSC by giving an analysis of the sense in which ascriptions of congruence (and of simultaneity) are conventional.*

In the present section, we shall examine the following corollary of Eddington's contention as elaborated by H. Putnam and P. K. Feyerabend:[38] GC must be a subthesis of TSC because GC has bona fide analogues in every branch of human inquiry, such that GC cannot be construed as an insight into the structure of space or time. As Eddington puts it: "The law of Boyle states that the pressure of a gas is proportional to its density. It is found by experiment that this law is only approximately true. A certain mathematical simplicity would be gained by conventionally redefining *pressure* in such a way that Boyle's law would be rigorously obeyed. But it would be high-handed to appropriate the word pressure in this way, unless it had been ascertained that the physicist had no further use for it in its original meaning."

P. K. Feyerabend has noted that what Eddington seems to have in mind here is the following: instead of revising Boyle's law $pv = RT$ in favor of van der Waals' law

$$\left(p + \frac{a}{v^2}\right)(v - b) = RT,$$

we could preserve the statement of Boyle's law by merely redefining "pressure"—now to be symbolized by "P" in its new usage—putting

[38] Professor Feyerabend no longer endorses Eddington's position. I cite his name in this context merely to acknowledge his further *articulation* of the Eddington-Putnam thesis.

* This critique of Eddington is extended in §3 of Chapter III on the basis of that chapter's §2.10.

$$P =_{\text{Def}} \left(p + \frac{a}{v^2}\right)\left(1 - \frac{b}{v}\right).$$

In the same vein, H. Putnam maintains that instead of using phenomenalist (naive realist) color words as we do customarily in English, we could adopt a new usage for such words—to be called the "Spenglish" usage—as follows: we take a white piece of chalk, for example, which is moved about in a room, and we lay down the rule that depending (in some specified way) upon the part of the visual field which its appearance occupies, its color will be called "green," "blue," "yellow," etc., rather than "white" under constant conditions of illumination.

It is a fact, of course, that whereas actual scientific practice in the GTR, for example, countenances and uses remetrizational procedures based on *non*customary congruence definitions,[39] scientific practice does *not* contain any examples of Putnam's "Spenglish" *space-dependent* (or *time-dependent*) use of phenomenalist (naive realist) *color predicates* to denote the color of a given object in various places (or at various times) under like conditions of illumination. According to Eddington and Putnam, the existence of *non*customary usages of "congruent" in the face of there being no such usages of color predicates is no more than a fact about the linguistic behavior of the members of our linguistic community. We saw that the use of linguistic alternatives in the specifically geochronometric contexts reflects fundamental *structural properties* of the facts to which these alternative descriptions pertain. And we must now show that the alleged Eddington-Putnam analogues of GC are *pseudo*-analogues.

The essential point in assessing the cogency of the purported

[39] The proponents of ordinary language usage in science, to whom the "ordinary man" seems to be the measure of all things, may wish to rule out *non*customary congruence definitions as linguistically illegitimate. But they would do well to remember that it is no more incumbent upon the scientist (or philosopher of science) to use the *customary scientific* definition of congruence in every geochronometric description than it is obligatory for, say, the student of mechanics to be bound by the familiar *common-sense* meaning of "work," which *contradicts* the mechanical meaning as given by the space integral of the force.

analogues is the following: do the domains from which they are drawn (e.g., phenomenalist or naive realist color properties, and pressure phenomena) exhibit *structural* counterparts to (a) those factual properties of the world postulated by relativity which entail the nonuniqueness of topological simultaneity, and (b) the postulated topological properties of physical space and time which make for the nonuniqueness of the congruence axioms by assuring the nonexistence of an intrinsic metric in these manifolds? *Or* are the examples cited by Eddington and Putnam analogues of the conventionality of metrical simultaneity or of congruence *only* in the impoverished, trivial sense that they feature linguistically alternative equivalent descriptions *while lacking* the following decisive property of the geochronometric cases: the alternative metrizations are the linguistic renditions or reverberations, as it were, of the *structural properties* assuring the aforementioned two kinds of nonuniqueness enunciated by GC? If the examples given are analogues only in the superficial, impoverished sense—as indeed I shall show them to be—then what have Eddington and Putnam accomplished by their examples? In that case they have merely provided unnecessary illustrations of the correctness of TSC *without* proving their examples to be on a par with the geochronometric ones. In short, their examples will then have served in no way to make good their claim that GC is a subthesis of TSC.

We shall find presently that their examples fail* because (a) the domains to which they pertain do *not* exhibit structural counterparts to those features of the world which make the definitions of topological simultaneity and the axioms for spatial or temporal congruence *nonunique*, and (b) Putnam's example in "Spenglish" is indeed an illustration *only* of the trivial conventionality of all language: *no* structural property of the domain of phenomenal color (e.g., in the appearances of chalk) is rendered by the feasibility of the Spenglish description.

* An elaboration of the statement which is to follow is given in §§4, 6, and 7 of Chapter III.

To state my objections to the Eddington-Putnam thesis, I call attention to the following two sentences:

(A) Person X does not have a gall bladder.
(B) The platinum-iridium bar in the custody of the Bureau of Weights and Measures in Paris (Sèvres) is 1 meter long everywhere rather than some other number of meters (after allowance for "differential forces").

I maintain that there is a *fundamental difference* between the senses in which each of these statements can possibly be held to be conventional, and I shall refer to these respective senses as "A-conventional" and "B-conventional": in the case of statement (A), what is conventional is *only* the use of the given *sentence* to render the *proposition* of X's not having a gall bladder, *not* the factual *proposition* expressed by the sentence. This A-conventionality is of the trivial weak kind affirmed by TSC. On the other hand, (B) is conventional not merely in the *trivial* sense that the English sentence used could have been replaced by one in French or some other language but in the much *stronger* and *deeper* sense that it is *not* a factual proposition that the Paris bar has everywhere a length *unity* in the meter scale even *after* we have specified what sentence or string of noises will express this proposition. In brief, in (A), semantic conventions are *used,* whereas, in (B), a semantic convention is *mentioned.* Now I maintain that the alleged analogues of Eddington and Putnam illustrate conventionality only in the sense of *A*-conventionality and therefore cannot score against my contention that geochronometric conventionality is *non*trivial by having the character of *B*-conventionality. Specifically, I assert that statements about phenomenalist colors are empirical statements pure and simple in the sense of being *only* A-conventional and *not* B-conventional, while an important class of statements of geochronometry possesses a different, deeper conventionality of their own by being B-conventional. What is it that is conventional in the case of the color of a given piece of chalk, which appears white in various parts of the visual field? I answer: only our customary decision to use the *same* *word* to refer to the various qualitatively same white chalk appearances in different parts of the visual field. But it is *not* conventional whether the various chalk appearances do have the same phenomenal color property (to within the precision allowed by vagueness) and thus are "*color congruent*" to one another or not! Only the *color words* are conventional, *not* the *obtaining* of specified color properties and of color congruence. And the *obtaining* of color congruence is *non*conventional quite independently of whether the various occurrences of a particular shade of color are denoted by the same color word or not.

In other words, there is *no* convention in whether two objects or two appearances of the same object under like optical conditions have the same phenomenal color property of whiteness (apart from vagueness) but only in whether the *noise* "white" is applied to both of these objects or appearances, to one of them and not to the other (as in Putnam's chalk example), or to neither. And the alternative color descriptions *do not render any structural facts of the color domain* and are therefore purely trivial. Though failing in this decisive way, Putnam's chalk color case is falsely given the *semblance* of being a bona fide analogue to the spatial congruence case by the device of laying down a rule which makes the use of color names *space dependent:* the rule is that *different noises* (color names) will be used to refer to the same *de facto* color property occurring in different portions of visual space. But this stratagem cannot overcome the fact that while the assertion of the possibility of assigning a space-dependent length to a transported rod reflects linguistically the objective nonexistence of an intrinsic metric, the space-dependent use of color names does *not* reflect a corresponding property of the domain of phenomenal colors in visual space. In short, the phenomenalist color of an appearance is an intrinsic, objective property of that appearance (to within the precision allowed by vagueness), and color congruence is an objective relation. But the length of a body and the congruence of noncoinciding intervals are *not* similarly *non*conventional. And we saw in our critique of Whitehead that this conclusion cannot be invalidated by the fact that two noncoinciding intervals can *look* spatially congruent no less than two color patches can appear color congruent.

Next consider Eddington's example of the preservation of the language of Boyle's law to render the *new facts* affirmed by van der Waals' law by the device of giving a new meaning to the word "pressure" as explained earlier. The customary concept of pressure has geochronometric ingredients (force, area), and any alterations made in the geochronometric congruence definitions will, of course, issue in changes as to what pressures will be held to be equal. But the conventionality of the geochronometric ingredients is *not* of course at issue, and we ask: Of what *structural feature* of the domain of pressure phenomena does the possibility of Eddington's above linguistic transcription render testimony? The answer clearly is *of none*. Unlike GC, the thesis of the "conventionality of pressure," if put forward on the basis of Eddington's example, concerns *only* A-conventionality and is thus merely a special case of TSC. We observe, incidentally, that two pressures which are equal on the customary definition will also be equal (congruent) on the suggested redefinition of that term: *apart* from the distinctly geochronometric ingredients *not* here at issue, the domain of pressure phenomena does not present us with any structural property as the counterpart of the lack of an intrinsic metric of space which would be reflected by the alternative definitions of "pressure."

The absurdity of likening the conventionality of spatial or temporal congruence to the conventionality of the choice between the two above meanings of "pressure" or between English and Spenglish color discourse becomes patent upon considering the expression given to the conventionality of congruence by the Klein-Lie group-theoretical treatment of congruences and metric geometries. For their investigations likewise serve to show, as we shall now indicate, how far removed from being a semantically uncommitted noise the term "congruent" is while still failing to single out a unique congruence class of intervals, and how badly amiss it is for Eddington and Putnam to maintain that this nonuniqueness is merely a special example of the semantical nonuniqueness of all uncommitted noises.

Felix Klein's Erlangen Program (1872) of treating geometries from the point of view of groups of spatial transformations was rooted in the following two observations: (1) the properties in virtue of which *spatial congruence* has the logical status of an *equality* relation depend upon the fact that displacements are given by a *group of transformations*, and (2) the congruence of two figures consists in their being intertransformable into one another by means of a certain transformation of points. Continuing Klein's reasoning, Sophus Lie then showed that, in the context of this group-theoretical characterization of metric geometry, the conventionality of congruence issues in the following results: (i) all the continuous groups in space having the property of displacements in a bounded region fall into three types which respectively characterize the geometries of Euclid, Lobachevski-Bolyai, and Riemann [3, p. 153], and (ii) for *each* of these metrical geometries, there is *not* one but an *infinitude* of different congruence classes ([99, p. 49]; cf. §3(iii), where this latter result is obtained *without* group-theoretical devices). On the Eddington-Putnam thesis, Lie's profound and justly celebrated results no less than the relativity of simultaneity and the conventionality of temporal congruence must be consigned absurdly to the limbo of trivial semantical conventionality along with Spenglish color discourse!*

These objections against the Eddington-Putnam claim that GC has bona fide analogues in every empirical domain are not intended to deny the existence of one or another genuine analogue but to deny only that GC may be deemed to be trivial on the strength of such relatively few bona fide analogues as may obtain. Putnam has given one example which does seem to qualify as a bona fide analogue. This example differs from his color case in that not merely the *name* given to a property but the sameness of the property named is dependent on spatial position as follows: when two bodies are at

* The point of the preceding two paragraphs concerning Eddington and Lie is elaborated in §3.4 of Chapter III.

essentially the same place, their sameness with respect to a certain property is a matter of fact, but when they are (sufficiently) apart spatially, no objective relation of sameness or difference with respect to the given property obtains between them. And, in the latter case, therefore, it becomes a matter of convention whether sameness or difference is ascribed to them in this respect. Specifically, suppose that we do not aim at a definition of mass adequate to classical physical theory and thus ignore Mach's definition of mass, which we discussed earlier (cf. §2(iii)). Then we can consider Putnam's hypothetical definition of "mass equality," according to which two bodies balancing one another on a suitable scale at what is essentially the same place in space have equal masses. Whereas on the Machian definition mass equality obtains between two bodies *as a matter of fact* independently of the extent of their spatial separation, on Putnam's definition such separation leaves the relation of mass equality at a distance *indeterminate.* Hence, on Putnam's definition, it would be a matter of convention whether (a) we would say that two masses which balance at a given place remain equal to one another in respect to mass after being spatially separated, or (b) we would make the masses of two bodies space dependent such that two masses that balance at one place would have different masses when separated, as specified by a certain function of the coordinates. The conventionality arising in Putnam's mass example is *not* a consequence of GC but is logically independent of it. For it is *not* spatial *congruence* of noncoinciding intervals but spatial *position* that is the source of conventionality here.

In conclusion, we must persist therefore with Poincaré, Einstein, Reichenbach, and Carnap in attaching a very different significance to alternative metric geometries or chronometries as equivalent descriptions of the same facts than to alternate types of visual color discourse as equivalent descriptions of the same phenomenal data. By the same token, we must attach much greater significance to being able to render *factually different* geochronometric states of affairs by the *same* geometry or chronometry, coupled with appropriately *different* congruence definitions, than to formulating both Boyle's law and van der Waals' law, which *differ* in factual content, by the *same* law statement coupled with appropriately different semantical rules. In short, there is an important respect in which physical geochronometry is less of an empirical science than all or almost all of the nongeochronometric portions (ingredients) of other sciences.

6. The Bearing of Alternative Metrizability on the Interdependence of Geochronometry and Physics

(i) The Fundamental Difference between the LINGUISTIC *Interdependence of Geometry and Physics Affirmed by the Conventionalism of H. Poincaré and Their* EPISTEMOLOGICAL *(Inductive) Interdependence in the Sense of P. Duhem.**

The central theme of Poincaré's so-called conventionalism is essentially an elaboration of the thesis of alternative metrizability whose fundamental justification we owe to Riemann, and *not* [32, Section 5] the *radical* conventionalism attributed to him by Reichenbach [72, p. 36].

Poincaré's much-cited and often misunderstood statement concerning the possibility of always giving a Euclidean description of any results of stellar parallax measurements [63, pp. 81–86] is a less lucid statement of exactly the same point made by him with magisterial clarity in the following passage: "In space we know rectilinear triangles the sum of whose angles is equal to two right angles; but equally we know curvilinear triangles the sum of whose angles is less than two right angles. . . . To give the name of straights to the sides of the first is to adopt Euclidean geometry; to give the name of straights to the sides of the latter is to adopt the non-Euclidean geometry. So that to ask what geometry it is proper to adopt is to ask, to what line is it proper to give the name straight? It is evident that experiment can not settle such a question" [63, p. 235].

Now, the equivalence of this contention to Riemann's view of

* See Chapter II, §2(i) for an elaboration of the material on interdependence presented here.

congruence becomes evident the moment we note that the legitimacy of identifying lines which are curvilinear in the usual geometrical parlance as "straights" is vouchsafed by the warrant for our choosing a new definition of congruence such that the previously curvilinear lines become geodesics of the new congruence. And we note that whereas the *original* geodesics in space exemplified the formal relations obtaining between Euclidean "straight lines," the *different* geodesics associated with the new metrization embody the relations prescribed for straight lines by the formal postulates of *hyperbolic* geometry. Awareness of the fact that Poincaré begins the quoted passage with the words "In [physical] space" enables us to see that he is making the following assertion here: the same *physical* surface or region of three-dimensional *physical* space admits of *alternative* metrizations so as to constitute a physical realization of either the formal postulates of Euclidean geometry or one of the non-Euclidean abstract calculi. To be sure, syntactically, this alternative metrizability involves a formal intertranslatability of the *relevant portions* of these incompatible geometrical calculi, the "intertranslatability" being guaranteed by a "dictionary" which pairs off with one another the *alternative names* (or descriptions) of each physical path or configuration. But the essential point made here by Poincaré is *not* that a purely formal translatability obtains; instead, Poincaré is emphasizing here that a *given* physical surface or region of physical 3-space can indeed be a model of one of the *non*-Euclidean geometrical calculi no less than of the Euclidean one. In *this* sense, one can say, therefore, that Poincaré affirmed the conventional or definitional status of *applied* geometry.

Hence, we must reject the following wholly syntactical interpretation of the above citation from Poincaré, which is offered by Ernest Nagel, who writes: "The thesis he [Poincaré] establishes by this argument is simply the thesis that choice of notation in formulating a system of pure geometry is a convention" [55, p. 261]. Having thus misinterpreted Poincaré's conventionalist thesis as pertaining only to formal intertranslatability, Nagel fails to see that Poincaré's avowal of the conventionality of physical or *applied* geometry is none other than the assertion of the *alternative metrizability of physical space* (or of a portion thereof). And, in this way, Nagel is driven to give the following unfounded interpretation of Poincaré's conception of the status of applied (physical) geometry: "Poincaré also argued for the definitional status of *applied* as well as of *pure* geometry. He maintained that, even when an interpretation is given to the primitive terms of a pure geometry so that the system is then converted into statements about certain physical configurations (for example, interpreting 'straight line' to signify the path of a light ray), no experiment on physical geometry can ever decide against one of the alternative systems of physical geometry and in favor of another" [55, p. 261]. But far from having claimed that the geometry is still conventional even *after* the provision of a particular physical interpretation of a pure geometry, Poincaré merely reiterated the following thesis of alternative metrizability in the passages which Nagel [55, pp. 261–262] then goes on to quote from him: suitable alternative semantical interpretations of the term "congruent" (for line segments and for angles), and correlatively of "straight line," etc., can readily demonstrate that, subject to the restrictions imposed by the existing topology, it is always a live option to give *either* a Euclidean *or* a *non*-Euclidean description of any given set of *physico*-geometric facts. And since alternative metrizations are just as legitimate epistemologically as alternative systems of units of length or temperature, one can always, in principle, *reformulate* any physical theory based on a given metrization of space—or, as we saw in §4 above, of time—so as to be based on an *alternative* metric.

There is therefore no warrant at all for the following caution expressed by Nagel in regard to the feasibility of what is merely a reformulation of physical theory on the basis of a new metrization: ". . . even if we admit universal forces in order to retain Euclid . . . we must incorporate the assumption of universal forces into the rest of our physical theory, rather than introduce such forces piecemeal subsequent to each observed 'deformation' in bodies. It is by no means self-evident, however, that physical theories can in fact always be devised that have built-in provisions for such universal forces" [55, pp. 264–265]. Yet, precisely that fact is self-evident, and its self-evidence is obscured from view by the logical havoc created by the statement of a remetrization issuing in Euclidean geometry in terms of "universal forces." For that metaphor seems to have misled Nagel into imputing the status of an *empirical* hypothesis to the use of a *non*standard spatial metric merely because the latter metric is described by saying that we "assume" appropriate universal forces. In fact, our discussion in §4(i), has shown mathematically for the one-dimensional case of time how Newtonian mechanics is to be recast via suitable transformation equations, such as equation (4) there, so as to implement a remetrization given by $T = f(t)$, which can be described metaphorically by saying that all clocks are "accelerated" by "universal forces."

Corresponding remarks apply to Poincaré's contention that we can always preserve Euclidean geometry in the face of any data obtained from stellar parallax measurements: if the paths of light rays are geodesics on the customary definition of congruence, as indeed they are in the Schwarzschild procedure cited by Robertson [77], and if the paths of light rays are found parallactically to sustain non-Euclidean relations on that metrization, then we need only choose a different definition of congruence such that these *same* paths will no longer be geodesics and that the geodesics of the newly chosen congruence are Euclideanly related. From the standpoint of synthetic geometry, the latter choice effects a *renaming* of optical and other paths and thus is merely a *recasting of the same factual content in Euclidean language rather than a revision of the extra-linguistic content of optical and other laws.* The *remetrizational retainability* of Euclideanism affirmed by Poincaré therefore *involves a merely linguistic interdependence of the geometric theory of rigid solids and the optical theory of light rays.* And since Poincaré's claim here is a straightforward elaboration of the metric amorphousness of the continuous manifold of space, it is not clear how H. P.

Robertson [77, pp. 324–325] can reject it as a "pontifical pronouncement" and even regard it as being in contrast with what he calls Schwarzschild's "sound operational approach to the problem of physical geometry." For Schwarzschild had rendered the question concerning the prevailing geometry *factual* only by the adoption of a particular spatial metrization based on the travel times of light, which does indeed turn the direct light paths of his astronomical triangle into geodesics.

Poincaré's interpretation of the parallactic determination of the geometry of a stellar triangle has also been obscured by Ernest Nagel's statement of it. Apart from encumbering that statement with the metaphorical use of "universal forces," Nagel fails to point out that the crux of the preservability (retainability) of Euclidean geometry lies in (i) the denial of the *geodesicity* (*straightness*) of optical paths which are found parallactically to sustain *non*-Euclidean relations on the *customary* metrization of line segments and angles, *or* at least in the rejection of the customary congruence for *angles* (cf. §3(iii)),[40] and (ii) the ability to guarantee the existence of a suitable new metrization whose associated geodesics are paths which *do* exhibit the formal relations of Euclidean straights. For Nagel characterizes the retainability of Euclidean geometry come what may by asserting that the latter's retention is effected "only by maintaining that the sides of the stellar triangles are not really Euclidean [*sic*] straight lines, and he [the Euclidean geometer] will therefore adopt the hypothesis that the optical paths are deformed by some fields of force" [55, p. 263]. But apart from the obscurity of the notion of the deformation of the optical *paths*, the unfortunate inclusion of the word "Euclidean" in this sentence of Nagel's obscures the very point which the advocate of Euclid is concerned to make in this context

[40] Under the assumed conditions as to the parallactic findings, the optical paths might still be interpretable as Euclideanly-related geodesics on a *new* metrization, but only if the customary *angle* congruences were abandoned and changed suitably as part of the remetrization (cf. §3(iii)). In that case, the paths of light rays would be straight lines *even in the Euclidean description* obtained by the new metrization, but the optical laws involving angles would have to be suitably restated.

in the interests of his thesis. And this point is *not*, as Nagel would have it, that the optical paths are not really *Euclidean* straight lines, a fact whose admission (assuming the customary congruence for angles) provided the starting point of the discussion. Instead, what the proponent of Euclid is concerned to point out here is that the legitimacy of alternative metrizations enables him to offer a metric such that the optical paths do *not* qualify as *geodesics* (straights) from the outset. For it is by denying altogether the geodesicity of the optical paths that the advocate of Euclid can uphold his thesis successfully in the face of the admitted prima-facie non-Euclidean parallactic findings.

The invocation of the conventionality of congruence to carry out remetrizations is not at all peculiar to Poincaré. For F. Klein's relative consistency proof of hyperbolic geometry via a model furnished by the interior of a circle on the Euclidean plane [3, pp. 164–175], for example, is based on one particular kind of possible remetrization of the circular portion of that plane, projective geometry having played the heuristic role of furnishing Klein with a suitable definition of congruence. Thus what from the point of view of synthetic geometry appears as intertranslatability via a dictionary appears as alternative metrizability from the point of view of differential geometry.

There are two respects, however, in which Poincaré is open to criticism in this connection: (i) He maintained [63, p. 81] that it would always be regarded as most convenient to preserve Euclidean geometry, even at the price of remetrization, on the grounds that this geometry is the simplest analytically [63, p. 65]. Precisely the opposite development materialized in the general theory of relativity: Einstein forsook the simplicity of the geometry itself in the interests of being able to maximize the simplicity of the definition of congruence. He makes clear in his fundamental paper of 1916 that had he insisted on the retention of Euclidean geometry in a gravitational field, then he could *not* have taken "one and the same rod, independently of its place and orientation, as a realization of the same interval" [21, p. 161]. (ii) Even if the simplicity of the geometry itself were the sole determinant of its adoption, that simplicity might be judged by criteria other than Poincaré's analytical simplicity. Thus, Menger has urged that from the point of view of a criterion grounded on the simplicity of the undefined concepts used, hyperbolic and not Euclidean geometry is the simplest.

On the other hand, if Poincaré were alive today, he could point to an interesting recent illustration of the sacrifice of the simplicity and accessibility of the congruence standard on the altar of maximum simplicity of the resulting theory. Astronomers have recently proposed to remetrize the time continuum for the following reason: as indicated in §4(i), when the mean solar second, which is a very precisely known fraction of the period of the earth's rotation on its axis, is used as a standard of temporal congruence, then there are three kinds of discrepancies between the actual observational findings and those predicted by the usual theory of celestial mechanics. The empirical facts thus present astronomers with the following choice: either they retain the rather natural standard of temporal congruence at the cost of having to bring the principles of celestial mechanics into conformity with observed fact by revising them appropriately, or they remetrize the time continuum, employing a less simple definition of congruence so as to preserve these principles intact. Decisions taken by astronomers in the last few years were exactly the reverse of Einstein's choice of 1916 as between the simplicity of the standard of congruence and that of the resulting theory. The mean solar second is to be supplanted by a unit to which it is nonlinearly related: the sidereal year, which is the period of the earth's revolution around the sun, due account being taken of the irregularities produced by the gravitational influence of the other planets [13].

We see that the implementation of the requirement of descriptive simplicity in theory construction can take alternative forms, because agreement of astronomical theory with the evidence now available is achievable by revising *either* the definition of temporal congruence *or* the postulates of celestial mechanics. The existence of this alternative likewise illustrates that for an axiomatized physical theory containing a geochronometry, it is *gratuitous* to single out the postulates of the theory as having been prompted by *empirical* findings in contradistinction to deeming the *definitions of congruence* to be wholly a priori, or vice versa. This conclusion bears out geochronometrically Braithwaite's contention [4] that there is an important sense in which axiomatized physical theory does not lend itself to compliance with Heinrich Hertz's injunction to "distinguish thoroughly and sharply between the elements . . . which arise from the necessities of thought, from experience, and from arbitrary choice" [38, p. 8]. The same point is illustrated by the possibility of characterizing the factual innovation wrought by Einstein's abandonment of Euclidean geometry in favor of Riemannian geometry in the GTR in several ways as follows: (i) *Upon using the customary definition of spatial congruence*, the geometry near the sun is *not* Euclidean, contrary to the claims of pre-GTR physics. (ii) The geometry near the sun is *not* Euclidean on the basis of the *customary* congruence, but it *is* Euclidean on a suitably modified congruence definition which makes the length of a rod a specified function of its position and orientation.[41] (iii) *Within the confines* of the requirement of giving a *Euclidean* description of the nonclassical facts postulated by the GTR, Einstein recognized the *factually dictated* need to *abandon* the *customary* definition of congruence, which had yielded a Euclidean description of the classically assumed facts. Thus, the revision of the Newtonian theory made necessary by the discovery of relativity can be formulated as *either* a change in the postulates of geometric theory *or* a change in the correspondence rule for congruence.

We saw that Poincaré's remetrizational retainability of Euclidean geometry or of some other particular geometry involves a merely *linguistic* interdependence of the geometric theory of rigid solids and the optical theory of light rays. Preparatory to clarifying the important difference between Duhem's and Poincaré's conceptions of

[41] The function in question is given in [8, p. 58].

the interdependence of geometry and physics, we first give a statement of Duhem's view of the falsifiability and confirmability of an isolated explanatory hypothesis in science.*

We must distinguish the following two forms of Duhem's thesis: (i) The logic of every disconfirmation, no less than of every confirmation, of a presumably empirical hypothesis H is such as to *involve at some stage or other* an entire network of interwoven hypotheses in which H is ingredient rather than the separate testing of the component H. (ii) The falsifiability of (part of) *an explanans* is unavoidably inconclusive in every case: *no one constituent hypothesis* H can ever be *extricated* from the ever-present web of collateral assumptions so as to be *open* to *decisive refutation* by the evidence as part of an *explanans* of that evidence, just as no such isolation is achievable for purposes of verification.

Duhem seems to think that the latter contention is justified by the following twofold argument or schema of unavoidably inconclusive falsifiability: (a) It is an elementary fact of *deductive* logic that if certain observational consequences O are entailed by the *conjunction* of H and a set A of auxiliary assumptions, then the *failure* of O to materialize entails *not* the falsity of H by itself but only the weaker conclusion that H and A cannot *both* be true; the falsifiability of H is therefore *inconclusive* in the sense that the *falsity* of H is *not deductively inferable* from the premise $[(H \cdot A) \rightarrow O] \cdot \sim O$. (b) The actual observational findings O', which are incompatible with O, *allow* that H be true while A is false, because they permit the theorist to preserve H with impunity as part of an *explanans* of O' by so modifying A that the *conjunction* of H and the *revised* version A' of A does explain (entail) O'. This preservability of H is to be understood as a retainability in *principle* and does not depend on the ability of scientists to propound the required set A' of collateral assumptions at any given time.

* For a fuller statement, which distinguishes the Duhemian *legacy* from Duhem's own views, see my essay "The Falsifiability of a Component of a Theoretical System," in *Mind, Matter, and Method*, edited by Paul K. Feyerabend and Grover Maxwell (Minneapolis: University of Minnesota Press, 1966).

Thus, *according to Duhem, there is an inductive (epistemological) interdependence and inseparability between H and the auxiliary assumptions.* And there is claimed to be an ingression of a kind of a priori choice into physical theory: at the price of suitable compensatory modifications in the remainder of the theory, *any one* of its *component* hypotheses *H* may be retained in the face of seemingly contrary empirical findings as part of an *explanans* of these very findings. According to Duhem, this quasi a priori preservability of *H* is sanctioned by the far-reaching theoretical *ambiguity* and flexibility of the logical constraints imposed by the observational evidence [19].[42]

Duhem would point to the fact that in a sense to be specified *in detail* in §7, the physical laws used to correct a measuring rod for substance-specific distortions presuppose a geometry and comprise the laws of optics. And hence he would *deny*, for example, that either of the following kinds of *independent* tests of geometry and optics are feasible:

1. Prior to and independently of knowing or presupposing the geometry, we find it to be a law of optics that the paths of light coincide with the geodesics of the congruence defined by rigid bodies.

Knowing this, we then use triangles consisting of a geodesic base line in the solar system and the stellar light rays connecting its extremities to various stars to determine the geometry of the system of rigid-body geodesics: stellar parallax measurements will tell us whether the angle sums of the triangles are 180° (Euclidean geometry), less than 180° (hyperbolic geometry), or in excess of 180° (spherical geometry).

[42] Duhem's explicit disavowal of both decisive falsifiability and crucial verifiability of an *explanans* will *not* bear K. R. Popper's reading of him [66, p. 78]: Popper, who is an exponent of decisive falsifiability [66], misinterprets Duhem as allowing that tests of a hypothesis may be decisively falsifying and as denying only that they may be crucially verifying. Notwithstanding Popper's *exegetical* error, we shall see in §7 that his thesis of the feasibility of decisively falsifying tests can be buttressed by a telling counterexample to Duhem's categorical denial of that thesis.

Feigl [28] has outlined a defense of the claim that isolated parts of physical theory can be confirmed.

If we thus find that the angle sum is different from 180°, then we shall know that the geometry of the rigid-body geodesics is *not* Euclidean. For in view of our prior independent ascertainment of the paths of light rays, such a non-Euclidean result could *not* be interpreted as due to the failure of optical paths to coincide with the rigid-body geodesics.

2. Prior to and independently of knowing or presupposing the laws of optics, we ascertain what the geometry is relatively to the rigid-body congruence.

Knowing this we then find out whether the paths of light rays coincide with the geodesics of the rigid-body congruence by making a parallactic or some other determination of the angle sum of a light-ray triangle.

Since we know the geometry of the rigid-body geodesics independently of the optics, we know what the corresponding angle sum of a triangle whose sides are geodesics ought to be. And hence the determination of the angle sum of a light-ray triangle is then decisive in regard to whether the paths of light rays coincide with the geodesics of the rigid-body congruence.

In place of such independent confirmability and falsifiability of the geometry and the optics, Duhem affirms their *inductive* (epistemological) inseparability and interdependence. Let us consider the interpretation of observational parallactic data to articulate, in turn, the differences between Duhem's and Poincaré's conceptions of (a) the feasibility of alternative geometric interpretations of observational findings, and (b) the retainability of a particular geometry as an account of such findings.

(a) *The feasibility of alternative geometric interpretations of parallactic or other observational findings.* The Duhemian conception envisions scope for alternative geometric accounts of a given body of evidence only to the extent that these geometries are associated with alternative *non*equivalent sets of physical laws that are used to compute corrections for substance-specific distortions. On the other hand, the range of alternative geometric descriptions of given evidence affirmed by Poincaré is far wider and rests on very different grounds: instead of invoking the Duhemian *inductive* latitude, Poincaré bases the possibility of giving *either* a Euclidean *or* a non-Euclidean description of the same spatio-physical facts on alternative metrizability. For Poincaré tells us [63, pp. 66–80] *that quite apart from any considerations of substance-specific distorting influences and even after correcting for these in some way or other*, we are at liberty to define congruence—and thereby to fix the geometry appropriate to the given facts—either by calling the solid rod equal to itself everywhere or by making its length vary in a specified way with its position and orientation. The particular case of the interpretation of certain parallactic data will give concrete meanings to these assertions.

The attempt to explain parallactic data yielding an angle sum different from 180° for a stellar light-ray triangle by different geometries which constitute live options *in the inductive sense of Duhem* would presumably issue in the following alternative between two theoretical systems.* Each of these theoretical systems comprises a geometry G and an optics O which are epistemologically inseparable and which are inductively interdependent in the sense that the combination of G and O must yield the observed results:

> (a) G_E: the geometry of the rigid body geodesics is Euclidean, and
> O_1: the paths of light rays do *not* coincide with these geodesics but form a non-Euclidean system,

or

> (b) $G_{\text{non-}E}$: the geodesics of the rigid-body congruence are *not* a Euclidean system, and
> O_2: the paths of light rays *do* coincide with these geodesics, and thus they form a non-Euclidean system.

To contrast this Duhemian conception of the feasibility of alternative geometric interpretations of the assumed parallactic data with that of Poincaré, we recall that the physically interpreted alternative

* A refined statement of this alternative is given in §2(i) of Chapter II.

geometries associated with two (or more) different metrizations *in the sense of Poincaré* have precisely the same total factual content, as do the corresponding two sets of optical laws. For an alternative metrization in the sense of Poincaré affects only the *language* in which the facts of optics and the coincidence behavior of a transported rod are described: the two geometric descriptions respectively associated with two alternative metrizations are *alternative representations of the same factual content*, and so are the two sets of optical laws corresponding to these geometries. Accordingly, Poincaré is affirming a *linguistic* interdependence of the geometric theory of rigid solids and the optical theory of light rays. By contrast, in the *Duhemian* account, G_E and $G_{\text{non-}E}$ not only *differ* in factual content but are logically incompatible, and so are O_1 and O_2. And on the latter conception, there is sameness of factual content *in regard to the assumed parallactic data* only between the *combined* systems formed by the two conjunctions (G_E and O_1) and ($G_{\text{non-}E}$ and O_2).[43] Thus, the need for the combined system of G and O to yield the empirical facts, coupled with the avowed epistemological (inductive) *inseparability* of G and O lead the Duhemian to conceive of the *interdependence* of geometry and optics as *inductive* (epistemological).

Hence whereas Duhem construes the interdependence of G and O inductively such that the geometry by itself is *not* accessible to empirical test, Poincaré's conception of their interdependence allows for an empirical determination of *G by itself*, *if we have renounced* recourse to an alternative metrization in which the length of the rod is held to vary with its position or orientation. This is not, of course, to say that Duhem regarded alternative metrizations as such to be illegitimate.

It would seem that it was Poincaré's discussion of the interdependence of optics and geometry by reference to stellar parallax measurements which has led many writers such as Einstein [26],

[43] These combined systems do *not*, however, have the same *over-all* factual content.

Eddington [20, p. 9], and Nagel [55, p. 262] to regard him as a proponent of the Duhemian thesis. But this interpretation appears untenable not only in the light of the immediate context of Poincaré's discussion of his astronomical example, but also, as we shall see briefly in §6(ii), upon taking account of the remainder of his writings. An illustration of the widespread conflation of the linguistic and inductive kinds of interdependence of geometry and physics (optics) is given by D. M. Y. Sommerville's discussion of what he calls "the inextricable entanglement of space and matter." He says:

A . . . "vicious circle" . . . arises in connection with the astronomical attempts to determine the nature of space. These experiments are based upon the received laws of astronomy and optics, which are themselves based upon the euclidean assumption. It might well happen, then, that a discrepancy observed in the sum of the angles of a triangle could admit of an explanation by some modification of these laws, or that even the absence of any such discrepancy might still be compatible with the assumptions of non-euclidean geometry.

Sommerville then quotes the following assertion by C. D. Broad:

All measurement involves both physical and geometrical assumptions, and the two things, space and matter, are not given separately, but analysed out of a common experience. Subject to the general condition that space is to be changeless and matter to move about in space, we can explain the same observed results in many different ways by making compensatory changes in the qualities that we assign to space and the qualities we assign to matter. Hence it seems theoretically impossible to decide by any experiment what are the qualities of one of them in distinction from the other.

And Sommerville's immediate comment on Broad's statement is the following:

It was on such grounds that Poincaré maintained the essential impropriety of the question, "Which is the true geometry?" In his view it is merely a matter of convenience. Facts are and always will be most simply described on the euclidean hypothesis, but they can still be described on the non-euclidean hypothesis, with suitable modifications of the physical laws. To ask which is the true geometry is then just as unmeaning as to ask whether the old or the metric system is the true one [84, pp. 209–210].

(b) *The retainability of a particular geometry as an account of observational findings.* The key to the difference between the geometric conventionalism of H. Poincaré and the geometrical form of the conventionalism of P. Duhem is furnished by the distinction between preserving a particular geometry (e.g., the Euclidean one) by a remetrizational change in the congruence definition, on the one hand, and intending to retain a particular geometry *without* change in that definition (or in other semantical rules) by an alteration of the factual content of the auxiliary assumptions, on the other.* More specifically, whereas Duhem's affirmation of the retainability of Euclidean geometry in the face of any observational evidence and the associated interdependence of geometry and the remainder of physics are *inductive* (epistemological), the preservability of that geometry asserted by Poincaré is *remetrizational:* Poincaré's conventionalist claim regarding geometry is that if the customary definition of congruence on the basis of the coincidence behavior common to all kinds of solid rods does not assure a particular geometric description of the facts, then such a description can be guaranteed remetrizationally, i.e., by merely choosing an appropriately different noncustomary congruence definition which makes the length of every kind of solid rod a specified *non*constant function of the *independent* variables of position and orientation.

(ii) Exegetical Excursus: Poincaré's Philosophy of Geometry.

Einstein [22, 26] and Reichenbach [71, 73] have interpreted Poincaré to have been a proponent of the Duhemian conception of the interdependence of geometry and physics. As evidence *against* this interpretation, I now cite the following two passages from Poincaré, the first being taken from his rejoinder to Bertrand Russell and the second from *Science and Hypothesis:*

* For my rebuttal to the contention that this kind of distinction is spurious, see p. 143 in my *Philosophical Problems of Space and Time* (New York: Knopf, 1963).

The term "to preserve one's form" has no meaning by itself. But I confer a meaning on it by *stipulating* that certain bodies will be said to preserve their form. These bodies, thus chosen, can henceforth serve as instruments of measurement. But if I say that these bodies preserve their form, it is because *I choose to do so* and not because experience obliges me to do so.

In the present context I choose to do so, because by a series of observations ("constatations") . . . *experience has proven to me* that their movements form a Euclidean group. I have been able to make these observations in the manner just indicated *without having any preconceived idea concerning metric geometry*. And, having made them, I judge that the convention will be convenient and I adopt it [65; the italics in the second of these paragraphs are mine].[44]

No doubt, in our world, natural solids . . . undergo variations of form and volume due to warming or cooling. But we neglect these variations in laying the foundations of geometry, because, besides their being very slight, they are irregular and consequently seem to us accidental [63, p. 76].

But one might either contest my interpretation here or conclude that Poincaré was inconsistent by pointing to the following passage by him:

Should we . . . conclude that the axioms of geometry are experimental verities? . . . If geometry were an experimental science, it would not be an exact science, it would be subject to a continual revision. Nay, it would from this very day be convicted of error, since we know that there is no rigorously rigid solid.

The axioms of geometry therefore are . . . conventions . . . Thus it is that the postulates can remain rigorously true even though the experimental laws which have determined their adoption are only approximative [63, pp. 64–65].[45]

The only way in which I can construe the latter passage and others like it in the face of our earlier citations from him is by assuming that Poincaré maintained the following: there are *practical* rather than *logical* obstacles which frustrate the complete elimination of

[44] For a documentation of the fact that Poincaré espoused a similarly empiricist conception of the three-dimensionality of space, see [32, Sec. 5].

[45] Similar statements are found in [63, pp. 79, 240].

perturbational distortions, and the resulting vagueness (spread) as well as the finitude of the empirical data provide scope for the exercise of a certain measure of convention in the adoption of a metric tensor.

This non-Duhemian reading of Poincaré accords with the interpretation of him in L. Rougier's *La Philosophie Géométrique de Henri Poincaré.* Rougier writes:

> The conventions fix the language of science which can be indefinitely varied: once these conventions are accepted, the facts expressed by science necessarily are either true or false. . . . Other conventions remain possible, leading to other modes of expressing oneself; but the truth, thus diversely translated, remains the same. One can pass from one system of conventions to another, from one language to another, by means of an appropriate dictionary. The very possibility of a translation shows here the existence of an invariant. . . . Conventions relate to the variable language of science, not to the invariant reality which they express [78, pp. 200–201].

7. The Empirical Status of Physical Geometry

(*i*) *Einstein's Duhemian Espousal of the Interdependence of Geometry and Physics.*

Our statement in §6 of the difference between Duhem's and Poincaré's conceptions of the interdependence of geometry and the remainder of physics calls for a critical appraisal of Duhem's thesis that the falsification of a hypothesis on the geometry of physical space is unavoidably inconclusive in isolation from the remainder of physics.*

Since Duhem's argument was *articulated* and endorsed by Einstein a decade ago, I shall state and then examine Einstein's version of it.†

We have noted throughout that physical geometry is usually conceived as the system of metric relations exhibited by transported solid bodies *independently of their particular chemical composition.*

* The 1962 statement given here is amplified in my 1966 essay in *Mind, Matter, and Method*, edited by Feyerabend and Maxwell.

† For a statement of my subsequent doubts that Einstein's articulation was also intended as an endorsement, see Chapter III, §9.1.

On this conception, the criterion of congruence can be furnished by a transported solid body for the purpose of determining the geometry by measurement only if the computational application of suitable "corrections" (or, ideally, appropriate shielding) has assured rigidity in the sense of essentially eliminating inhomogeneous thermal, elastic, electromagnetic, and other perturbational influences. For these influences are "deforming" in the sense of producing changes of *varying degree* in different kinds of materials. Since the existence of perturbational influences thus issues in a dependence of the coincidence behavior of transported solid rods on the latter's *chemical composition*, and since physical geometry is concerned with the behavior common to all solids apart from their substance-specific idiosyncrasies, the discounting of idiosyncratic distortions is an essential aspect of the logic of physical geometry. The demand for the computational *elimination* of such distortions as a prerequisite to the experimental determination of the geometry has a thermodynamic counterpart: the requirement of a means for measuring temperature which does not yield the discordant results produced by expansion thermometers at other than fixed points when different thermometric substances are employed. This thermometric need is fulfilled successfully by Kelvin's thermodynamic scale of temperature. But attention to the implementation of the corresponding prerequisite of physical geometry has led Einstein to impugn the empirical status of that geometry. He considers the case in which congruence has been defined by the diverse kinds of transported solid measuring rods *as corrected for their respective idiosyncratic distortions* with a view to then making an empirical determination of the prevailing geometry. And Einstein's thesis is that the very logic of computing these corrections precludes that the geometry itself be accessible to experimental ascertainment *in isolation from* other physical regularities.[46] Specifically, he states his case in the form of a dialogue [26, pp. 676–678] in which he attributes his own

[46] For a very detailed treatment of the relevant computations, see [94, 60, 85, 45].

Duhemian view to Poincaré and offers that view in opposition to Hans Reichenbach's conception, which we discussed in §3. But, as we saw in §6(ii), Poincaré's text will *not* bear Einstein's interpretation. For we noted that in speaking of the variations which solids exhibit under distorting influences, Poincaré says, "we neglect these variations in laying the foundations of geometry, because, besides their being very slight, they are irregular and consequently seem to us accidental" [63, p. 76]. I am therefore taking the liberty of replacing the name "Poincaré" in Einstein's dialogue by the term "Duhem and Einstein." *With this modification*, the dialogue reads as follows:

Duhem and Einstein: The empirically given bodies are not rigid, and consequently cannot be used for the embodiment of geometric intervals. Therefore, the theorems of geometry are not verifiable.
Reichenbach: I admit that there are no bodies which can be *immediately* adduced for the "real definition" [i.e., physical definition] of the interval. Nevertheless, this real definition can be achieved by taking the thermal volume-dependence, elasticity, electro- and magneto-striction, etc., into consideration. That this is really and without contradiction possible, classical physics has surely demonstrated.
Duhem and Einstein: In gaining the real definition improved by yourself you have made use of physical laws, the formulation of which presupposes (in this case) Euclidean geometry. The verification, of which you have spoken, refers, therefore, not merely to geometry but to the entire system of physical laws which constitute its foundation. An examination of geometry by itself is consequently not thinkable. Why should it consequently not be entirely up to me to choose geometry according to my own convenience (i.e. Euclidean) and to fit the remaining (in the usual sense "physical") laws to this choice in such manner that there can arise no contradiction of the whole with experience? [26, pp. 676–678.]

By speaking here of the "real definition" (i.e., the coordinative definition) of "congruent intervals" by the corrected transported rod, Einstein is ignoring that the actual and potential physical meaning of congruence in physics *cannot* be given exhaustively by any *one* physical criterion or test condition. But here as elsewhere throughout this essay (cf. §5(ii)), we can safely ignore this open cluster character of the concept of congruence. For our concern as well as Einstein's is merely to single out *one* particular congruence class from among an infinitude of such alternative classes. And as long as our specification of that one chosen class is *unambiguous*, it is wholly immaterial that there are also *other* physical cirteria (or test conditions) by which it could be specified.

Einstein is making two major points here: (1) In obtaining a physical geometry by giving a physical interpretation of the postulates of a formal geometric axiom system, the specification of the physical meaning of such theoretical terms as "congruent," "length," or "distance" is *not* at all simply a matter of giving an operational definition in the strict sense. Instead, what has been variously called a "rule of correspondence" (Margenau and Carnap), a "coordinative definition" (Reichenbach), an "epistemic correlation" (Northrop), or a "dictionary" (N. R. Campbell) is provided here *through the mediation of hypotheses and laws* which are collateral to the geometric theory whose physical meaning is being specified. Einstein's point that the physical meaning of congruence is given by the transported rod *as corrected theoretically* for idiosyncratic distortions is an illuminating one and has an abundance of analogues throughout physical theory, thus showing, incidentally, that strictly operational definitions are a rather simplified and limiting species of rules of correspondence (cf. §2(ii)). (2) Einstein's second claim, which is the cardinal one for our purposes, is that the role of collateral theory in the physical definition of congruence is such as to issue in the following *circularity*, from which there is no escape, he maintains, short of acknowledging the existence of an a priori element *in the sense of the Duhemian ambiguity:* the rigid body is not even defined without first *decreeing* the validity of Euclidean geometry (or of some other particular geometry). For *before* the *corrected* rod can be used to make an empirical determination of the *de facto* geometry, the required corrections must be computed via laws, such as those of elasticity, which involve Euclideanly calculated areas and volumes [83, 89]. But clearly the warrant for thus introducing Euclidean geometry *at this stage* cannot be empirical.

In the same vein, H. Weyl endorses Duhem's position as follows:

> Geometry, mechanics, and physics form an inseparable theoretical whole [96, p. 67]. . . . Philosophers have put forward the thesis that the validity or non-validity of Euclidean geometry cannot be proved by empirical observations. It must in fact be granted that in all such observations essentially physical assumptions, such as the statement that the path of a ray of light is a straight line and other similar statements, play a prominent part. This merely bears out the remark already made above that it is only the whole composed of geometry and physics that may be tested empirically [96, p. 93].

If Einstein's Duhemian thesis were to prove correct, then it would have to be acknowledged that there is a sense in which physical geometry *itself* does not provide a geometric characterization of physical reality. For by this characterization we understand the articulation of the system of relations obtaining between bodies and transported solid rods quite apart from their substance-specific distortions. And to the extent to which physical geometry is a priori in the sense of the Duhemian ambiguity, there is an ingression of a priori elements into physical theory to take the place of distinctively geometric gaps in our knowledge of the physical world.

(*ii*) *Critique of Einstein's Duhemian Thesis.*

I shall set forth my doubts regarding the soundness of Einstein's contention in three parts consisting of the following: (1) A critique of the general Duhemian schema of the logic of falsifiability, as presented in §6(i), in the form of assertions (a) and (b). (2) An analysis of the status of the Einstein-Duhem argument in the special case in which effectively no deforming influences are present in a certain region whose geometry is to be ascertained. (3) An evaluation of Einstein's version of Duhem as applied to the empirical determination of the geometry of a region which *is* subject to deforming influences.

1. Referring to §6(i), let us now consider the two parts (a) and (b) of the schema which the *stronger* form (ii) of the Duhemian thesis claims to be the *universal paradigm* of the logic of falsifiability in empirical science. Clearly, part (a) is valid, being a *modus tollens* argument in which the antecedent of the conditional premise is a conjunction. But part (a) utilizes the *de facto* findings O' only to the extent that they are *incompatible* with the observational expectations O derived from the conjunction of H and A. And part (a) is *not at all sufficient to show that the falsifiability of H as part of an explanans of the actual empirical facts O′ is unavoidably inconclusive.* For neither part (a) nor other general logical considerations can guarantee the deducibility of O' from an *explanans* constituted by the conjunction of H and some *nontrivial* revised set A' of the auxiliary assumptions which is logically incompatible with A under the hypothesis H.[47]

[47] The requirement of nontriviality of A' requires clarification. If one were to allow O' itself, for example, to qualify as a set A', then, of course, O' could be deduced trivially, and H would not even be needed in the *explanans*. Hence a *necessary* condition for the *nontriviality* of A' is that H be required in addition to A' for the deduction of the *explanandum*. But, as N. Rescher has pointed out to me, this necessary condition is not also sufficient. For it fails to rule out an A' of the trivial form $\sim H \vee O'$ (or $H \supset O'$) from which O' could *not* be deduced without H.

The unavailability of a formal sufficient condition for nontriviality is not, however, damaging to the critique of Duhem presented in this paper. For surely Duhem's illustrations from the history of physics as well as the whole tenor of his writing indicate that *he intends his thesis to stand or fall on the existence of the kind of A′ which we would all recognize as nontrivial in any given case.* Any endeavor to save Duhem's thesis from refutation by invoking the kind of A' which no scientist would accept as admissible would turn Duhem's thesis into a most *unenlightening* triviality that no one would wish to contest. Thus, I have no intention whatever of denying the following compound formal claim: "if H and A entail O, the falsity of O does not entail the falsity of H, *and* there will always be *some kind of A′* which, in conjunction with H, will entail O'."

Interestingly, G. Rayna has pointed out to me that there is *equivalence* between the following two claims:

(i) Duhem's thesis that for *any* H and O',

$(\exists A')[(H \cdot A') \rightarrow O']$, where A' is nontrivial,

and

(ii) the seemingly stronger assertion that there exist infinitely many significantly different $A'_1, A'_2, A'_3, \ldots$ such that

How then does Duhem propose to assure that there exists such a *nontrivial* set A' for any one component hypothesis H *independently* of the domain of empirical science to which H pertains? It would seem that such assurance *cannot* be given on general logical grounds at all but that the existence of the required set A' needs *separate* and *concrete* demonstration for each particular case. In short, even in contexts to which part (a) of the Duhemian schema is applicable—which is *not* true for *all* contexts, as we shall see—neither the premise

$$[(H \cdot A) \rightarrow O] \cdot \sim O$$

nor other general logical considerations entail that

$$(\exists A')[(H \cdot A') \rightarrow O'],$$

where A' is nontrivial in the sense discussed in footnote 47. And hence Duhem's thesis that the falsifiability of an *explanans H* is unavoidably inconclusive is a *non sequitur.**

That the Duhemian thesis is not only a *non sequitur* but actually

---

$$(H \cdot A_i') \rightarrow O', \qquad i = 1, 2, 3, \ldots$$

To prove this equivalence, we consider any infinite set $B_1, B_2, B_3, \ldots$ of pairwise incompatible hypotheses of the kind that would qualify as a constituent of A'. Upon treating the conjunction $(H \cdot B_i)$ as one hypothesis to which Duhem's thesis can now be applied, we obtain

$$(\exists B_i')[\{(H \cdot B_i) \cdot B_i'\} \rightarrow O'] \qquad i = 1, 2, 3, \ldots$$

But this can be written as

$$(\exists B_i')[\{H \cdot (B_i \cdot B_i')\} \rightarrow O'].$$

Let $A_i' = B_i \cdot B_i'$ by definition. In view of the previously granted existence of the B_i, we then have

$$(\exists A_i')[(H \cdot A_i') \rightarrow O'], \qquad i = 1, 2, 3, \ldots$$

where the A_i' are significantly distinct, since the B_i, B_j were chosen incompatible for $i \neq j$. Q.E.D.

* Here and in the next paragraph, I should have stated the issue between Duhem and myself as pertaining to *separate* falsifiability and *not* to "conclusive" falsifiability. The degree to which the separate falsification of a particular H will be "conclusive" will clearly depend on the degree of our *inductive confidence* in the empirical premises of the deductive argument on which we rest the conclusion that H is false. The mere fact that $\sim H$ is *deductively* inferable does not justify my use of the term "conclusive" in the text. For further details, see my essay, "The Falsifiability of a Component of a Theoretical System," in *Mind, Matter, and Method*, edited by Feyerabend and Maxwell, pp. 285–289.

false is borne out, as we shall now see, by the case of testing the hypothesis that a certain *physical geometry* holds, a case of conclusive falsifiability which yields an important counterexample to Duhem's stronger thesis concerning falsifiability but which does justify the *weaker* form (i) of his thesis stated in §6(i).

This counterexample will serve to show that (1) by denying the feasibility of conclusive falsification, the Duhemian schema is a serious *misrepresentation* of the actual logical situation characterizing an important class of cases of falsifiability of a purported *explanans*, and that (2) the plausibility of Duhem's thesis derives from the false supposition that part (a) of the schema *is* always applicable *and* that its formal validity guarantees the applicability of part (b) of the schema.

2. If we are confronted with the problem of the falsifiability of the geometry ascribed to a region which is effectively free from deforming influences, then the *correctional* physical laws play no role as auxiliary assumptions, and the latter reduce to the claim that the region in question is, in fact, effectively *free* from deforming influences. And *if* such freedom can be affirmed *without* presupposing collateral theory, then the geometry alone rather than only a wider theory in which it is ingredient will be falsifiable. On the other hand, if collateral theory *were* presupposed here, then Duhem and Einstein might be able to adduce its modifiability to support their claim that the geometry *itself* is *not* conclusively falsifiable. The question is therefore whether freedom from deforming influences can be asserted and ascertained independently of (sophisticated) collateral theory. My answer to this question is Yes.* For quite independently of the conceptual elaboration of such physical magnitudes as temperature whose constancy would characterize a region free from deforming influences, the absence of perturbations is certifiable for

* For an important amplification of the reply to this question which is about to be given here, see my essay, "The Falsifiability of a Component of a Theoretical System," in *Mind, Matter, and Method*, edited by Feyerabend and Maxwell, pp. 285–289.

the region as follows: two solid rods of very *different* chemical constitution which coincide at one place in the region will also coincide everywhere else in it independently of their paths of transport. It would *not* do for the Duhemian to object here that the certification of two solids as quite *different chemically* is theory-laden to an extent permitting him to uphold his thesis of the inconclusive falsifiability of the geometry. For suppose that observations were so ambiguous as to permit us to assume that two solids which *appear* strongly to be chemically *different* are, in fact, chemically identical in all relevant respects. If so rudimentary an observation were thus ambiguous, then no observation could ever possess the required univocity to be incompatible with an observational consequence of a *total theoretical system.* And if that were the case, Duhem could hardly avoid the following conclusion: "observational findings are always so unrestrictedly ambiguous as *not* to permit even the refutation of any given *total theoretical system.*" But such a result would be tantamount to the absurdity that *any* total theoretical system can be espoused as true a priori. Thus, it would seem that if Duhem is to maintain, as he does, that a *total theoretical system is* refutable by confrontation with observational results, then he must allow that the coincidence of diverse kinds of rods independently of their paths of transport is certifiable observationally. Accordingly, the absence of deforming influences is ascertainable *independently* of any assumptions as to the geometry and of other (sophisticated) collateral theory.

Let us now employ our earlier notation and denote the geometry by "H" and the assertion concerning the freedom from perturbations by "A." Then, once we have laid down the congruence definition and the remaining semantical rules, the physical geometry H becomes conclusively falsifiable as an *explanans* of the posited empirical findings O'. For the actual logical situation is characterized *not* by part (a) of the Duhemian schema but instead by the schema $[\{(H \cdot A) \rightarrow O\} \cdot \sim O \cdot A] \rightarrow \sim H$.

It will be noted that we identified the H of the Duhemian schema with the geometry. But since a geometric theory, at least in its synthetic form, can be axiomatized as a conjunction of logically independent postulates, a particular axiomatization of *H* could be decomposed logically into various sets of component subhypotheses. Thus, for example, the hypothesis of Euclidean geometry could be stated, if we wished, as the conjunction of two parts consisting respectively of the Euclidean parallel postulate and the postulates of absolute geometry. And the hypothesis of hyperbolic geometry could be stated in the form of a conjunction of absolute geometry and the hyperbolic parallel postulate.

In view of the logically compounded character of a geometric hypothesis, Professor Grover Maxwell has suggested that the Duhemian thesis may be tenable in this context if we construe it as pertaining *not* to the falsifiability of a geometry as a whole but to the falsifiability of its component subhypotheses in any given axiomatization. There are two ways in which this proposed interpretation might be understood: (1) as an assertion that *any one component sub*hypothesis eludes conclusive refutation on the grounds that the empirical findings can falsify the set of axioms only as a whole, or (2) in any given axiomatization of a physical geometry there exists *at least one component sub*hypothesis which eludes conclusive refutation.

The first version of the proposed interpretation will not bear examination. For suppose that *H* is the hypothesis of Euclidean geometry and that we consider absolute geometry as one of its subhypotheses and the Euclidean parallel postulate as the other. If now the empirical findings were to show that the geometry is hyperbolic, then indeed absolute geometry would have eluded refutation. But if, on the other hand, the prevailing geometry were to turn out to be spherical, then the mere replacement of the Euclidean parallel postulate by the spherical one could not possibly save absolute geometry from refutation. For absolute geometry alone is logically incompatible with spherical geometry and hence with the posited empirical findings.

If one were to read Duhem according to the very cautious *second* version of Maxwell's proposed interpretation, then our analysis of the logic of testing the geometry of a *perturbation-free* region could not be adduced as having furnished a counterexample to so mild a form of Duhemism. And the question of the validity of this highly attenuated version is thus left open by our analysis without any detriment to that analysis.

We now turn to the critique of Einstein's Duhemian argument as applied to the empirical determination of the geometry of a region which is subject to deforming influences.

3. There can be no question that when deforming influences *are* present, the laws used to make the corrections for deformations involve areas and volumes in a fundamental way (e.g., in the definitions of the elastic stresses and strains) and that this involvement presupposes a geometry, as is evident from the area and volume formulae of differential geometry, which contain the square root of the determinant of the components g_{ik} of the metric tensor [27, p. 177]. Thus, the empirical determination of the geometry involves the joint assumption of a geometry and of certain collateral hypotheses. But we see already that this assumption *cannot* be adequately represented by the conjunction $H \cdot A$ of the Duhemian schema where H represents the geometry.

Now suppose that we begin with a set of Euclideanly formulated physical laws P_0 in correcting for the distortions induced by perturbations and then use the thus Euclideanly corrected congruence standard for *empirically* exploring the geometry of space by determining the metric tensor. *The initial stipulational affirmation of the Euclidean geometry G_0 in the physical laws P_0 used to compute the corrections in no way assures that the geometry obtained by the corrected rods will be Euclidean!* If it is *non*-Euclidean, then the question is, What will be required by Einstein's fitting of the physical laws to preserve Euclideanism and avoid a contradiction of the theoretical system with experience? Will the adjustments in P_0 necessitated by the retention of Euclidean geometry entail merely a change in the dependence of the length assigned to the transported rod on such *nonpositional* parameters as temperature, pressure, and magnetic field? Or could the putative empirical findings compel that the length of the transported rod be likewise made a nonconstant function of its *position* and *orientation* as *independent* variables in order to square the coincidence findings with the requirement of Euclideanism? The possibility of obtaining *non*-Euclidean results by measurements carried out in a spatial region uniformly characterized by standard conditions of temperature, pressure, electric and magnetic field strength, etc., shows it to be *extremely doubtful*, as we shall now show, that the preservation of Euclideanism could *always* be accomplished short of introducing *the dependence of the rod's length on the independent variables of position or orientation.*

Suppose that, relatively to the customary congruence standard, the geometry prevailing in a given region when *free* from perturbational influences is that of a strongly *non*-Euclidean space of spatially and temporally constant curvature. Then what would be the character of the alterations in the *customary* correctional laws which Einstein's thesis would require to assure the *Euclideanism* of that region relatively to the customary congruence standard under *perturbational* conditions? The required alterations would be *independently falsifiable*, as will now be demonstrated, because they would involve affirming that such coefficients as those of linear thermal expansion *depend on the independent variables of spatial position.* That such a space dependence of the correctional coefficients might well be necessitated by the exigencies of Einstein's Duhemian thesis can be seen as follows by reference to the law of linear thermal expansion.* In the usual version of physical theory, the first approximation of that law is given by $L = L_0(1 + \alpha \cdot \Delta T)$. If Einstein is to guarantee the Euclideanism of the region under discussion by means of logical devices that are consonant with his thesis, and if our region is subject only to *thermal* perturbations for some time, then,

* The thermal argument against the Duhemian position to be sketched here is developed in great detail in §§9.1 and 9.2 of Chapter III.

unlike the customary law of linear thermal expansion, the revised form of that law needed by Einstein will have to bear the twin burden of effecting *both* of the following *two* kinds of superposed corrections: (1) the *changes* in the lengths ascribed to the transported rod in different positions or orientations which would be required even if our region *were* everywhere at the standard temperature, merely for the sake of rendering Euclidean its otherwise *non*-Euclidean geometry, and (2) corrections compensating for the effects of the *de facto* deviations from the standard temperature, these corrections being the sole onus of the *usual* version of the law of linear thermal expansion. What will be the consequences of requiring the *revised* version of the law of thermal elongation to implement the *first* of these two kinds of corrections in a context in which the deviation ΔT from the standard temperature is *the same* at *different* points of the region, that temperature deviation having been measured in the manner chosen by the Duhemian? Specifically, what will be the character of the coefficients α of the *revised* law of thermal elongation under the posited circumstances, if Einstein's thesis is to be implemented by effecting the *first* set of corrections? Since the new version of the law of thermal expansion will then have to guarantee that the lengths L assigned to the rod at the various points of *equal* temperature T *differ* appropriately, it would seem clear that logically possible empirical findings could compel Einstein to make the coefficients α of solids *depend* on the *space coordinates.**

But such a spatial dependence is *independently falsifiable:* comparison of the thermal elongations of an aluminum rod, for example, with an invar rod of essentially zero α by, say, the Fizeau method *might* well show that the α of the aluminum rod is a characteristic of aluminum which is *not* dependent on the space coordinates. And even if it *were* the case that the α's are found to be space dependent, how could Duhem and Einstein assure that this space dependence

* In the case of points for which $\Delta T = 0$, the assumption of the space-dependence of α is unavailing, since the formula shows that L has the constant value L_0 at all such points independently of what α is assumed to be, whereas the lengths L are required to be different.

would have the particular functional form required for the success of their thesis?

We see that the required resort to the introduction of a spatial dependence of the thermal coefficients might well *not* be open to Einstein. Hence, in order to retain Euclideanism, it would then be necessary to *remetrize* the space in the sense of abandoning the customary definition of congruence, entirely apart from any consideration of idiosyncratic distortions and even after correcting for these in some way or other. But this kind of remetrization, though entirely admissible in *other* contexts, does *not* provide the requisite support for Einstein's Duhemian thesis! For Einstein offered it as a criticism of Reichenbach's conception. And hence it is the avowed onus of that thesis to show that the geometry *by itself* cannot be held to be empirical, i.e., falsifiable, even when, with Reichenbach, we have sought to assure its empirical character by choosing and then adhering to the usual (standard) definition of spatial congruence, which *excludes* resorting to such remetrization.

Thus, there may well obtain observational findings O', expressed in terms of a particular definition of congruence (e.g., the *customary* one), which are such that there does *not* exist any nontrivial set A' of auxiliary assumptions capable of preserving the Euclidean H in he face of O'. And this result alone suffices to invalidate the Einsteinian version of Duhem's thesis to the effect that any geometry, such as Euclid's, can be preserved in the face of any experimental findings which are expressed in terms of the customary definition of congruence.

It might appear that my geometric counterexample to the Duhemian thesis of unavoidably inconclusive falsifiability of an *explanans* is vulnerable to the following criticism: "To be sure, Einstein's geometric articulation of that thesis does not leave room for saving it by resorting to a remetrization in the sense of making the length of the rod *vary* with position or orientation even *after* it has been corrected for idiosyncratic distortions. But why saddle the Duhemian thesis as such with a restriction peculiar to Einstein's particular version of it? And thus why not allow Duhem to save his thesis by countenancing those *alterations in the congruence definition* which are *remetrizations?*"

My reply is that to deny the Duhemian the invocation of such an alteration of the congruence definition *in this context* is *not* a matter of gratuitously requiring him to justify his thesis within the confines of Einstein's particular version of that thesis; instead, the imposition of this restriction is entirely legitimate here, and the Duhemian could hardly wish to reject it as unwarranted. For it is of the essence of Duhem's contention that H (in this case, Euclidean geometry) can always be preserved *not* by tampering with the semantical rules (interpretive sentences) linking H to the observational base but rather by availing oneself of the alleged *inductive latitude* afforded by the ambiguity of the experimental evidence to do the following: (a) leave the factual commitments of H *unaltered* by retaining both the statement of H and the semantical rules linking its terms to the observational base,* and (b) replace the set A by a set A' of auxiliary assumptions *differing in factual content* from A such that A and A' are logically incompatible under the hypothesis H. Now, the factual content of a geometrical hypothesis can be changed either by preserving the original statement of the hypothesis while changing one or more of the semantical rules or by keeping all of the semantical rules intact and suitably changing the statement of the hypothesis (cf. §6(i)). We can see, therefore, that the retention of a Euclidean H by the device of changing through remetrization the semantical rule governing the meaning of "congruent" (for line segments) effects a retention *not* of the *factual commitments* of the original Euclidean H but only of its *linguistic trappings*. That the thus "preserved" Euclidean H actually repudiates the factual commitments of the

* In the geometric example here at issue, the *essential* restriction imposed by (a) is to adhere to the claim that the differentially undeformed, self-congruent transported rod yields a metric issuing in the stipulated Euclidean H, i.e., that the chosen metric $ds^2 = g_{ik}\,dx^i\,dx^k$ is realized physically by such a self-congruent rod. For a statement of a refinement of (a), which precludes the impugning of its feasibility, see my *Philosophical Problems of Space and Time*, p. 143.

original one is clear from the following: the *original* Euclidean *H* had asserted that the coincidence behavior common to all kinds of solid rods is Euclidean, *if* such transported rods are taken as the physical realization of congruent intervals; but the Euclidean *H* which survived the confrontation with the posited empirical findings only by dint of a *remetrization* is predicated on a denial of the very assertion that was made by the original Euclidean *H*, which it was to "preserve." It is as if a physician were to endeavor to "preserve" an a priori diagnosis that a patient has acute appendicitis in the face of a negative finding (yielded by an exploratory operation) as follows: he would redefine "acute appendicitis" to denote the healthy state of the appendix!

Hence, the confines within which the Duhemian must make good his claim of the preservability of a Euclidean *H* do *not* admit of the kind of change in the congruence definition which alone would render his claim tenable under the assumed empirical conditions. Accordingly, the geometrical critique of Duhem's thesis given in this paper does *not* depend for its validity on restrictions peculiar to Einstein's version of it.

Even apart from the fact that Duhem's thesis precludes resorting to an alternative metrization to save it from refutation in our geometrical context, the very feasibility of alternative metrizations is vouchsafed *not* by any general Duhemian considerations pertaining to the logic of falsifiability but by a property peculiar to the subject matter of geometry (and chronometry): the latitude for *convention* in the ascription of the spatial (or temporal) *equality* relation to intervals in the continuous manifolds of physical space (or time).*

But what of the possibility of actually *extricating* the unique underlying geometry (to within experimental accuracy) from the network of hypotheses which enter into the testing procedure?

That contrary to Duhem and Einstein, the geometry itself may

* The defense of this conclusion given in §§2(i) and 5(i) above will be substantially strengthened in Chapter III, §§2.9, 2.10, 2.11, 3.2, 4, and 7.

well be empirical, once we have renounced the kinds of alternative congruence definitions employed by Poincaré, is seen from the following possibilities of its successful empirical determination. After assumedly obtaining a non-Euclidean geometry G_1 from measurements with a rod corrected on the basis of Euclideanly formulated physical laws P_0, we can revise P_0 so as to conform to the non-Euclidean geometry G_1 just obtained by measurement. This retroactive revision of P_0 would be effected by recalculating such quantities as areas and volumes on the basis of G_1 and changing the functional dependencies relating them to temperatures and other physical parameters. Let us denote by "P_1'" the set of physical laws P resulting from thus revising P_0 to incorporate the geometry G_1. Since various physical magnitudes ingredient in P_1' involve lengths and durations, we now use the set P_1' to correct the rods (and clocks) with a view to seeing whether rods and clocks thus corrected will reconfirm the set P_1'. If not, *modifications* need to be made in this set of laws so that the functional dependencies between the magnitudes ingredient in them reflect the new standards of spatial and temporal congruence defined by P_1'-corrected rods and clocks. We may thus obtain a new set of physical laws P_1.

Now we employ this set P_1 of laws to correct the rods for perturbational influences and then determine the geometry with the thus corrected rods. Suppose the result is a geometry G_2 different from G_1. Then if, upon repeating this two-step process several more times, there is convergence to a geometry of constant curvature we must continue to repeat the two-step process an additional finite number of times until we find the following: the geometry G_n ingredient in the laws P_n providing the basis for perturbation corrections is indeed the same (to within experimental accuracy) as the geometry obtained by measurements with rods that have been corrected via the set P_n. *If* there is such convergence at all, it will be to the same geometry G_n even if the physical laws used in making the *initial* corrections are not the set P_0, which presupposes Euclidean geometry, but a different set P based on some non-Euclidean geometry or other.* That there can exist only one such geometry of constant curvature G_n would seem to be guaranteed by the identity of G_n with the unique underlying geometry G_t characterized by the following properties: (i) G_t would be exhibited by the coincidence behavior of a transported rod if the *whole* of the space were actually free of deforming influences, (ii) G_t would be obtained by measurements with rods corrected for distortions on the basis of physical laws P_t presupposing G_t, and (iii) G_t would be found to prevail in a given relatively small, perturbation-free region of the space quite independently of the assumed geometry ingredient in the correctional physical laws. Hence, *if* our method of successive approximation does converge to a geometry G_n of *constant* curvature, then G_n would be this unique underlying geometry G_t.† And, in that event, we can claim to have found empirically that G_t is indeed the geometry prevailing in the entire space which we have explored.

But what if there is no convergence? It might happen that whereas convergence would obtain by starting out with corrections based on the set P_0 of physical laws, it would *not* obtain by beginning instead with corrections presupposing some particular *non*-Euclidean set P or vice versa: just as in the case of Newton's method of successive approximation [15, p. 286], there are conditions, as A. Suna has pointed out to me, under which there would be no convergence. We might then nonetheless succeed as follows in finding the geometry G_t empirically, *if* our space is one of constant curvature.

The geometry G_r resulting from measurements by means of a corrected rod is a single-valued function of the geometry G_a assumed in the correctional physical laws, and a Laplacian demon having sufficient knowledge of the facts of the world would know this function $G_r = f(G_a)$. Accordingly, we can formulate the problem

* This statement is too strong and is amended in Chapter II, §2(ii). There it is also noted in what sense a sequence of geometries can meaningfully converge to a particular geometry of *variable* curvature.

† *Ibid.*

of determining the geometry empirically as the problem of finding the point of intersection between the curve representing this function and the straight line $G_r = G_a$. That there exists one and only one such point of intersection follows from the existence of the geometry G_t defined above, provided that our space is one of constant curvature.* Thus, what is now needed is to make determinations of the G_r corresponding to a number of geometrically different sets of correctional physical laws P_a, to draw the most reasonable curve $G_r = f(G_a)$ through this finite number of points (G_a, G_r), and then to find the point of intersection of this curve and the straight line $G_r = G_a$.

Whether this point of intersection turns out to be the one representing Euclidean geometry or not is beyond the reach of our conventions, *barring* a remetrization. And thus the least that we can conclude is that since empirical findings can greatly narrow down the range of uncertainty concerning the prevailing geometry, there is no assurance of the *latitude* for the choice of a geometry which Einstein takes for granted. Einstein's Duhemian position would appear to be inescapable *only* if our proposed method of determining the geometry by itself empirically *cannot* be generalized in some way to cover the general relativity case of a space of variable curvature and if the latter kind of theory turns out to be true.[48]

It would seem therefore that, contrary to Einstein, the logic of eliminating distorting influences prior to stipulating the rigidity of a solid body is *not* such as to provide scope for the ingression of conventions over and above those acknowledged in Riemann's analysis

[48] The extension of our method to the case of a geometry of variable curvature is not simple. For in that case, the geometry G is no longer represented by a single scalar given by a Gaussian curvature, and our graphical method breaks down. Whatever the answer to this open question of the extensibility of our method, I have argued in another publication [31] that it is wholly misconceived to suppose with J. Maritain [49] that there are *supra*-scientific philosophical means for ascertaining the underlying geometry, if *scientific* procedures do not succeed in unraveling it.

* This assertion of uniqueness is likewise amended (weakened) in Chapter II, §2(ii).

of congruence, and trivial ones such as the system of units used. *Mutatis mutandis*, an analogous conclusion can be established in regard to the application of corrections to provide a standard of isochronism for clocks.

8. Summary

The present essay has endeavored to answer the following multifaceted question: In what sense and to what extent can the ascription of a particular metric geometry to physical space and the chronometry ingredient in physical theory be held to have an *empirical* warrant?

Our analysis of the logical status of the concept of a rigid body and of an isochronous clock leads to the conclusion that once the physical meaning of congruence has been stipulated by reference to a solid body and to a clock respectively for whose distortions allowance has been made computationally as outlined, then the geometry and the ascriptions of durations to time intervals is determined uniquely by the totality of relevant empirical facts. It is true, of course, that even apart from experimental errors, not to speak of quantum limitations on the accuracy with which the metric tensor *of space-time* can be meaningfully ascertained by measurement [103, 82], no *finite* number of data can uniquely determine the functions constituting the representations g_{ik} of the metric tensor in any given coordinate system. But the criterion of *inductive* simplicity which governs the free creativity of the geometer's imagination in his choice of a particular metric tensor here is the same as the one employed in theory formation in any of the nongeometrical portions of empirical science. And choices made on the basis of such inductive simplicity are in principle true or false, unlike those springing from considerations of descriptive simplicity, which merely reflect conventions.*

* The conclusion stated here will be defended further in some detail in §9 of Chapter III.

Bibliography

1. Barankin, E. W. "Heat Flow and Non-Euclidean Geometry," *American Mathematical Monthly*, 49:4–14 (1942).
2. Baruch, J. J. "Horological Accuracy: Its Limits and Implications," *American Scientist*, 46:188A–196A (1958).
3. Bonola, R. *Non-Euclidean Geometry*. New York: Dover Publications, 1955.
4. Braithwaite, R. B. "Axiomatizing a Scientific System by Axioms in the Form of Identification," in *The Axiomatic Method*, L. Henkin, P. Suppes, and A. Tarski, eds. Amsterdam: North Holland Publishing Company, 1959. Pp. 429–442.
5. Brouwer, D. "The Accurate Measurement of Time," *Physics Today*, 4:7–15 (1951).
6. Brown, F. A., Jr. "Biological Clocks and the Fiddler Crab," *Scientific American*, 190:34–37 (1954).
7. Brown, F. A., Jr. "The Rhythmic Nature of Animals and Plants," *American Scientist*, 47:147–168 (1959); "Living Clocks," *Science*, 130:1535–1544 (1959); and "Response to Pervasive Geophysical Factors and the Biological Clock Problem," *Cold Spring Harbor Symposia on Quantitative Biology*, 25:57–71 (1960).
8. Carnap, R. *Der Raum*. Berlin: Reuther and Reichard, 1922.
9. Carnap, R. Preface to H. Reichenbach's *The Philosophy of Space and Time*. New York: Dover Publications, 1958.
10. Clemence, G. M. "Astronomical Time," *Reviews of Modern Physics*, 29:2–8 (1957).
11. Clemence, G. M. "Dynamics of the Solar System," in *Handbook of Physics*, E. Condon and H. Odishaw, eds. New York: McGraw-Hill, 1958. P. 65.
12. Clemence, G. M. "Ephemeris Time," *Astronomical Journal*, 64:113–115 (1959) and *Transactions of the International Astronomical Union*, Vol. 10 (1958).
13. Clemence, G. M. "Time and Its Measurement," *American Scientist*, 40:260 (1952).
14. Clifford, W. K. *The Common Sense of the Exact Sciences*. New York: Dover Publications, 1955.
15. Courant, R. *Vorlesungen über Differential und Integralrechnung*, Vol. 1. Berlin: Springer, 1927.
16. D'Abro, A. *The Evolution of Scientific Thought from Newton to Einstein*. New York: Dover Publications, 1950.
17. Denbigh, K. G. "Thermodynamics and the Subjective Sense of Time," *British Journal for the Philosophy of Science*, 4:183–191 (1953).
18. Dingler, H. "Die Rolle der Konvention in der Physik," *Physikalische Zeitschrift*, 23:47–52 (1922).
19. Duhem, P. *The Aim and Structure of Physical Theory*. Princeton: Princeton University Press, 1954.
20. Eddington, A. S. *Space, Time and Gravitation*. Cambridge: Cambridge University Press, 1953.

21. Einstein, A. "The Foundations of the General Theory of Relativity," in *The Principle of Relativity, a Collection of Original Memoirs.* New York: Dover Publications, 1952. Pp. 111–164.
22. Einstein, A. *Geometrie und Erfahrung.* Berlin: Springer, 1921.
23. Einstein, A. "On the Electrodynamics of Moving Bodies," in *The Principle o- Relativity, a Collection of Original Memoirs.* New York: Dover Publicaf tions, 1952. Pp. 37–65.
24. Einstein, A. "On the Influence of Gravitation on the Propagation of Light," in *The Principle of Relativity, a Collection of Original Memoirs.* New York: Dover Publications, 1952. Pp. 97–108.
25. Einstein, A. "Prinzipielles zur allgemeinen Relativitätstheorie," *Annalen der Physik,* 55:241 (1918).
26. Einstein, A. "Reply to Criticisms," in *Albert Einstein: Philosopher-Scientist,* P. A. Schilpp, ed. Evanston: The Library of Living Philosophers, 1949. Pp. 665–688.
27. Eisenhart, L. P. *Riemannian Geometry.* Princeton: Princeton University Press, 1949.
28. Feigl, H. "Confirmability and Confirmation," *Revue Internationale de Philosophie,* 5:268–279 (1951). Reprinted in *Readings in Philosophy of Science,* P. P. Wiener, ed. New York: Scribner, 1953.
29. Fraenkel, A. A., and Y. Bar-Hillel. *Foundations of Set Theory.* Amsterdam: North Holland Publishing Company, 1958.
30. Goodhard, C. B. "Biological Time," *Discovery,* 18:519–521 (December 1957).
31. Grünbaum, A. "The *A Priori* in Physical Theory," in *The Nature of Physical Knowledge,* L. W. Friedrich, ed. Bloomington: Indiana University Press, 1960. Pp. 109–128.
32. Grünbaum, A. "Carnap's Views on the Foundations of Geometry," in *The Philosophy of Rudolf Carnap,* P. A. Schilpp, ed. La Salle, Ill.: Open Court, 1963.
33. Grünbaum, A. "A Consistent Conception of the Extended Linear Continuum as an Aggregate of Unextended Elements," *Philosophy of Science,* 19:288–306 (1952).
34. Grünbaum, A. "Logical and Philosophical Foundations of the Special Theory of Relativity," in *Philosophy of Science: Readings,* A. Danto and S. Morgenbesser, eds. New York: Meridian, 1960. Pp. 399–434.
35. Grünbaum, A. "Operationism and Relativity," *Scientific Monthly,* 79:228–231 (1954). Reprinted in *The Validation of Scientific Theories,* P. Frank, ed. Boston: Beacon Press, 1957. Pp. 84–94.
36. Grünbaum, A. "The Philosophical Retention of Absolute Space in Einstein's General Theory of Relativity," *Philosophical Review,* 66:525–534 (1957).
37. Grünbaum, A. "Relativity and the Atomicity of Becoming," *Review of Metaphysics,* 4:143–186 (1950).
38. Hertz, H. *The Principles of Mechanics.* New York: Dover Publications, 1956.
39. Hoagland, H. "Chemical Pacemakers and Physiological Rhythms," in *Colloid Chemistry,* Vol. V, J. Alexander, ed. New York: Reinhold, 1944. Pp. 762–785.

40. Hoagland, H. "The Physiological Control of Judgments of Duration: Evidence for a Chemical Clock," *Journal of General Psychology*, 9:267–287 (1933).
41. Hobson, E. W. *The Theory of Functions of a Real Variable*, Vol. 1. New York: Dover Publications, 1957.
42. Hood, P. *How Time Is Measured.* London: Oxford University Press, 1955.
43. Jeffreys, H. *The Earth.* 3rd ed.; Cambridge: Cambridge University Press, 1952.
44. Klein, F. *Vorlesungen über Nicht-Euklidische Geometrie.* Berlin: Springer, 1928.
45. Leclercq, R. *Guide Théorique et Pratique de la Recherche Expérimentale.* Paris: Gauthier-Villars, 1958.
46. Lyons, H. "Atomic Clocks," *Scientific American*, 196:71–82 (February 1957).
47. McVittie, G. C. "Distance and Relativity," *Science*, 127:501–505 (1958).
48. Margenau, H., and G. M. Murphy. *The Mathematics of Physics and Chemistry.* New York: Van Nostrand, 1943.
49. Maritain, J. *The Degrees of Knowledge.* New York: Scribner, 1959.
50. Milham, W. I. *Time and Timekeepers.* New York: Macmillan, 1929. A more recent edition appeared in 1941.
51. Milne, E. A. *Kinematic Relativity.* London: Oxford University Press, 1948.
52. Milne, E. A. *Modern Cosmology and the Christian Idea of God.* London: Oxford University Press, 1952.
53. Møller, C. *The Theory of Relativity.* London: Oxford University Press, 1952.
54. Newton, I. *Principia*, F. Cajori, ed. Berkeley: University of California Press, 1947.
55. Nagel, E. *The Structure of Science.* New York: Harcourt, Brace and World, 1961.
56. Northrop, F. S. C. *The Meeting of East and West.* New York: Macmillan, 1946.
57. Northrop, F. S. C. "Whitehead's Philosophy of Science," in *The Philosophy of Alfred North Whitehead*, P. A. Schilpp, ed. New York: Tudor, 1941. Pp. 165–207.
58. Page, L. *Introduction to Theoretical Physics.* New York: Van Nostrand, 1935.
59. Page, L., and N. I. Adams. *Electrodynamics.* New York: Van Nostrand, 1940.
60. Pérard, A. *Les Mesures Physiques.* Paris: Presses Universitaires de France, 1955.
61. Poincaré, H. *Dernières Pensées.* Paris: Flammarion, 1913. Chs. 2 and 3.
62. Poincaré, H. "Des Fondements de la Géométrie, à propos d'un Livre de M. Russell," *Revue de Métaphysique et de Morale*, 7:251–279 (1899).
63. Poincaré, H. *The Foundations of Science.* Lancaster, Pa.: Science Press, 1946.
64. Poincaré, H. "La Mesure du Temps," *Revue de Métaphysique et de Morale*, 6:1–13 (1898).
65. Poincaré, H. "Sur les Principes de la Géométrie, Réponse à M. Russell," *Revue de Métaphysique et de Morale*, 8:73–86 (1900).
66. Popper, K. R. *The Logic of Scientific Discovery.* London: Hutchinson, 1959.

67. Price, D. J. "The Prehistory of the Clock," *Discovery*, 17:153–157 (April 1956).
68. Reichenbach, H. "Discussion of Dingler's Paper," *Physikalische Zeitschrift*, 23:52–53 (1922).
69. Reichenbach, H. *Experience and Prediction*. Chicago: University of Chicago Press, 1938.
70. Reichenbach, H. *Philosophic Foundations of Quantum Mechanics*. Berkeley: University of California Press, 1948.
71. Reichenbach, H. "The Philosophical Significance of the Theory of Relativity," in *Albert Einstein: Philosopher-Scientist*, P. A. Schilpp, ed. Evanston: Library of Living Philosophers, 1949.
72. Reichenbach, H. *The Philosophy of Space and Time*. New York: Dover Publications, 1958.
73. Reichenbach, H. *The Rise of Scientific Philosophy*. Berkeley: University of California Press, 1951.
74. Reichenbach, H. "Über die physikalischen Konsequenzen der relativistischen Axiomatik," *Zeitschrift für Physik*, 34:35 (1925).
75. Riemann, B. "Über die Hypothesen, welche der Geometrie zu Grunde liegen," in *Gesammelte Mathematische Werke*, H. Weber, ed. New York: Dover Publications, 1953. Pp. 272–287.
76. Robertson, H. P. "The Geometries of the Thermal and Gravitational Fields," *American Mathematical Monthly*, 57:232–245 (1950).
77. Robertson, H. P. "Geometry as a Branch of Physics," *Albert Einstein: Philosopher-Scientist*, P. A. Schilpp, ed. Evanston: Library of Living Philosophers, 1949. Pp. 313–332.
78. Rougier, L. *La Philosophie Géométrique de Henri Poincaré*. Paris: Alcan, 1920.
79. Russell, B. *The Foundations of Geometry*. New York: Dover Publications, 1956.
80. Russell, B. *Our Knowledge of the External World*. London: Allen and Unwin, 1926.
81. Russell, B. "Sur les Axiomes de la Géométrie," *Revue de Métaphysique et de Morale*, 7:684–707 (1899).
82. Salecker, H., and E. P. Wigner. "Quantum Limitations of the Measurement of Space-Time Distances," *Physical Review*, 109:571–577 (1958).
83. Sokolnikoff, I. S. *Mathematical Theory of Elasticity*. New York: McGraw-Hill, 1946.
84. Sommerville, D. M. Y. *The Elements of Non-Euclidean Geometry*. New York: Dover Publications, 1958.
85. Stille, U. *Messen und Rechnen in der Physik*. Brunswick: Vieweg, 1955.
86. Struik, D. J. *Classical Differential Geometry*. Cambridge, Mass.: Addison-Wesley, 1950.
87. Tarski, A. "What Is Elementary Geometry?" in *The Axiomatic Method*, L. Henkin, P. Suppes, and A. Tarski, eds. Amsterdam: North Holland Publishing Company, 1959. Pp. 16–29.
88. Taylor, G. I. "Tidal Friction in the Irish Sea," *Philosophical Transactions*, *Royal Society of London*, Series A, 220:1–33 (1920).
89. Timoshenko, S., and J. N. Goodier. *Theory of Elasticity*. New York: McGraw-Hill, 1951.

90. Veblen, O., and J. H. C. Whitehead. *The Foundations of Differential Geometry*, No. 29 of Cambridge Tracts in Mathematics and Mathematical Physics. Cambridge: Cambridge University Press, 1932.
91. Von Helmholtz, H. *Schriften zur Erkenntnistheorie*, P. Hertz and M. Schlick, eds. Berlin: Springer, 1921.
92. Walker, A. G. "Axioms for Cosmology," in *The Axiomatic Method*, L. Henkin, P. Suppes, and A. Tarski, eds. Amsterdam: North Holland Publishing Company, 1959. Pp. 308–321.
93. Ward, F. A. B. *Time Measurement*. Part I, Historical Review. 4th ed.; London: Royal Stationery Office, 1958.
94. Weinstein, B. *Handbuch der Physikalischen Massbestimmungen*. Berlin: Springer, Vol. 1, 1886, and Vol. 2, 1888.
95. Weyl, H. *Philosophy of Mathematics and Natural Science*. Princeton: Princeton University Press, 1949.
96. Weyl, H. *Space-Time-Matter*. New York: Dover Publications, 1950.
97. Whitehead, A. N. *The Concept of Nature*. Cambridge: Cambridge University Press, 1926. Ch. VI.
98. Whitehead, A. N. *Essays in Science and Philosophy*. New York: Philosophical Library, 1947.
99. Whitehead, A. N. *The Principle of Relativity*. Cambridge: Cambridge University Press, 1922. Ch. III.
100. Whitehead, A. N. *The Principles of Natural Knowledge*. Cambridge: Cambridge University Press, 1955.
101. Whitehead, A. N. *Process and Reality*. New York: Macmillan, 1929.
102. Whyte, L. L. "Light Signal Kinematics," *British Journal for the Philosophy of Science*, 4:160–161 (1953).
103. Wigner, E. P. "Relativistic Invariance and Quantum Phenomena," *Review of Modern Physics*, 29:255–268 (1957).

II

GEOMETRY AND PHYSICS

1. The Physical Status of the Hypothesis That Everything Has Doubled in Size Overnight

The clash between Newton's and Riemann's conceptions of congruence, discussed in §2(i) of Chapter I and, *in much greater detail,* in §§2 and 3 of Chapter III, bears on the assessment of the physical import of the hypothesis of a universal nocturnal expansion. And I shall offer such an assessment in the light of that clash. But first I need to amend my articulation of Riemann's idea that in the case of a *discrete* manifold, the congruences (and the metric) can be based on the *number* of elements in parts thereof. In Chapter I (p. 12), I stated that for a discrete manifold, the "distance" between two elements can be defined in a rather natural way by the cardinality of the least number of "intervening elements." But this formulation of Riemann's assertion that for "definite parts of a manifold," the "quantitative comparison is effected by means of counting in the case of discrete magnitudes" (this volume, p. 11) undesirably yields a value of zero *both* for the "distance" between two consecutive elements and for the "distance" of an element from itself. Hence the count which is to yield the "distance" should be based instead on the cardinality of the least number of "steps" linking the two given elements of the discrete manifold. And Riemann's point had been that if physical space were granular (discrete, atomic, quantized), an extensional measure of each spatial "interval" would be *built into the interval itself* in the form of a function of the cardinal number of constituent space atoms. I shall therefore say, for brevity, that a granular space has an "intrinsic metric."

In anticipation of the discussion in Chapter III, §2.10, consider an interval AB in the mathematically continuous physical space of, say, a given blackboard, and also an interval T_0T_1 in the continuum of instants constituted by, say, the movement of a classical particle. In contrast to the situation in an atomic space, neither the cardinality of AB nor any other property *built into* the interval provides a measure of its particular spatial extension, and similarly for the temporal extension of T_0T_1. For AB has the same cardinality as any of its *proper* subintervals and also as any other nondegenerate interval CD. Corresponding remarks apply to the time interval T_0T_1. If the intervals of physical space or time *did* possess a built-in measure or "intrinsic metric," then relations of *congruence* (and also of incongruence) would obtain among disjoint space intervals AB and CD on the strength of that intrinsic metric. And in that hypothetical case, neither the existence of congruence relations among disjoint intervals nor their epistemic ascertainment would logically involve the iterative application and transport of any length standard. But the intervals of mathematically continuous physical space and time are devoid of a built-in metric. And in the absence of such an intrinsic metric, the basis for the extensional measure of an interval of physical space or time must be furnished by a relation of the interval to a body or process which is applied to it from outside itself and is thereby "extrinsic" to the interval. Hence *the very existence* and not merely the epistemic ascertainment of relations of congruence (and of incongruence) among disjoint space intervals AB and CD of continuous physical space will depend on the respective relations sustained by such intervals to an extrinsic metric standard which is applied to them. Thus, *whether two disjoint intervals are congruent at all or not will depend on the particular coincidence behavior of the extrinsic metric standard under transport* and not only on the particular intervals AB and CD. Similarly for the case of disjoint time intervals T_0T_1 and T_2T_3, and the role of clocks.

Moreover, the failure of the intervals of physical space to possess an intrinsic metric—a failure which compels recourse to an extrinsic transported metric standard to begin with—has the consequence that the continuous structure of physical space *cannot* certify the self-congruence (rigidity) of any extrinsic standard *under transport*. And similarly for physical time and the uniformity (isochronism) of clocks. Thus, physical space and time can be said to be intrinsically metrically amorphous. For precisely this reason, the following two claims made at least implicitly in the first Scholium of Newton's *Principia* [1] are unsound: (1) The criterion for the adequacy of an extrinsic length standard is that it is self-congruent (rigid) under transport as a matter of spatial fact. (2) Of two extrinsic length standards which yield *incompatible* congruence findings for disjoint intervals, there might be one which is *truly* spatially self-congruent (rigid) under transport. It should therefore not be misleading to render our verdict of unsoundness in regard to these two claims as follows: an extrinsic metric standard is self-congruent under transport *as a matter of convention* and not as a matter of spatial fact, although any *concordance* between its congruence findings and those of another such standard is indeed a matter of *fact*. If the term "transport" is suitably generalized so as to pertain as well to the extrinsic metric standards applied to time intervals, then this claim of conventionality holds for physical time as well.

Let us now combine this conclusion with our earlier one that the very *existence* of congruences among disjoint intervals depends on their empirically ascertainable relations to the conventionally self-congruent transported standard. Then it becomes evident that the existence of congruence relations among disjoint intervals is (1) a matter of convention precisely to the extent that the self-congruence of the extrinsic metric standard under transport is conventional, and (2) a matter of fact precisely to the extent to which the respective relations of the intervals to the applied extrinsic metric standard are matters of fact.

We can now turn to the analysis of the hypothesis of a universal nocturnal doubling.[1]

G. Schlesinger has maintained that "a set of circumstances can be conceived under which one would have to conclude that overnight everything has doubled in size" and that "in the absence of these circumstances we are entitled to claim that it is *false* that overnight everything has doubled in size."[2] And Schlesinger went on to contend that the only way of actually rendering the hypothesis of nocturnal doubling physically vacuous is as follows: construe this hypothesis so as to make it a singular instance of the tautology that no unverifiable proposition can be verifiable, a triviality which does not qualify as an interesting ascription of physical vacuousness to the *particular* hypothesis of nocturnal doubling.

In order to appraise Schlesinger's contention, we must note at the outset that he takes quite insufficient cognizance of two relevant facts: (1) the hypothesis of nocturnal doubling—hereafter called the "ND-hypothesis"—can be construed in several different ways, and (2) the charge of physical meaninglessness, inherent nonfalsifiability, and unverifiability has been leveled against the ND-hypothesis in the philosophical literature on a number of quite distinct grounds.

For my part, I have always construed the ND-hypothesis as being predicated on the kind of conception of spatial and temporal congruence which is set forth by Newton in the Scholium of his *Principia* [1, pp. 6–12]. As we saw, on this kind of conception, container-space (and, *mutatis mutandis*, container-time) exhibits congruence relations which are intrinsic in the following sense: its structure is such

[1] The following analysis of this hypothesis draws on Sec. III of my essay "The Falsifiability of a Component of a Theoretical System," which appeared in [3, pp. 295–305].

[2] [8, p. 68.] Similar views were put forward earlier by B. Ellis [2]. The 1964 views which I attribute to Schlesinger are not found in their entirety in his own paper but were partly set forth by him in oral discussion at the meeting of Section L of the AAAS in December 1963, and also in a departmental colloquium at the University of Pittsburgh. The stimulus of Schlesinger's remarks was very helpful to me in articulating my views.

that the *existence*—as distinct from the epistemic ascertainment—of congruence relations between nonoverlapping intervals is *built into* these intervals themselves, and therefore cannot depend in any way on the particular coincidence behavior of any kind of physical object under transport. This *transport-independence* of all congruence relations existing among intervals in principle permits the invocation of these intrinsic congruences to *authenticate* as a congruence standard any standard which is "operational" in the following *special sense:* the "operational" standard's performance of its metrical function is extrinsic and hence *transport-dependent* by involving the iterative application of a transported length standard. Given the context of the Newtonian assumption of the existence of a container-space having intrinsic congruences between different intervals at the same time and between a given interval *and itself at different times,* it was therefore possible to construe the ND-hypothesis as asserting the following: relatively to the congruences intrinsic to container-space there has been a nocturnal doubling of all extended physical objects contained in space, and *also* a doubling of *all* "operational," i.e., extrinsic and hence transport-dependent, congruence standards. Because of the legacy of Newton's metrical philosophy, it was this nontrivial version of the ND-hypothesis which I appraised logically in a 1963 book [5, Chapter 1, especially pp. 42–43]. And when I denied the falsifiability of this interpretation of the hypothesis, I did *not* rest my denial on Schlesinger's triviality that no unverifiable proposition is verifiable; instead, my denial was based on the following weighty presumed fact: the *actual failure* of physical space to possess the kind of structure which is endowed with intrinsic congruences. For this (presumed) failure deprives space of the very type of metric on whose existence the ND-hypothesis depends for its physical significance and hence falsifiability. But, one might ask: Is it *logically* possible that there be a world whose space does have a structure which exhibits intrinsic congruences?

Like Riemann, I shall answer this question in the affirmative, and I shall give an adumbration of such a world. First I wish to recall, however, that Newtonian mechanics and the nineteenth-century aether theory of light propagation each cut the ground from under their own thesis that their absolute space possesses intrinsic congruences, and they did so by postulating the mathematical *continuity* of their absolute space. For, as Riemann pointed out in his Inaugural Lecture, if physical space is mathematically continuous, the *very existence* (as distinct from epistemic ascertainment) of congruences among its intervals is not built into them but transport-dependent, *and nothing in the continuous structure of space can vouch for the rigidity (self-congruence) of the extrinsic congruence standard under transport.*[3] Thus, if Newton's space is continuous, it is devoid of intrinsic congruences. And we shall see in Chapter III, §3.3, that the absoluteness of Newton's rotational or other *accelerations* and the absoluteness of *velocity* in the aether theory of wave optics both allow the *non*existence of intrinsic congruences among any intervals of space or time. Given, therefore, that the continuous absolute space of Newtonian mechanics and nineteenth-century wave optics cannot be adduced to prove the *logical* possibility of a space endowed with intrinsic congruences, can a kind of world be imagined which is such as to establish this logical possibility?

In now confronting this question, rigorous care must be exercised lest the challenge to furnish this proof of logical possibility be turned into the following very different second challenge: to specify empirical conditions which would justify that we *reinterpret* our present information about the physical space of our *actual* world so as to attribute a structure endowed with intrinsic congruences to that space. This second challenge has to be faced *not* by my impending characterization of an appropriate imaginary world, but rather by contemporary speculations of space and time *quantization* going back, according to at least one historian, to Democritus' *mathe-*

[3] Mr. Peter Woodruff has pointed out to me that David Hume anticipated Riemann's insight in his *Treatise of Human Nature*, Part II, Sec. IV. It should be recalled from Chap. I, §2(iii), however, that Riemann's thesis of the intrinsic metrical amorphousness of the continua of space and time does *not* apply to *all* kinds of continuous manifolds.

matical atomism, a kind of atomism which must be distinguished from his much better known physical atomism.[4] Moreover, we must guard against being victimized by the tacit acceptance of a challenger's implicit question-begging requirement as follows: his implicit demand that the spatial vocabulary in which I couch my compliance with the *first* of these challenges be assigned meanings which are predicated on the assumption of spatial continuity ingredient in present-day confirmed physical theories. For the acceptance of that question-begging requirement would indeed import into the description of the sought-after imaginary world inconsistencies which could then be adduced to maintain that a physical space endowed with intrinsic congruences is logically impossible. And this conclusion would then, in turn, enable Schlesinger to uphold his contention that my version of the ND-hypothesis cannot be falsified in *any* logically possible world and hence is only trivially physically meaningless.

Having cautioned against these pitfalls, I now maintain that just as the three-dimensionality of actual physical space cannot be vouchsafed on a priori grounds, so also it cannot be guaranteed a priori that in every kind of spatial world the congruences would be transport-dependent rather than intrinsic. For imagine a world whose space has elements (space atoms, lumps, or granules) such that (1) for any one space atom, there are a fixed finite number of space atoms which are next to it or contiguous with it (the magnitude of this finite number will determine the granulo-"dimensionality" of the atomic space by being two in the case of 1-space, eight for 2-space, etc.), (2) there are only a finite (though large) number of distinct space atoms all told, and (3) there are no physical foundations whatever for mathematically attributing proper (nonempty) parts to the space atom, as attested by the existence of only finitely many rest positions for any object. It is clear that the structure of this granular space is such as to endow it with a transport-independent metric: the congruences and metrical attributes of "intervals" are

[4] For details on this distinction, see [7].

intrinsic, being based on the cardinal *number* of space atoms, although it is, of course, trivially possible to introduce various other units each of which is some fixed *integral* multiple of one space atom.[5] To be an *extended* object in this kind of granular space is to comprise more than one space atom. And the metric intrinsic to the space permits the factual determination of the rigidity under transport of any object which is thereby to qualify as an "operational" congruence standard. By the same token, the sudden nocturnal doubling of *all* "operational" (i.e., transport-dependent) metric standards along with all extended physical objects could be said to obtain in such a granular space, if on some morning they each suddenly occupied or corresponded to twice as many space atoms as before. Accordingly, the ND-hypothesis formulated above is indeed falsifiable in the context of an atomic theory of space. But it is not falsifiable within the framework of any of the confirmed modern physical theories pertaining to our actual world: their affirmation of the mathematical continuity of physical space precludes that there be an intrinsic kind of metric, which is the very kind on whose existence the ND-hypothesis depends for its physical significance and hence falsifiability. Thus, we note that the content of the theory to which the ND-hypothesis is adjoined is relevant to its falsifiability. But this relevance of the remainder of the theoretical system does *not* detract from the fact that within the latter theoretical system the ND-hypothesis is itself in principle *non*falsifiable.

Ellis and Schlesinger mistakenly believe that they have corrected all previous treatments of the ND-hypothesis in the philosophical literature, because they have completely overlooked a fundamental logical fact which is an immediate corollary of my Riemannian justification for making *alternative* choices from among *incompatible* spatial and temporal congruences (see Chapter I, §§2(i), 3(i), and 4). This logical fact is that Riemann's doctrine of the intrinsic metrical

[5] The reason for the restriction to an *integral* multiple is that a proposed unit based on a *non*integral multiple of one space atom would have no physical realization.

amorphousness of the spatial continuum permits the deduction both of the *non*falsifiability of the ND-hypothesis *and* of the falsifiability of a certain modified version of the ND-hypothesis which I shall call the "ND-*conjecture*." The ND-conjecture differs from the ND-hypothesis by *not* attributing the doubling to *all* operational standards of length measurement but only to one of them, while exempting the others from the doubling. In order to demonstrate the falsifiability of the ND-*conjecture* on the basis of Riemann's metrical doctrine, I must first clarify the logical relation of that doctrine to the view that the physical concepts of congruence and of length are multiple-criteria concepts or open-cluster concepts rather than single-criterion concepts.

Riemann's thesis that the spatial and temporal continua are each intrinsically amorphous metrically sanctions a choice among alternative metrizations of these continua corresponding to *incompatible* congruence classes of intervals and thus to incompatible congruence relations. By thus espousing the nontrivial ingredience of a convention both in the *self*-congruence of a given interval *at different times* and in the equality of two different intervals at the same time, Riemann denies that the existence or nonexistence of these congruence relations is a matter of spatial or temporal fact. But clearly, Riemann's conception of congruence fully allows that any *one* set of congruence relations be specified physically by each of several different operational criteria rather than by merely one such criterion! For example, this conception countenances the specification of one and the same congruence class of intervals in inertial systems by the equality of the travel times required by light to traverse these intervals no less than by the possibility of their coincidence with an unperturbed transported solid rod. Thus, as I have pointed out elsewhere [5, pp. 14–15], Riemann's conventionalist conception allows that congruence (and length) is an open, multiple-criteria concept in the following sense: no one physical criterion, such as the one based on the solid rod, can exhaustively render the actual and potential physical meaning of congruence or length in physics, there being a potentially growing multiplicity of compatible physical criteria rather than only one criterion by which any one spatial congruence class can be specified to the exclusion of every other congruence class. And hence the conventionalist conception of congruence is not at all committed to the crude operationist claim that some *one* physical criterion renders "*the* meaning" of spatial congruence in physical theory.

A whole cluster of physical congruence criteria can be accepted by the proponent of Riemann's conventionalist view during a given time period, provided that the members of this cluster form a *compatible* family during this time period in the following twofold sense: (1) the various criteria yield *concordant* findings during the time period in question on the self-congruence or rigidity of any one interval at different times, and (2) one and the same congruence class of separated spatial intervals must be specified by each member of the cluster of congruence criteria. It is a matter of *empirical* law whether the concordance required for the compatibility and interchangeable use of the members of the cluster obtains during a given time period or not. And if that concordance does obtain, an explanation of it, and thereby of the physical geometry associated with each member of the cluster, may be attempted along the lines of the GTR. As we shall see in detail in Chapter III, §2.12, this theory sought to implement the very vision to which Riemann had been led in his Inaugural Lecture by his conception of congruence [5, pp. 426–427]. But if there is no concordance, an explanation may be sought of the spatial geometry obtaining with respect to *one* congruence criterion no less than of the spatial geometry obtaining with respect to a concordant family of such criteria. Hence, Riemann's conception of congruence can be upheld entirely without prejudice to the multiple-criterion character of congruence and is highly germane to the quest for an explanation of the spatial geometry obtaining with respect to one or several compatible congruence criteria.

Now imagine a hypothetical empirical situation relevant to the falsifiability of the ND-*conjecture:* up to one fine morning, there has been concordance among the various congruence criteria in regard to the self-congruence of particular space intervals in time, but after that particular morning, that concordance gives way to the following kind of *discordance:* intervals which, *in the metric furnished by metersticks,* remain spatially congruent to those states of themselves that antedated the momentous morning thereafter have *twice* their pre-morning sizes in the metric furnished by *all the other* congruence (length) criteria, such as the travel times of light. In that case, the requirement of logical consistency precludes our continuing to use metersticks interchangeably with the other congruence criteria after the fateful morning to certify continuing self-congruence in time. And it is perfectly clear that in the hypothetical eventuality of a sudden discordance between previously concordant criteria of self-congruence, the thesis of the intrinsic metrical amorphousness of the spatial continuum sanctions our choosing *either one* of the two incompatible congruences as follows: *either* we discard the meterstick from the cluster of criteria for self-congruence of space intervals in time *or* we retain the meterstick as a criterion of self-congruence to the exclusion of all the other criteria which had been interchangeable with it prior to the fateful morning.

The two alternatives of this choice give rise to the following two physical descriptions of nature, the *first* of which states the ND-*conjecture:*

Description I: Since the fateful morning, all metersticks and extended bodies have doubled in size relatively to all the *other* length criteria belonging to the original cluster, which are being retained after the fateful morning, and the laws of nature are unchanged, provided that (i) the lengths l ingredient in these laws are understood as referring to the latter retained length criteria and (ii) the instant of doubling during the fateful morning is excluded from the temporal range of validity of the laws in the following sense: the doubling itself is *not* deducible from the conjunction of these laws with the boundary conditions prevailing before that instant; *or*

Description II: Relatively to the retention of the meterstick as a length standard after the fateful morning and the discarding of all other previously used length criteria, there is the following *unitary* change in those functional dependencies or laws of nature which involve variables ranging over lengths: all magnitudes which were correlated with and hence measures of a length l *prior* to the fateful morning will be functionally correlated with the new length $l/2$ *after* the fateful morning. Given that the lengths symbolized by l are referred to and measured by metersticks after the fateful morning no less than before, this mathematical change in the equations means *physically* that after the fateful morning, the same lengths will no longer correspond to the previous values of other magnitudes but to new values specified by the new laws. Thus, if L is the length of a path traversed by light, and c the velocity of light, then the time T required by light for the traversal of the distance L before the momentous morning is given by $T = L/c$, whereas the time T' which corresponds to the *same* length L after the fateful morning is $T' = 2L/c$. Again, if L is the length of a simple pendulum (of small amplitude) and g the acceleration of gravity, then the period P corresponding to L before the fateful morning is given by $P = 2\pi\sqrt{L/g}$, whereas the period P' which corresponds to the *same* length L after the momentous morning is $P' = 2\sqrt{2}\pi\sqrt{L/g}$. It is to be clearly understood that although the numerical constants "c" and "g" occur in these new laws, after the fateful morning they no longer respectively represent the light velocity and the terrestrial acceleration of gravity.

Thus the new light velocity c' in Description II is given by $c' = L/T' = L/2T = c/2$. But the light velocity is also the product of the wave length and the frequency. By hypothesis, in Description II the wave length changes from λ to $\lambda/2$ while the light velocity changes from c to $c/2$. And since the light velocities c and $c' = c/2$ must be respectively yielded by the products of λ and $\lambda/2$ with the frequency, there is no change in the frequency. Similarly, the pre-morning law of gravitational acceleration is

$$\frac{d^2R}{dt^2} = -G\frac{M}{R^2}$$

and yields the value $g = -G\,M_E/R_E^2$ for the terrestrial acceleration, where M_E and R_E are respectively the mass and radius of the earth. After the momentous morning, this same acceleration g prevails in Description II on a planet which has only *half* the radius of the earth but the same mass M_E. Hence the new post-morning law of gravitational acceleration is

$$\frac{d^2R}{dt^2} = -\frac{1}{4}G\frac{M}{R^2},$$

which yields the required acceleration g for a planet of smaller radius $R = R_E/2$ and mass M_E. But the post-morning radius of the *earth* remains R_E in Description II, and therefore the new post-morning terrestrial acceleration of gravity is

$$g' = -\frac{1}{4}G\frac{M_E}{R_E^2} = \frac{g}{4}.$$

Clearly we can express the new period P' in terms of the new acceleration g' by replacing g by $4g'$ in the new pendulum law, obtaining $P' = \sqrt{2}\pi\sqrt{L/g'}$ in place of the old law $P = 2\pi\sqrt{L/g}$.

It should be noted that the nocturnal change from P to P', which Description II characterizes as a change in the law relating the period to the fixed length L, is described equivalently in Description I as a nocturnal doubling of the length of the pendulum *not* involving any change in the *law* relating the period to the length. By the same token, Description II renders the post-morning change in the coincidence relations between certain points on a solid rod and successive crests (or troughs) of a monochromatic light wave by saying that the pre-morning wave length λ was reduced to the post-morning value $\lambda/2$; but Description I characterizes that very same change as a doubling of the rod, because any portion of the rod which originally coincided with the span of one complete cycle of light now coincides with the span of *two* such cycles. Similarly, the change in the terrestrial acceleration of gravity from g to $g' = g/4$, which Description II characterizes as due to a change in the law of gravitational acceleration, is attributed by Description I to a change in the earth's radius from R_E to $2R_E$ amid the constancy of the law. And while II speaks of a change in the pre-morning velocity c of light to a post-morning value of $c' = c/2$, I asserts that the light velocity remains numerically unchanged. For let us see how Description I determines the light velocity both as the quotient of distance and transit time, and as the product of wave length and frequency.

According to I, on the momentous morning the distance between two fixed points A and B on the path of a direct light ray changes from $L = cT$ to $L' = 2L = 2cT = cT'$ not only as measured by the transit time of light but also as measured by any other length criterion *except for a solid rod.* Hence the quotient L'/T', which yields the post-morning light velocity, will have the unchanged value c. Also, the standard of length measurement constituted by the span of one cycle of a given kind of monochromatic light remains concordant after the fateful morning with all the other length standards *except for solid rods*, which are then no longer in use as length standards. Hence the pre-morning wave length λ of any such light does not change. Nor does the frequency n change in Description I, just as it does not change in Description II. For the frequency n is a number yielded by a single stationary clock and does not depend on the paths traversed by light *in given times,* so that a nocturnal change in these paths leaves the frequency unchanged, independently of whether this path change is described as in I or as in II. Since both L'/T' and λn remain unchanged, we see that in Description I the pre-morning light velocity does not change after the fateful morning.

A number of important points need to be noted regarding the two descriptions I and II.

1. Descriptions I and II are *logically equivalent* in the context of

the physical theory in which they function. And, contrary to Schlesinger's allegation, they incontestably have the same explanatory and predictive import or scientific legitimacy. Descriptions I and II each enunciate an unexplained regularity in the form of a time-dependence: I asserts the unexplained new fact that all metersticks and extended bodies have expanded by doubling, while II enunciates an unexplained change in a certain set of equations or laws according to a unitary principle.

2. Schlesinger mistakenly believes that he has shown on methodological grounds that the doubling of the metersticks asserted by I is the uniquely correct account as contrasted with the change in the laws of nature affirmed by II. Specifically, he supposes that he can disqualify II as a valid equivalent alternative to I on the grounds that I explains all the phenomena under consideration by the *single* assumption of doubling, whereas II allegedly can claim no comparable methodological merit but is a linguistically unfortunate and futile attempt to evade admitting I. And he derives this supposition from the fact that in Description II, the laws describing nature after the fateful morning can be derived from the previously valid *different* laws by the mathematical device of replacing the length l in the original laws by $2l$. But the error of Schlesinger's inference lies in having misconstrued the mere computational stratagem of replacing l by $2l$ as a warrant for claiming that the only *physically admissible* account of the phenomena is that all metersticks have doubled. For he overlooks the fact that the alternative metrizability of the spatial continuum allows us to *retain* the meterstick as our congruence standard after the fateful morning to the exclusion of the other criteria in the original cluster, no less than it alternatively permits us to preserve the others while jettisoning the meterstick. Thus, Description II is based on the convention that the meterstick does not change but is itself the standard to which changes are referred, since the lengths l are measured by the meterstick after the fateful morning just as they were before then.

In this connection, I should remark that I find Schlesinger's own computations misguided when he uses the *actual known laws of nature* (e.g., the conservation laws for angular momentum, etc.) as a basis for calculating whether the hypothetical changes postulated by the ND-conjecture would be accompanied by other compensatory changes so as to render the detection of the hypothetical changes *physically* impossible. For, as I noted in the statement of Description I, the actual laws and boundary conditions of the standard physics disallow the ND-conjecture, since they provide no basis for generating the forces that would be required to effect the conjectured observable doubling. Thus, what is at issue here in the context of the assumed continuity of physical space is only the *logical* rather than the physical possibility of an observational falsification (refutation) of the ND-conjecture. The issue is whether we can specify logically possible observable occurrences relevant to the falsifiability of the ND-conjecture, *not* the compatibility of such occurrences with the time-invariance of the actual laws of nature.

3. The sudden discordance between previously concordant criteria of self-congruence in time is indeed falsifiable. Being a proponent of Riemann's view of congruence, I countenance Description I no less than Description II to formulate that hypothetical discordance. And I therefore maintain that if the language of Description I is used, the falsifiability of the discordance assures the falsifiability of the ND-conjecture. By contrast, suppose that a narrowly operationist defender of the single-criterion conception of length *were* to declare that he is wedded to the meterstick once and for all even in the event of our hypothetical discordance. The putative defender of such a position would then have to insist on Description II while rejecting I as meaningless, thereby denying the falsifiability of the ND-conjecture.

4. Schlesinger is not at all entitled to claim that II is false while I is true on the basis of the following argument: if one were to countenance II as veridically equivalent to I as a description of our hypothetical nocturnal changes, then one would have no reason to reject analogously equivalent descriptions of phenomena of temperature

change and of other kinds of change. This argument will not do. For suppose that one could construct analogues of my Riemannian justification of the equivalence of Descriptions I and II and thus could show that temperature changes also lend themselves to equivalent descriptions. How could this result then serve to detract from either the truth or the interest of my contention that II is no less true than I? Indeed, we shall see below that it does not.

As is shown by the treatment of nocturnal expansion in my earlier book [5, pp. 42–43], I have not denied the physical meaningfulness of the kind of doubling (expansion) which results from a sudden discordance between previously concordant criteria of self-congruence in time. And it is puzzling to me that Reichenbach's writings and mine both failed to convey our awareness to Ellis and Schlesinger that the very considerations which serve to exhibit the physical vacuousness of the ND-hypothesis also allow the deduction of the falsifiability of the ND-*conjecture*. It is with regard to the ND-hypothesis but not with regard to the ND-conjecture that I have maintained and continue to maintain the following: on the assumption of the continuity of physical space, this hypothesis can have no more empirical import than a trivial change of units consisting in abandoning the Paris meter in favor of a new unit, to be called a "napoleon," which corresponds to a semi-meter.[6]

[6] These considerations readily allow the correction of an inept formulation of a sound idea given in §3(i) of Chap. I, p. 42 above, and in [5, p. 438]: quantities g'_{ik} pertaining to meters are subtracted there illicitly from quantities g_{ik} which refer to centimeters. But the quantities g'_{ik} and g_{ik} ingredient in the subtractions can all be held to refer to meters by first stipulating in the sense of the ND-hypothesis that there is an instant t such that *all* objects *alike* fail to be spatially self-congruent as between times before and after t in the following respect: any object whose size was 1 meter up to the instant t—prior to which the metric tensor was g'_{ik}—thereafter "expands" to 100 meters, so that after the instant t the metric tensor acquires new components g_{ik} which are proportional to g'_{ik} by a factor of 10,000. Given this stipulation, the quantities $g_{ik}-g'_{ik}$ are reinterpreted such that all the terms ingredient in the subtraction refer to the meter unit, although no one object can provide a physical realization of that unit both before and after the instant t. (I am indebted to Mr. Richard Burian for pointing out the need to modify my original interpretation of the subtractions involved here.)

Schlesinger [9] has recently published a rejoinder to my views on the ND-hypothesis and on the ND-conjecture as part of his inquiry into the question What does the denial of absolute space and time mean? I shall therefore proceed to examine the most recent arguments offered by him in that rejoinder.

It will be recalled that I formulated the ND-*hypothesis* in the context of the Newtonian theory of absolute space as follows: relatively to the congruences which that theory presumes to be intrinsic to its container-space, there has been a nocturnal doubling of all extended physical objects contained in space, and *also* a doubling of *all transport-dependent* congruence standards. Unfortunately, Schlesinger overlooks my explicit formulation and considers instead a sketchy and quite different statement given by Schlick half a century ago [10]. To see that Schlesinger's treatment is substantively inadequate, let us contrast Schlick's statement with my formulation in respect to their success in rendering *spatially informative* Poincaré's claim that a universal nocturnal expansion is physically vacuous. And it will then be apparent that my version of Poincaré's thesis can hardly be indicted as *spatially uninformative* on the strength of such criticisms as Schlesinger can validly level against Schlick's version.

For simplicity but without loss of generality, I shall deal again with the case of a twofold increase of all lengths rather than with Schlick's example of a hundredfold increase. Then Schlick's sketch of the physically vacuous doubling hypothesis can be articulated by interpreting him to have offered *an open-ended conjunction of sub-hypotheses* in which we find the following, among others: (1) since the period of a simple pendulum is given by $P = 2\pi\sqrt{l/g}$ and since the acceleration of gravity on the earth's surface is given by $g = GM/r^2$, there has been an increase in the periods of all simple pendula in virtue of the nocturnal doubling of their lengths and of the radius of the earth, (2) the rates of material clocks have changed by a factor whose magnitude is such that, after the eventful night, these clocks continue to yield the same duration numbers for the periods of simple pendula, which have increased as stated by the preceding subhypothesis, (3) since there has been a nocturnal doubling of the length L of any given path traversed by an electromagnetic pulse in the course of a *round*-trip ABA, the pre-nocturnal round-trip time $T = L/c$ increases post-nocturnally by a factor of two, and (4) any incompatibility between the respective changes of the rates of material clocks required by the last two preceding subhypotheses is precluded by appropriate changes in other physical magnitudes (e.g., the velocity of light) which are, in principle, specifiable by additional subhypotheses. The fourth of these subhypotheses indicates that if *each* of Schlick's conjuncts is to be *physically specific* in regard to the changes required by it, the conjunction of his subhypotheses must be open-ended. For our knowledge of the laws of nature is incomplete, and Schlick characterizes the totality of the changes which he postulates along with the nocturnal doubling by the requirement that *no observable change whatever* result from any or all of the consequences of the universal expansion.

It would appear that Schlick's recourse to this requirement of neutralizing changes or "counter"-events *in his very statement of the doubling hypothesis* makes him vulnerable to the following charge of Schlesinger's: instead of having presented an interesting denial of the absoluteness of physical *space*, Schlick offered only the very *general* principle that no completely unverifiable proposition can have physical content. Moreover, Schlick's formulation is infelicitous precisely because it fails to articulate the spatially informative content that Schlesinger did not discern in Poincaré's thesis. To see just how Schlick's statement fails to render Poincaré's point, let us inquire in what sense the open-ended conjunction of Schlick's subhypotheses can be true under the assumed condition that there is no observable change. Then we find that (1) *taken singly*, each of Schlick's *physically specific* subhypotheses is individually incompatible with the *observed absence* of the kind of change asserted by it and hence cannot itself be literally true, and (2) even if a Laplacean demon succeeded in closing the open end of Schlick's conjunction by rendering its fourth subhypothesis specific, the individual conjuncts in Schlick's statement would singly still be literally false. Thus, if it is to be true at all, Schlick's open-ended conjunction *cannot* be true in the sense that each of its conjuncts is literally true individually. And indeed it appears that, under the assumed conditions, his conjunction is true *only* if taken as *metaphorically* asserting a change of units as follows: after the night in question, all length numbers are to be based on a new unit which corresponds to one half of *whatever unit* was the basis of length numbers prior to that night.

On this construal, Poincaré considers it spatially informative to claim that whatever the initial unit, such a change of unit is always feasible with empirical impunity, and he rests this claim on the assumption that space is a mathematical continuum of like point-elements. Schlesinger cannot validly level his charge of spatial uninformativeness against the above claim of the universal feasibility of a change of the unit of length. Specifically, as I shall now argue, it is unavailing to Schlesinger's case that corresponding changes of the unit of duration and of other physical magnitudes are feasible as well. And the *unrestricted* theoretical feasibility of such changes of unit is not at all trivial *either* in the case of length *or* in the case of any of the other physical magnitudes for which it obtains. That there is no triviality in the case of length becomes evident the moment we realize the following: if physical space were quantized (atomic), the pre-nocturnal use of one atom of space as a unit would not allow the physically meaningful post-nocturnal use of a new unit which corresponds to one half of one space quantum, since the latter unit would have no physical realization, although it would, of course, be feasible to use a new unit which is any *integral* multiple of one space atom. Thus, the unrestricted theoretical feasibility of a change of the unit of length rules out a quantized space. How can Schlesinger hope to gainsay the spatial informativeness of the assertion of this feasibility by noting that corresponding assertions hold with respect to the unit of time and with respect to the units of other physical magnitudes that presuppose the units of length or of duration? By the same token, if both space and mass turned out to be quantized, one could hardly establish the spatial uninformativeness of the *denial* of the unrestricted feasibility of a change of the unit of length by noting that a corresponding denial holds for mass! It does not lessen the interest or specificity of the ascription of genius to Newton to make a like affirmation concerning Einstein.

I have maintained that Poincaré should be interpreted as making the above feasibility claim with respect to length *and* as regarding it to be a consequence of the assumed mathematical continuity of physical space. As a mathematical continuum of like punctal elements, physical space fails to be absolute *in the sense relevant here* by lacking intrinsic, transport-independent congruences. On the other hand, if physical space were quantized, it would be endowed with such congruences. Thus Poincaré's denial of the absoluteness of physical space in this context is to be understood as the denial of the existence of intrinsic congruences between different intervals at the same time and between a given interval and itself at different times. And on my construal of the ND-hypothesis, the *non*falsifiability of that hypothesis is tantamount—in the context of the physical assumptions of the remainder of the theory—to the nonexistence of such intrinsic congruences: in the original publication cited by Schlesinger, I formulated the ND-hypothesis as asserting the nocturnal doubling of all extended physical objects *and* of *all* transport-dependent congruence standards with respect to the intrinsic congruences of an assumedly absolute physical space. Thus I rested my denial of the physical significance (and hence falsifiability) of the ND-hypothesis on the presumed nonexistence of intrinsic congruences in physical space. It was therefore negligent on Schlesinger's part to have disregarded my explicitly *spatial* formulation of the ND-hypothesis in favor of Schlick's misleading metaphorical statement. No wonder that Schlesinger then arrived at the incorrect verdict that I failed to exhibit the bearing of the nonfalsifiability of the ND-hypothesis on the denial of the absoluteness of physical space.

Specifically, Schlesinger attacks the straw man whose easy demolition issues in this *ignoratio elenchi* by giving his version of the two senses in which the hypothesis of a universal (nocturnal) expansion can be understood. Referring to the statement of a universal expansion as "the UE-statement," he considers the following two versions of the hypothesis:

Version (1): The falsifiable claim to which I have referred both as stating the "ND-*conjecture*" and as "Description I" of a hypothetical sudden discordance of previously concordant criteria in regard to the self-congruence of space intervals in time.

Version (2): The assertion that along with the UE-process, "innumerably many other *compensatory* changes have occurred, the combined effect of all of which was to eliminate all the traces of UE, giving rise to a universe which appears the same in all respects today as it appeared yesterday" [9, p. 55].

He then proceeds to invoke *these* two versions as a basis for indicting the following claim of mine: if space is atomic (and hence is endowed with intrinsic congruences), then the ND-hypothesis is indeed falsifiable. In an endeavor to show that the nonfalsifiability of the ND-hypothesis does *not* preclude the atomicity of physical space, Schlesinger writes: "If . . . the UE-statement is made in the second way, into which we have built the condition that all the traces of the UE have been obliterated, then, even if space were atomic, the statement would remain vacuous as before. For, on this way of asserting the UE-statement, we must either take it that the size of the ultimate space quanta have doubled too, or if they did not, the fact that all bodies occupy more space quanta will not be detectable since appropriate changes to disguise *this too* have accompanied the UE" [9, p. 56]. But Schlesinger's charge that "even if space were atomic, the [UE-]statement would remain vacuous as before" applies to his Version (2) but manifestly *not* to my formulation of the ND-hypothesis. For "the condition that all the traces of the UE have been obliterated" has indeed been built into his Schlickean Version (2) but *not* into my statement of the ND-hypothesis.

In a further unsuccessful attempt to demonstrate that "introducing the question of the atomicity of space is of no help whatever," he first recalls my thesis that the ND-hypothesis would be confirmable in an atomic space by the fact that all extended physical objects and all transport-dependent congruence standards would correspond to more space quanta than before. And he believes himself to be posing a difficulty for my thesis by then proceeding to argue as follows: "If the UE-statement is asserted in the first way then, irrespective of the ultimate nature of space, the statement may be confirmed or disconfirmed depending on whether or not the various length-dependent properties of physical systems have changed in the manner required by the form in which they are functionally related to the length of their relevant elements" [9, p. 56].

That Schlesinger is again attacking a straw man is evident from my account of the falsifiability status of the ND-CONJECTURE: far from being incompatible with that account, it follows from it that the ND-*conjecture* is falsifiable both in a continuous and in an atomic space, whereas the ND-hypothesis is falsifiable in an atomic but not in a continuous space. I had *not* asserted that the continuity (and hence non-atomicity) of physical space is *necessary* for the physical significance (falsifiability or confirmability) of the ND-conjecture;[7] what I had claimed is that in the context of the remainder of the theory, spatial continuity is *sufficient* for the testability of the ND-conjecture. Hence Schlesinger cannot score against my thesis by noting that the ND-*conjecture* is falsifiable in an atomic space no less than in a continuous one. And it is now clear that he has misconstrued my account of the relevance of the atomicity of space to the physical significance of the UE-statement.

It will be instructive for our subsequent purposes to examine next Schlesinger's argument that Poincaré's thesis is *spatially* uninforma-

[7] Nor did I assert that the continuity of physical space is *necessary* for the *non*-falsifiability of the ND-hypothesis.

tive on the grounds that it has analogues in such other domains of physics as thermometry. He writes:

Innumerable other statements taken from any domain you wish can be stripped of meaning in exactly the same way as Schlick has stripped the talk about the uniform expansion of the universe of meaning. Consider 'the temperature of everything has dropped by 100°F last night.' Is this statement verifiable? Obviously, if we take it that all that happened is what is asserted by the previous statement and what is entailed by it in the context of present day physics, the next morning, in addition to feeling very cold, we shall be faced with a very strange universe containing countless clues of the general decrease in temperature. If, however, the statement is taken to be accompanied with the qualification that in addition suitable adjustments have also taken place so as to conceal from us the change in the temperature of everything, (e.g. that our physiology has changed in such manner that we now feel quite pleasant at −45°F; that the freezing point of all liquids has dropped by 100°F; that the coefficients of thermal expansion of all substances have undergone changes so as to affect appropriately all expansion thermometers, etc.) then the change will be unnoticeable because we have stipulated so. But so taken, the meaninglessness of the statement reflects nothing special about the nature of temperature but a very general principle (to which we may or may not agree) about the meaninglessness of statements completely unverifiable. . . .

It would . . . be a mistake to assume that something of special interest can be conveyed about spatial properties by drawing attention to the fact that a uniform expansion . . . will go undetected. The reason, after all, why the meter-rod method cannot reveal a uniform and world wide change in length is because it is a particular case of a differential method of measurement. It is a well known and general truth that a differential method of measurement can disclose either partial or non-uniform change in the quantities it measures. The meter rod method of measuring distances is one of the countless differential methods we have in physics. The thermocouple method of measuring temperature (i.e. the method which exploits the fact that electric current flows in a circuit formed by two different metals when a difference in temperature is maintained between their two junctions and that this current flow varies with the temperature difference) is another one. Consequently one might just as well point at the case of a uniform temperature drop . . . that manifestation of the temperature drop which is measurable by the thermocouple method disappears when the temperature of everything is affected [9, pp. 46–48].

My earlier statement of the sense in which Schlick's open-ended conjunction of subhypotheses can be true in the case of the ND-hypothesis now enables us to exhibit the defects of Schlesinger's treatment of his example from thermometry.

I argued above that Schlick's conjunction is true only if taken as *metaphorically* asserting a change to a new unit of length which corresponds to one half of whatever unit was the original basis of length numbers. For precisely analogous reasons, I now maintain that under the assumed condition that there is no observable change whatever, the hypothesis of a universal drop in temperature (the "UDT"-hypothesis) can be true only if taken as metaphorically asserting the following: the original temperature scale, constituted by the assignment of real numbers T to the various states of heat intensity, has been replaced by a different scale T' based on a different kind of state as the zero of temperature such that the two scales are related by the equation $T' = T - 100$. How can the legitimacy of such a transformation of temperature scale possibly invalidate my demonstration of the spatial informativeness of the feasibility of Poincaré's unrestricted change of the units of length?

Schlesinger believes that his thermometric example of the physically vacuous UDT-hypothesis can serve to exhibit the *spatial* triviality of the physical vacuousness of Poincaré's UE-hypothesis. His further reason is that reference to a standard body, process, or state enters essentially into both length and temperature measurement, each of which he therefore calls "differential." To this I reply:

1. The fact that *all* methods of spatial measurement inevitably involve a *transport-dependent* standard body or process is indeed spatially informative, and is tantamount to the physical vacuousness of my statement of the ND-hypothesis. And the *spatial* informativeness of this fact can hardly be gainsaid by pointing to the employ-

ment of standard bodies or processes in the measurement of other physical attributes.

2. What Schlesinger calls the "differential" character of temperature measurement (by thermocouple, for example) results from the following facts of the relevant thermodynamic theory: (a) the temperature of a body or system is a property which determines whether or not the system is *in thermal equilibrium with the standard body* (*system*), and (b) the zero of temperature of the standard system is arbitrary, subject to the restriction of being consistent with the existence of the minimum temperature at −460°F [10, pp. 7–9].

Is it not evident, therefore, that such similarities as obtain between the measurements of length and of temperature cannot possibly serve to establish Schlesinger's contention that the nonfalsifiability of the UDT-hypothesis exhibits Poincaré's thesis to be devoid of any distinctively *spatial* interest?

Schlesinger goes on to offer what he considers a constructive proposal of his own, aimed at articulating the valid core of Poincaré's allegedly infelicitous denial of the absoluteness of space:

> One can envisage ways in which the repudiation of Absolute Space and Time through the denial of the meaningfulness of a sentence like "Overnight everything has uniformly expanded" might be construed as having unique significance. . . . By emphasizing the meaninglessness of this sentence one might for instance desire to combat the view according to which a completely counteracted universal expansion is yet not totally erased for its occurrence involves larger amounts of the continuum being occupied . . . [9, p. 57].

But as a sequel to my earlier discussion of UE, this proposal plainly carries coals to Newcastle: the notion of "larger amounts of the continuum ['really'] being occupied" is precisely the one whose defects I made apparent by noting the absence of (a substratum of) intrinsic congruences in the spatial continuum. In saying that Schlesinger did not contribute a new gloss which renders *Poincaré*'s account of the status of the ND-hypothesis spatially informative, I do *not* intend to disparage the merit of his articulation of the ND-*conjecture* as a falsifiable hypothesis!

I now turn to the examination of Schlesinger's views concerning the interpretation (hypothesis, description) which is appropriate to an observable sudden discordance of previously concordant congruence criteria which is such that the sizes of extended objects remain the same only in the metric furnished by solid rods (meter or yard sticks). Schlesinger's thesis is that what I have called "Description I" is *uniquely* true vis-à-vis Description II, i.e., he maintains that under the posited circumstances, we would "be forced to conclude that a uniform increase in the size of everything has taken place" [9, p. 48]. By contrast, I have argued for the complete equivalence or co-legitimacy of Descriptions I and II with regard to both truth-value *and* inductive (though not *descriptive!*) simplicity. Hence we must now consider Schlesinger's most recent arguments against the co-legitimacy of Description II.

To begin with, Schlesinger misstates my claim of the co-legitimacy of Description II in two places by attributing to me a view which I reject no less than I reject Schlesinger's: the thesis that Description II is uniquely legitimate vis-à-vis Description I. By upholding the co-legitimacy of I and II, I reject *alike* the claim that I is unique (Schlesinger) *and* the claim that II is unique (made by the *hypothetical* defender of the single-criterion conception of length who is putatively wedded to the meterstick once and for all). Referring to Description II as "the alternative hypothesis," Schlesinger writes:

According to the alternative hypothesis the vast number of changes observed are due to the equally vast number of causes such as the change in the universal gravitational constant, a change in the velocity of light and so on. The obvious complaint, that the alternative hypothesis is inferior to the one postulating a universal increase in length which accounts singlehanded or with the aid of at most one or two additional assumptions for everything observed, is met by pointing out that these various causes need not be spelled out separately. A unitary principle connects these causes. The principle is this: all the various events that have taken place during the night have occurred in such a manner that their combined effects give rise to a situation which could obtain if everything had increased in size. In other words, all material objects have not in fact increased in size, but everything is as if they increased in size.

We shall have to agree with Professor Grünbaum that even though his alternative hypothesis stipulates innumerably many changes, their effects can be described precisely in as simple a way as the effects of a universal expansion. Thus the two hypotheses may be claimed to have equal descriptive simplicity. This, however, does not necessarily indicate that they are of equal simplicity in all essential respects. According to our hypothesis the innumerably many changes observed are the inevitable consequences of a single event: the increase in the size of the universe. According to Grünbaum, however, these are caused by a vast number of independent physical occurrences. Thus his hypothesis postulates a vast number of unexplained events and is therefore inferior to ours which postulates only one unexplained happening [9, pp. 49–50].

Schlesinger then emphasizes that, in his view, the difference between I and II is a difference between describing reality and mere appearance:

The compelling argument is best brought out by . . . considering a fictitious situation in which we arise one bright morning and find ourselves in a vastly transformed world, being confronted with an immense number of strange phenomena, an endless number of physical constants and familiar laws of nature *seemingly* having undergone various changes during the previous night. It is assumed that the apparent changes have to be assigned an untold number of independent causes unless we are prepared to stipulate that all the apparent changes are due to the fact that during the night everything, and the distances between them, have doubled in size. . . . However, by stipulating a universal doubling of all sizes we succeed in accounting for all the seemingly diverse anomalies by a single hypothesis [9, p. 55].

As I see it, Schlesinger's reasoning is vitiated by the following several mistakes:

1. His statement of Description II is a travesty. I had stated the pertinent unitary principle of change in the laws involving variables ranging over lengths by writing [6, p. 302]: "all magnitudes whice were correlated with and hence measures of a length *l prior* to thh

fateful morning will be functionally correlated with the new length $l/2$ *after* the fateful morning. Given that the lengths symbolized by l are referred to and measured by metersticks after the fateful morning no less than before, this mathematical change in the equations means *physically* that after the fateful morning, the same lengths will no longer correspond to the previous values of other magnitudes but to new values specified by the new laws." Schlesinger paraphrased this statement of mine by saying: "all material objects have not in fact increased in size, but everything is as if they increased in size." To see what is wrong with Schlesinger's paraphrase, let us be mindful of the fact that a length of 100 inches corresponds to (i.e., is equivalent to) a length of 254 centimeters and then first consider as an example the assumedly true statement "this table is 100 inches long." Now suppose that someone proposed to paraphrase the latter statement by saying "this table does *not in fact* have a length of 254 but its extensional properties *are as if* it *had* a length of 254." Then if the tacitly presupposed units of length are either the centimeter or the inch, and if the *first* part of this conjunction is to be true, the number 254 in it must refer to INCHES, *not* to centimeters. But if the number 254 in the *second* conjunct likewise refers to inches, the second conjunct is *false:* the extensional properties of a 100-inch or 254-centimeter table are *not* AS IF the table were 254 INCHES long! Thus, if both conjuncts are to be true, the two occurrences of the number 254 in the conjunction *cannot refer to the same unit* (i.e., inches or centimeters), and the 254 of the second conjunct must refer to centimeters; but in that case, the phrase "as if" is quite misleading or, at best, otiose.

Precisely analogous difficulties beset Schlesinger's paraphrase of my Description II. The phrase "increased in size" tacitly refers to a congruence criterion. And Schlesinger's conjunction "all material objects have not in fact increased in size, but everything is as if they increased in size" simultaneously invokes two *different congruence criteria*, just as the above paraphrase of the statement about the table simultaneously invokes two different *units* of length. Specifically, under the posited hypothetical circumstances, the two *incompatible* congruence standards are given by solid rods (metersticks), on the one hand, and by the members of a family F of other congruence criteria, on the other. And they are respectively the counterparts of inches and centimeters in the paraphrase of the sentence concerning the 100-inch table. For if the conjunction in Schlesinger's paraphrase is to be true, then the expression "have not in fact increased in size" in his first conjunct must be predicated on the solid rod congruences, whereas the sentence "everything is as if they increased in size" is predicated on the *different* congruences in F. Hence what Schlesinger's purported paraphrase of my Description II asserts is the following: the sizes of material objects have not in fact increased with respect to the solid rod congruences, but everything is as if it were the case that their sizes had increased relatively to the different congruences in F.

Here Schlesinger glosses over the important fact that the length variable l in Description I refers physically to the congruences in F, whereas in Description II that variable pertains only to the different congruences furnished by solid rods. And his paraphrase of Description II thereby insinuates that Description II draws on Description I for its very statement. In stating (the unitary principle of change relevant to) Description II, I had not employed the congruences in F but only the solid body congruences. And clearly if Schlesinger's second conjunct is to be predicated on the same congruences as the first (i.e., on solid rods) and is to be true under the posited circumstances, then it must be formulated in terms of the *change in the laws of nature* specified by my Description II. Here I must remind the reader of what I said above concerning the *physical significance* which is to be assigned to the *mathematical stratagem* of obtaining the new laws in II by replacing the length variable l in the old laws by $2l$. This mathematical stratagem in no way presupposes that II is tacitly parasitic upon the congruences in F employed by I.

Schlesinger's paraphrase is erroneous, because it rests on the falsehood that Description II draws on Description I for its very statement. And, in this way, he prepares the ground for his defense of the contention that I is *inductively simpler* than II, a defense whose fallaciousness I shall now demonstrate.

2. He claims that II invokes a "vast number" of independent unexplained physical occurrences or causes to account for the posited observed changes. But, according to him, I—which he espouses as uniquely true vis-à-vis II—interprets these changes as "the inevitable consequences of a single event: the increase in the size of the universe." And he concludes that II is inductively inferior to I, because the latter "postulates only one unexplained happening." But, as we shall see presently, any plausibility possessed by this argument derives entirely from (1) a *pseudo*-contrast between the *number* of causally independent happenings invoked by I and II respectively or (2) Schlesinger's careless use of the terms "single event" and "single hypothesis."

In the first place, in Schlesinger's cherished Description I, all of the doublings are clearly causally independent of one another, and there will be as many such causally independent events as there are extended bodies in the universe. Description I specifies no cause or set of causes for the doublings either severally or collectively. Indeed, the posited circumstances allow that the universe contains infinitely many extended bodies, and if it does, then the number of causally independent occurrences (doublings) invoked by I is vast to the point of being infinite! By a question-begging use of the term "vast number," to characterize the causally independent events in II, Schlesinger miscontrasts the number of causally independent events which are implicated in the changes that Description I calls "doublings" with the *equal* number implicated in *these very changes* as rendered in Description II by the assertion of a specified kind of unitary change in the members of a designated class of laws of nature. For example, one and the same change in the period of a simple pendulum is described by I as a change in the latter's length in accord with the pendulum law, and by II as a change in the law relating the unchanged length to the period. And similarly for each of the causally independent changes.

In the second place, by describing all of the many doublings as "a single event," Schlesinger uses the term "single event" in a collective sense which is question-begging in this context, since he will not allow that the identical, equally many changes which II describes as a change in certain laws of nature also be dubbed "a single event." Indeed, having lumped the many doublings together misleadingly in the "single event" of there being an "increase in the size of the universe," he feels entitled to render Description I by saying that the size of the universe has doubled. That this is a significant mistake in this context becomes evident from the fact that my statement of Description I applies meaningfully to a spatially infinite universe, whereas Schlesinger's rendition of Description I does not, since it does not make sense to say of an infinite universe that its size has doubled. In other words, the universe might be spatially infinite and yet allow that the ND-conjecture be true even though the doubling *of the universe* would then be meaningless. Thus, only his specious version of Description I allows Schlesinger to claim that I offers a "single hypothesis" in the honorific sense of greater inductive simplicity as contrasted with the rival Description II! Incidentally, unlike Schlesinger, I would say that I is *descriptively* simpler than II. But this fact is, of course, unavailing to his thesis, since the difference in descriptive simplicity cannot detract from the truth-equivalence of I and II, and hence cannot make for a difference in inductive simplicity.

In a further attempt to show that II is not co-legitimate with I, Schlesinger writes:

> But the most important objection is best made clear with the aid of an example. Suppose we observe that all water, mercury and oil have frozen; we feel severe cold and all the thermometers register a fall of 100°F in the temperature of the surroundings. A simple way to account for this would be to say that the temperature of our

surroundings has dropped 100°F. Imagine, however, someone coming along with an alternative hypothesis claiming that there has been no drop in temperature but numerous changes in the properties of matter have taken place which make it appear as if there were such a drop: the freezing point of all liquids has risen by 100°F; our physiology has changed in such a manner that we now feel intolerably cold at 55°F; the coefficient of thermal expansion of all substances has undergone a change in such a manner as to affect appropriately all expansion thermometers etc. He might also claim that his hypothesis which he regards as significantly different from the one postulating a general lowering of temperatures is no more complicated than it for we need not spell out separately all the changes that have occurred. A unitary principle governs them all: in all equations which represent functional dependencies or laws of nature that involve variables ranging over temperatures, substitute instead of θ (representing temperature in °F) the number $\theta - 100$. His position would be as good as Grünbaum's.

Consequently, it is seen that if we persist in saying that no matter what happened we would never be forced to concede that everything has expanded in size—and defended his thesis in the manner just described, we could not thereby demonstrate anything of significance about the nature of space. For such a claim amounts to a pronouncement in general that any observation *O* which can be accounted for by postulating the occurrence of *C*, can also always be accounted for by an alternative hypothesis; a hypothesis which need not be spelled out in all its details and of which one may say that certain things have occurred whose effect was to bring about observation *O*. In other words, Grünbaum's suggestion, if tenable, amounts to saying that every empirical hypothesis has a counter-hypothesis to it according to which the events stipulated by the hypothesis have not actually occurred but what has in fact taken place makes things appear as if they had occurred.

What has been said so far is sufficient to make one feel at quite a loss to understand how anyone might be tempted to insist that nothing could happen that would impose on us the conjecture of universal expansion . . . [9, pp. 50–51].

Schlesinger's argument here is in three steps: (a) he offers what he considers to be a thermometric counterpart to my spatial Description II; (b) he invokes his earlier mistaken paraphrase to argue that if I and II are co-legitimate in the spatial case, then so also are their thermometric counterparts, a conclusion which he considers absurd; (c) he adduces the co-legitimacy of the purported thermometric counterparts as providing a *reductio ad absurdum* of my claim that the two spatial descriptions are co-legitimate and that their co-legitimacy is a consequence of a significant property of physical space, i.e., of its being a mathematical continuum of like elements. And my objections to these contentions of his are the following:

1. He states the alternative hypothesis which is the purported thermometric counterpart of my Description II by saying: "there has been no drop in temperature but numerous changes in the properties of matter have taken place which make it appear as if there was such a drop: the freezing point of all liquids has risen by 100°F; our physiology has changed in such a manner that we now feel intolerably cold at 55°F . . . etc." [9, p. 50]. Just as his incorrect paraphrase of my Description II had invoked two incompatible congruence criteria simultaneously, so here his purported thermometric counterpart of my II tacitly appeals simultaneously to two different temperature scales (Scale 1 and Scale 2) each of which is graduated in degrees Fahrenheit but which differ by a downward shift of the zero by 100°F. For let us recall our discussion of Schlesinger's earlier temperature example apropos of Schlick's version of the ND-hypothesis. Then it becomes apparent that the difference between Schlesinger's earlier thermometric example and the present one is the following: in the earlier example, a change in the zero of the temperature scale was used to assert a *change* of temperature in the *absence* of an observable change, whereas in the present example, such a scale change is used to assert a *constancy* of temperature in the face of an observable change. And hence we see that in the face of the observable changes posited in the present example, the statement "there has been no drop in temperature" can be true only in the sense that the *number* furnished by *Scale 2* for the temperature at the time of the observably new state of affairs is the *same* as the number previously furnished by *Scale 1* for the temperature

prevailing during the observably different earlier state of affairs. Moreover, in the statement "numerous changes in the properties of matter have taken place which make it appear as if there was such a drop," the term "drop" refers to the original Scale 1. But the assertions which Schlesinger then offers as partial explications of the latter statement make simultaneous references to Scale 1 and Scale 2. Thus, the statement that "the freezing point of all liquids has risen by 100°F" is itself literally true under the posited circumstances when applied to a particular kind of liquid *only* in the following sense: a liquid whose freezing point is at T_1°F relatively to Scale 1, which was in use at the earlier time, has a freezing point at $(T_1 + 100)$°F relatively to Scale 2, which is used at the later time. For the phenomena posited in Schlesinger's example here differ observably in fundamental ways from those in which liquids would exhibit a *time-dependent* freezing point behavior relatively to the SAME scale, say as furnished by a helium gas thermometer: under the circumstances which he posits here, the freezing point of water, for example, would *not* be observed to have risen from 32°F to 132°F on a fixed scale furnished by such a thermometer! Instead, what is involved in Schlesinger's present example is that the temperature numbers to be associated with the unfrozen and frozen states of a liquid at times t_1 and t_2 respectively are to be the same on the strength of referring the temperature at time t_1 to Scale 1 and the temperature at time t_2 to Scale 2. By the same token, in the statement "whereas we used *not* to feel intolerably cold at 55°F, we now do feel that way at 55°F," the successive occurrences of "55°F" refer to Scale 1 and Scale 2 respectively.

If it is explicitly recognized as such, a change of scale involving a shift in the zero of temperature, far from being absurd is entirely permissible and would indeed be reflected in the law statements of the freezing points of liquids, etc., by adding (not subtracting as Schlesinger has it) 100°F to the Fahrenheit numbers. Hence, we can have two equivalent thermometric Descriptions I and II of the posited phenomena in the sense that I is based on Scale 1 alone, while II involves an avowed scale shift. But the conclusions which Schlesinger tries to ground on this feasibility in steps (b) and (c) of his present argument are *non sequiturs*, as we shall now see.

2. There are separate reasons for the co-legitimacy of I and II in the spatial case, on the one hand, and in the thermometric case, on the other. In the spatial case, the co-legitimacy of I and II arises, as I have argued, from the alternative metrizability of space WITH RESPECT TO ITS CONGRUENCE CLASSES OF INTERVALS. But in the thermometric case, the co-legitimacy of the thermometric alternatives I and II arises from the option of choosing different ZEROS of temperature. Therefore, it is wrong for Schlesinger to claim that the reasons for the two co-legitimacies are relevantly the same. And hence, he did not establish the scientific uniqueness of Description I vis-à-vis II in the spatial case by a *reductio* based on the indictment of the thermometric co-legitimacy as absurd. Nor did he establish the spatial uninformativeness of the co-legitimacy of the spatial versions of I and II. On the contrary, the recognition of the different kinds of reasons that make for the respective co-legitimacies exhibits the wrongheadedness of Schlesinger's attempt to base a *reductio* argument on the assimilation of the reason for the one to that of the other.

There is a thermometric example which is similar to the spatial one with respect to generating two equivalent, co-legitimate descriptions for a *like kind of reason*. This case arises from the following alternative metrizability of the temperature continuum with respect to the congruences among temperature changes (i.e., differences or intervals): the temperature-interval congruences furnished by an expansion thermometer employing one kind of thermometric substance do not, in general, accord with those furnished by such a thermometer when employing a different thermometric substance. And if we do not avail ourselves of Kelvin's so-called thermodynamic scale of temperature, which eliminates this discordance among the congruences, then we can base co-legitimate thermometric Descriptions I and II on two such alternative congruences. In that case, the resulting I and II are co-legitimate for a reason similar to the one pertinent to the spatial case. But the existence of this kind of similarity between a spatial and a thermometric example hardly shows that the respective properties of the two domains from which it arises are not each significant in their own right. Yet we shall see in Chapter III, §4, that there is an important residual dissimilarity between these two examples, as shown by the comparison there of *spatial* congruences with those among pressure changes.

Referring to W. C. Salmon's defense of my spatial thesis of the co-legitimacy of I and II against Schlesinger's earlier critique, Schlesinger first quotes Salmon as follows:

The issue, I take it, is this. Grünbaum insists upon the metrical amorphousness of space. He therefore asserts the vacuity of saying that physical objects have doubled in size in relation to this amorphous continuum. It makes good sense, however, to assert that a length measured by one method has doubled in relation to that same length measured by another method. But he insists this is perfectly equivalent to saying that an extension measured by the second method has halved in relation to that same length measured by the first method. We can indeed talk about doubling or halving of lengths with respect to any physical standard we choose but the metrically amorphous space is not itself such a physical standard. It strikes me that the only ground on which Schlesinger can accept one of Grünbaum's descriptions and reject the other is to maintain that all lengths *really* doubled—that is doubled with respect to space itself. But the amorphousness of space rules out this possibility" [9, p. 51].

And then Schlesinger recalls my emphasis on the failure of physical space to possess intrinsic congruences and comments:

This emphasis by the way constitutes one of the most baffling features of his argument since the nature of space, its continuousness or discontinuousness, the denumerability or non-denumerability of its constituents and altogether its metrizability or non-metrizability, is entirely irrelevant to the credibility of the nocturnal-doubling-conjecture. Accepting then Salmon's reading of Grünbaum we now know what aspect of the situation needs emphasis. What we have to stress is that in the circumstances encountered in the previous section we have definitely not been wishing to say that the length of everything has increased relative to space itself (whatever that might mean), but that it has increased *relative to the length it had before* [9, pp. 51–52].

Note Salmon's statement "the only ground on which Schlesinger can accept one of Grünbaum's descriptions [i.e., Description I] and reject the other [i.e., Description II] is to maintain that all lengths *really* doubled—that is doubled with respect to space itself." What Salmon is telling us here is that Schlesinger's denial of my thesis of the co-legitimacy of I and II rests on invoking a length increase with respect to space itself in the following sense: nonexistent intrinsic congruences of space itself need to be invoked to authenticate the *uniqueness* of the verdicts of the congruence criteria (in class **F**) which yield Description I *as against* the rival congruences furnished by solid rods, which yield Description II. I ask Schlesinger: Given my demonstration above of the *failure* of his inductive simplicity argument for the uniqueness of I vis-à-vis II, how else can he exclude II except by claiming falsely that solid rods cannot serve as congruence standards, because they have doubled *with respect to space itself?*

It will be noted that Salmon did *not* say at all that the credibility of Description I (the ND-conjecture) depends on any tacit metrical appeal to space itself but only that Schlesinger's claim of *uniqueness* for I *to the exclusion* of II rests on such an appeal! To this Schlesinger replies irrelevantly that the ND-conjecture (Description I) does not depend on any metrical appeal to space itself, a fact on which we are all agreed.

Furthermore, he again attacks a straw man when expressing puzzlement about the relevance of the continuity of space: as the reader will recall, I had never asserted that the continuity of physical space is *necessary* for the physical significance or credibility of the ND-conjecture. Finally, he errs most obviously in his statement here when saying of physical space "altogether its metrizability or non-metrizability, is entirely irrelevant to the credibility of the nocturnal-doubling conjecture." To this I reply by asking: without the metrizability of space, how, pray tell, can one even meaningfully assert, let alone believe, that any extended object has *doubled* its *size?* It is now evident that here Schlesinger mistakenly believes himself to have refuted the co-legitimacy of I and II, which I rested on the continuity of space, by pointing out that Description I itself could be credible without any invocation of that continuity.

Schlesinger goes on to cite a case of a different hypothetical world for which, he agrees, each of two spatial Descriptions I and II would be equally appropriate. And he believes that the difference between that hypothetical world (World B) and the one which has been at issue so far (World A) is such as to allow two co-legitimate descriptions in the former but not in the latter case. He writes:

> Salmon's argument about the automatic reversibility of any claim about the increase of X and the possibility of turning it into a claim about the decrease of non-X would be relevant to a situation in which a *discrepancy* was discovered between the length of different kinds of physical systems as measured by the meter rod while other methods of measurement have not, or for some reason could not, have been employed. Suppose one morning we rise to find by the meter rod method of measurement that all non-metallic bodies and the distances between all bodies have doubled in size relative to all metallic bodies. Suppose also that a group of philosophers, for some metaphysical reasons, regarded metals immutable and consequently viewed meter rods made of metal as truer standards of length. Such philosophers would insist that the whole of the non-metallic universe must have increased in size. Against them, Grünbaum would be fully justified in arguing that an alternative account attributing no change to anything except to metallic bodies which have contracted to half of their previous size is equally acceptable. Incidentally, assuming for a moment that in principle no other methods were available to decide the issue, it would then be altogether wrong to insist that either description constitutes the correct account of what took place . . .
>
> But it would be a confusion to compare such a situation to the one dealt with in the previous section. For in that situation no one wishes to maintain that one group of physical systems increased in size

while another group remained unchanged. Everything, without exception, has changed in size relative to what it was yesterday—that is what is claimed. And there is no conflict between the various standards of length measurement as to the validity of this claim . . . the issue is not whether one group of systems has increased its size or the other group has decreased its size. It is whether a change in a given property (namely size) of all systems has occurred or whether a change in a great number of laws, having a combined effect as if the former change occurred, has taken place [9, pp. 52–53].

It is agreed, of course, that two equivalent descriptions are appropriate to Schlesinger's World B. But the fallacies in Schlesinger's reasoning here that two descriptions are not *also* appropriate to our former World A in the sense of my Descriptions I and II seem to me to be the following:

1. It is clear, of course, that in World B there is a discordance between metallic solid (meter) rods and nonmetallic ones in regard to their congruence behavior, whereas in World A there is concordance among the congruence verdicts of all solid rods. But in World A there is discordance between the like congruence verdicts furnished by the criteria in Class F, on the one hand, and the like congruence verdicts yielded by solid rods (of whatever constitution), on the other. Why, I must ask, should the discordance in World B make for the admissibility of alternative descriptions of that world while the discordance in World A allegedly does not confer co-legitimacy on my Descriptions I and II? How does the fact that *all* solid rods exhibit a concordant congruence behavior in World A disqualify them from serving alike as a congruence standard which yields Description II? It is clear that Schlesinger has no case here, since the dissimilarity between Worlds A and B to which he points is irrelevant, while the relevant similarity is the obtaining of a discordance with which alternative, equivalent descriptions can be associated in *each* of the two worlds. And in either world the justification for the co-legitimacy of I and II alike involves the obtaining of a discordance.

2. He cannot now fall back on his earlier argument from inductive

simplicity, since the latter is altogether untenable. And in the light of the failure of that argument, it is misconceived on his part to suggest in the present quotation that if World B contained congruence criteria other than either metallic or nonmetallic solids, the further criteria could "decide the issue" in the sense of conferring *uniqueness* on the congruence verdicts of one of the two discordant subsets of solids.

3. Given the demonstrated failure of his inductive simplicity argument for the *uniqueness* of Description I of World A and his disavowal of a metrical appeal to space itself, his espousal of that uniqueness is as unsound as the position of the hypothetical philosophers in World B who insist that metallic solid rods are "truer standards of length" than nonmetallic ones.

Schlesinger gives an additional argument in an attempt to show that the presence in World B of congruence criteria like some of those in class F could "decide the issue" between the metallic and nonmetallic rods after all. He now assumes that there is a class F′ of further congruence criteria which yield the *same* congruences as the *metallic* solid rods and hence disagree with the nonmetallic rods. In that case, a Description I′ of the enriched World B (which is the counterpart of Description I in World A) would be associated with the congruences furnished alike by the members of F′ *and* the metallic solids: according to Description I′, the nonmetallic bodies have undergone a length change by the same factor but the laws of nature hold as before. But a Description II′ of the enriched World B (which is the counterpart of my Description II of World A) would be associated with the congruences furnished alike by the nonmetallic rods, and in Description II′, the lengths of metallic solids *and* the laws of nature will be said to have changed. And Schlesinger's argument for the uniqueness of I′ vis-à-vis II′ is the following:

It stands to reason that in a situation like this Grünbaum would agree that there is no room left for any doubt and we are entitled to claim to know which group of things has undergone a change of size and which group remained constant. For no matter how narrow an operationist one wishes to remain, one cannot get away from the fact that there is a disagreement among those very standards to which one so faithfully wishes to remain 'wedded' and it is not possible to maintain that none of the meter-rods in the world have altered size. If we agree that it is the non-metallic component of the universe which has increased in size, no more changes need to be stipulated, whereas if we insisted that the non-metallic part of the universe preserved its size while the metallic has halved in size, then we must in addition stipulate a change in a vast number of nature's laws, which, although governed by a unitary principle, could altogether be avoided [9, pp. 53–54].

My reasons for finding this argument fully as unconvincing as all of Schlesinger's earlier ones are the following:

1. Schlesinger uses the fact that I′ is *descriptively* much simpler than II′ to infer fallaciously that I′ is also *inductively* simpler than II′ and hence not equivalent in truth value. Only this fallacious inference or an appeal to his earlier unsuccessful arguments against II of World A enables him to speak here of there being "no room left for any doubt" and of being "entitled to claim" that we "know" that the nonmetallic rather than the metallic solids of the world are the ones which changed in size.

2. The discordance among the metallic and nonmetallic members of the class of solid rods is fully acknowledged by Description II′ and cannot possibly serve as an argument against II′. It is astonishing that Schlesinger should attribute to me a hypothetical kind of operationist position. For in the first place, under the *different* circumstances of World A in which there is concordance among *all* solids, the hypothetical single-criterion "operationist" claimed *uniqueness* for Description II while I claimed only co-legitimacy with I for II. And, in the second place, as Wilfrid Sellars has emphasized, the operationist's position is to insist on a single criterion *at a time* and hence allows him to use different single criteria at different times, and hence to adopt *either* the metallic rods *or* the nonmetallic ones after they had become discordant with one another.

3. Since I have affirmed the co-legitimacy of I′ and II′, I have

thereby asserted that II′ could be avoided by adopting Description I′. Hence Schlesinger's claim that II′ "could altogether be avoided" is hardly an argument against my thesis.

Having presented his argument for the uniqueness of I′ as a description of the enriched World B, Schlesinger's final argument is based on a modified enriched World B, which I shall call "World C." He writes:

> But now the utter untenability of the view according to which nothing could happen to make us assent to the conjecture that overnight everything has increased in size can fully be exposed. All we have to do is to modify our present example and, instead of having a sudden single change, have the range of the group of systems whose size has altered, increase gradually. Imagine a process whereby greater and greater proportions of the constituents of the universe change in size relative to what they were before. On the first day we notice that—say—1 percent of the universe has doubled in size while the size of the rest remained constant (and we know it is not the other way round, i.e. that it is not the 1 percent of the universe which preserved its size while 99 percent of it has shrunk, by the testimony of all the functionally length dependent properties of physical systems). By the next day 2 percent and the day after 3 percent of the furniture of the universe has doubled in size, and so on. On the hundredth day our observations will differ basically from what they were before in that no longer will there be a relative discrepancy between the sizes of various systems since by then the process of doubling has extended over the whole universe. Surely it would be absurd to claim that this process of the spreading increase in the dimensions of the universe's constituents has continued almost to the very end and just before it affected everything it suddenly went into reverse and instantaneously everything has contracted back to its original size and the strange behaviour of the universe which is as if the process of doubling were completed and not reversed should be put down to the sudden change in all the laws of nature which are functionally length dependent. It is seen then that no matter what strange views one may hold on the nature of scientific method, one is forced to admit at least in this instance, that at the completion of the previous process there is strong enough evidence to force us to say that everything has increased in size [9, p. 54].

We see that Schlesinger's World C is one in which there is first a period of 99 days before zero hour during which the successive percentages of solid rods and extended bodies play the role of the nonmetallic solids in the enriched World B; thereafter, i.e., after zero hour, World C exhibits the discordance of congruences between the criteria in class F, on the one hand, and *all* solid rods, on the other, which is familiar from World A. Hence the difference between Worlds A and C lies in the following: In World A, there is *concordance* in the congruence behavior of *all* solid meterrods throughout all time before, after, and at the zero hour, whereas in World C, there is *discordance* in the congruence behavior of the metersticks for a period of 99 days prior to the zero hour. Clearly, therefore, in World C the various solid rods cannot all consistently serve interchangeably as a congruence standard *during this 99-day period.* Hence there are differences between Worlds A and C with respect to the range of co-legitimate alternative descriptions. I shall state *some* of the co-legitimate descriptions of World C. And it will then be clear from all my arguments concerning World A that nothing in World C supports Schlesinger's claim of uniqueness for Description I of World A. *Nor* could World C be adduced in any way to establish the uniqueness of the particular description of that world which is favored by Schlesinger.

The following are *some* of the alternative equivalent descriptions of World C:

i. On the first of the 100 days, 99 percent of all bodies shrank while the laws of nature changed, and on each of the remaining 99 days, 1 percent of the shrunken bodies return to their unshrunken size. (Here a particular 1 percent of all solid rods play the congruence role played by *all* solid rods in Description II of World A.) By the zero hour, all extended bodies are of the same size again as before the first day, but the change in the laws of nature that occurred on the first day remains after the zero hour.

ii. The description which Schlesinger erroneously claims to be *uniquely* true vis-à-vis the alternative envisioned by him: Within 100 days all the solids of the universe have doubled in size.

iii. The description which Schlesinger question-beggingly indicts as "absurd": Beginning with the first of the 100 days, 1 percent of all bodies double in size for *each* of the next 99 days, but on the 100th day, the 99 percent of the bodies that had expanded shrink back to their unexpanded size. (Here a particular 1 percent of all metersticks play the congruence role assumed by all metersticks in Description II of World A, but the membership of this 1 percent is *not* the same as in the preceding Description (i) of World C.) By the zero hour, all bodies are back to normal, but the laws of nature change at the zero hour.

I conclude that all of Schlesinger's arguments against my original thesis of the co-legitimacy of I and II of World A have failed. And hence I reject his imputation that I hold "strange views . . . on the nature of scientific method" [9, p. 54].

2. The Interdependence of Geometry and Physics, and the Empirical Status of the Spatial Geometry

The elaboration of §§6 and 7 of Chapter I which I included in an essay on Duhem's epistemological holism published in 1966 [6, Sections I and II] did not amplify the special case of alternative geometric interpretations of stellar parallax observations in the sense of Duhem. Nor did my 1966 essay contain mention of an error in §7 of Chapter I, p. 140 above, which was pointed out by Arthur Fine [4, pp. 159–160]. Subsections (i) and (ii) which are to follow are designed to remedy these two omissions.

(*i*) *The Duhemian Interdependence of Geometry and Physics in the Interpretation of Stellar Parallax Observations.*

I shall elaborate on pp. 116–118 of Chapter I by reference to the special case in which stellar light ray triangles are found to have angle sums of less than 180°. This finding permits an inductive latitude in the sense emphasized by Duhem for the following reason:

uncertainty as to the separate validity of the correctional physical laws makes for uncertainty as to what spatial paths are indeed geodesic ones, and this uncertainty, in turn, *allows* that the optical paths are slightly non-geodesic ("curved") in the small *and* markedly *non*-geodesic in the large. Consequently, there is inductive latitude to postulate either of the following two theoretical systems to explain the observed fact that stellar light ray triangles have angle sums of less than 180°:

(a) G_E: the geometry of the rigid body geodesics is Euclidean, and
O_1: the paths of light rays do *not* coincide with these geodesics, and light ray triangles have area-dependent angle sums of less than 180°,

or

(b) $G_{\text{non-}E}$: the geodesics of the rigid body congruence are a *non*-Euclidean system of hyperbolic geometry
O_2: the paths of light rays *do* coincide with these geodesics, and thus light ray triangles are non-Euclidean ones of hyperbolic geometry.

(ii) The Empirical Status of the Spatial Geometry.

My statement on p. 140 above that there can exist only one geometry G_n of constant curvature which the specified convergence procedure could yield is too strong. The reason which I gave for this conclusion was that any geometry of constant curvature G_n yielded by the convergence procedure will be identical with the unique underlying geometry satisfying the third (and first) of the three conditions characterizing the desired geometry G_t. But, as Arthur Fine has noted correctly [4, pp. 159–160], there is no guarantee that this identity must obtain: a G_n of constant curvature furnished by the convergence procedure and thereby satisfying the *second* of the conditions characterizing the desired G_t might well not be the geometry prevailing in a perturbation-free subregion of the space, i.e., might not satisfy the third (or first) of these conditions. And hence the existence of a G_t of spatially (and temporally) constant curvature in a perturbation-free subregion cannot exclude the following possibility: there are two convergent sequences $G_0, \ldots, G_1$, and $G_0', \ldots, G_1'$, which satisfy the second condition governing G_t and in which G_1 and G_1' are *different* geometries (no less than G_0 and G_0') though each being of constant curvature. Thus, my claim in Chapter I, p. 140, that "there can exist only one . . . geometry of constant curvature G_n" satisfying the condition of convergence should be weakened to state the following: It *may* be that there is only one G_n of constant curvature which satisfies the latter condition *and* that this unique G_n is also the geometry satisfying the first and third of the three conditions characterizing G_t; i.e., it *may* be that there exists a geometry of constant curvature which satisfies all three of the requirements governing G_t. And if there is such a G_n, the empirical convergence procedure which I outlined in §7 of Chapter I will serve to uncover the underlying geometry to within the usual inductive uncertainties.

I should add that a sequence of geometries can meaningfully converge to a geometry of variable curvature and that the concept of a convergent sequence of geometries therefore need not be restricted to geometries of constant curvature, each of which can be represented by a single number (the Gaussian curvature). For in the case of variable curvature, there can be convergence to a particular set of functions g_{ik} in the following sense: for each space point there is convergence to some particular numerical value for each of the g_{ik}.

Bibliography

1. Cajori, F., ed. *Sir Isaac Newton's Mathematical Principles of Natural Philosophy and His System of the World.* Berkeley: University of California Press, 1947.
2. Ellis, B. "Universal and Differential Forces," *British Journal for the Philosophy of Science*, 14:189–193 (1963).
3. Feyerabend, P. K., and G. Maxwell, eds. *Mind, Matter, and Method.* Minneapolis: University of Minnesota Press, 1966.
4. Fine, A. "Physical Geometry and Physical Laws," *Philosophy of Science*, 31:156–162 (1964).
5. Grünbaum, A. *Philosophical Problems of Space and Time.* New York: Knopf, 1963.

6. Grünbaum, A. "The Falsifiability of a Component of a Theoretical System," in *Mind, Matter, and Method,* P. K. Feyerabend and G. Maxwell, eds. Minneapolis: University of Minnesota Press, 1966.
7. Luria, S. "Die Infinitesimaltheorie der antiken Atomisten," in *Quellen und Studien zur Geschichte der Mathematik, Astronomie, und Physik.* Abteilung B: Studien, II. Berlin: Springer, 1933.
8. Schlesinger, G. "It Is False That Overnight Everything Has Doubled in Size," *Philosophical Studies,* 15:65–71 (1964).
9. Schlesinger, G. "What Does the Denial of Absolute Space Mean?" *Australasian Journal of Philosophy,* 45:44–60 (1967).
10. Schlick, M. *Space and Time in Contemporary Physics,* H. L. Brose, tr. Oxford: Oxford University Press, 1920.
11. Zemansky, M. W. *Heat and Thermodynamics.* 3rd ed.; New York: McGraw-Hill, 1951.

III

REPLY TO HILARY PUTNAM'S "AN EXAMINATION OF GRÜNBAUM'S PHILOSOPHY OF GEOMETRY"

1. Introduction

Referring to my philosophical account of physical geometry and chronometry, H. Putnam says: "Grünbaum has in my opinion failed to give a true picture of one of the greatest scientific advances of all time" [37, p. 211].[1] I have not found one single argument in Putnam's fifty pages that could serve to sustain this judgment. Nay, reflection on fundamental *mathematical* and *physical* errors on which he rests much of his case has enabled me to uncover substantial new support for my position from the general theory of relativity and from elsewhere in physics.

The present chapter will endeavor to set forth this new support and to elaborate on some of my earlier arguments as part of a detailed, point-by-point rebuttal of all Putnam's criticisms. It will turn out that his critique rests not only on mistakes in physics whose erroneousness is not a matter for argument, but also on numerous

NOTE: By prior arrangement, a somewhat different version of this chapter is also published in Volume V of the *Boston Studies in the Philosophy of Science*, edited by R. S. Cohen and M. W. Wartofsky (Dordrecht: D. Reidel Publishing Company, 1968), pp. 1–150.

[1] All but one of my subsequent references to Putnam will be to this essay of his, and the pages from which my citations of it are drawn will be specified within brackets in my text immediately following the quoted passages. For the one exception, in §3.2, the different reference is given.

grossly careless misreadings of my text.[2] And in one important case (see §8.0 below), he even adduces against me a result that I had explicitly set forth and emphasized in a publication which was readily available before Putnam wrote his essay.

2. Riemann's Philosophy of Geometry and Einstein's General Theory of Relativity

2.0. Riemann's conception of the congruence relation among intervals of a continuous manifold of point-like elements countenances alternative metrizations of the time continuum and of the space continuum of physics which yield incompatible congruence classes of intervals in these respective continua. In view of the pivotal role played by Riemannian geometry in Einstein's general theory of relativity (hereafter called "GTR"), it is appropriate to inquire whether the GTR actually confronts us with alternative congruences among intervals of physical time and space. I contend that Riemann's conception of congruence represents an insight into the structure of physical space and time *and* that this insight exhibits the physical co-legitimacy of the incompatible spatial and temporal congruences actually encountered in the GTR. By contrast, Putnam, who, as we shall see, rejects Riemann's view of congruence, is unaware of the role of such incompatible congruences in the GTR, and hence he denies the relevance of Riemann's view to the GTR. But as we shall see, Putnam is quite unaware of the *import* of the fact that the GTR assigns measures to three-dimensional space intervals (via the *spatial* metric tensor $\gamma_{\iota K}$ (ι, $K = 1, 2, 3$)[3]) and to one-dimensional time

[2] Putnam's critique is addressed mainly to what I wrote in my 1962 publication "Geometry, Chronometry, and Empiricism," which first appeared in H. Feigl and G. Maxwell, eds., *Scientific Explanation, Space, and Time* (Minneapolis: University of Minnesota Press, 1962), pp. 405–526, and which constitutes Chapter I of the present book. Unless otherwise specified, all references to previously published statements of mine that are criticized by Putnam are to this essay. But since the pagination of the 1962 text and of the present Chapter I is different, each reference will furnish both sets of page numbers, and the year of the publication will precede each page number citation to identify it.

[3] Cf. [33, p. 238].

intervals no less than to four-dimensional space-time intervals (via the *space-time* metric tensor g_{ik} (i, k = 1, 2, 3, 4)). Putnam writes:

> . . . we are *not* free in general relativity to employ any metric tensor other than g_{ik}. It is possible to choose the reference system arbitrarily and still arrive at the same covariant laws of nature in general relativity *only* because we are *not* allowed to "choose" a space-time metric. Grünbaum obscures this situation by asserting that in general relativity we sometimes *do* employ non-standard "definitions of congruence"; he further asserts that our freedom to do this is a consequence of the alleged "insight" that we discussed above. This seems puzzling in view of the fact that all observers employ the same g_{ik} tensor in general relativity, until one realizes how Grünbaum employs his terms. What he is pointing out is the fact that the customary rule of correspondence (employing a clock at rest in the reference system to determine local time) is not valid under exceptional circumstances (on a rotating disk). But correspondence rules always break down under some exceptional circumstances. Thus no special insight is needed to allow for the fact that the g_{ik} tensor of the theory is to be preserved over the readings of a clock situated on a rotating disk! In sum: our alleged "freedom" to *choose a different* g_{ik} *tensor* (a different space-time metric) *at the cost of complicating the laws of nature is in fact never employed in the general theory of relativity*. All observers are required to "choose" the *same* space-time metric [p. 242].

Preparatory to pointing out the fallacies and scientific errors contained in this passage, I wish to set forth a number of physical and mathematical results that will serve as a basis for the discussion.

2.1. When deducing the Lorentz transformations of the *special* theory of relativity ("STR"), one seeks those *linear* transformations between the coordinates x, y, z, t of system S and x', y', z', t' of the system S' which will assure the invariance of the velocity of light,[4] i.e., which will yield either of the following two equations from the other: $x^2 + y^2 + z^2 - c^2t^2 = 0$ and $x'^2 + y'^2 + z'^2 - c^2t'^2 = 0$. Let the system S' move with constant velocity v along the x axis of S such that the x and x' axes are coincident. Then the imposition of

[4] There are also *non*linear transformations between these coordinates which leave the light-velocity invariant. Cf. [40, pp. 172–175].

the requirement that the transformations be linear rather than nonlinear has the following important consequence: it assures that if two spatial intervals on the x' axis which are congruent in system S' are measured at instants which are simultaneous in S, then they will likewise be congruent space intervals in S. For note that for our case of simultaneous events, $dt = 0$ in the equation

$$dx' = \frac{dx - v\,dt}{\sqrt{1 - v^2/c^2}},$$

and then two S' space intervals of equal lengths dx' will likewise have equal lengths dx in S. In this sense, the linearity of the Lorentz transformation equation between quasi-Cartesian coordinates assures the invariance of spatial congruence as between S and S' for intervals on the x and x' axes [3, pp. 31n and 33]. By contrast, for $dt = 0$ a transformation between the coordinates x, y, z, t and x', y', z', t' which is *non*linear in x would issue in a corresponding *non*invariance of spatial congruence among such intervals as between S and S'. For in the latter nonlinear case, the magnitudes dx of the coordinate differentials in S which correspond to a given dx' would depend not only on the magnitudes of dx' but *also* on the particular coordinates x between which they are taken. But even in the case of the linear Lorentz transformations, a body of length dx' which is congruent in S' to a body of length dy' or dz' will *not* be congruent to either of them in S, since $dx \neq dx'$ whereas $dy = dy'$ and $dz = dz'$.

To be sure, there are *some non*linear coordinate transformations relating S to S' but involving at least one set of *NON*-rectangular coordinates under which spatial congruence among intervals on the x axis remains invariant. An example would be those nonlinear transformations that would replace the Lorentz transformations if in the frame S' we were to use polar (spherical) space coordinates r', θ' (and ϕ') instead of the rectangular ones x', y' (and z'). But when we go on to consider the class of nonlinear coordinate transformations countenanced by the GTR, which are generally nonlinear in both the space and time coordinates, then spatial congruence typically fails to be invariant and there is also the following very important *non*invariance of temporal congruence: of any three events A, B, and C which pairwise *invariantly* sustain a time-like separation or are connectible by a light ray, if the time intervals $\overline{AB}$ and $\overline{BC}$ are congruent in one reference system S, then these intervals will generally be *incongruent* in the other. It is incontestable that we are here confronting alternative metrizations in the sense countenanced by Riemann's conception of congruence, a fact that will become further apparent from a typical illustration to be given below. And the complete compatibility of this *non*invariance of the *time* congruences with the invariance of the congruences among the infinitesimal *space-time* intervals of the GTR will become apparent after we have distinguished three different kinds of intervals, which we shall now proceed to do.

2.2. In addition to the space-time manifold of events, there are the submanifolds of time and space. Hence there are three different kinds of infinitesimal intervals whose respective metrics or interval measures are the following in the GTR:

(i) the invariant measure $ds_4^2 = g_{ik}\,dx^i\,dx^k$ $(i, k = 1, 2, 3, 4)$ of the four-dimensional space-time interval between two events;

(ii) the noninvariant measure $ds_3^2 = \gamma_{\iota K}\,dx^\iota\,dx^K$ $(\iota, K = 1, 2, 3)$ of the three-dimensional space interval between the spatial locations of two events in a given reference system. (While each system of coordinates determines a unique system of reference, in any given system of reference one can always introduce an infinite number of different space-time coordinate systems such that the space coordinates of any one set C of them are functions of only the space coordinates of the others while the time coordinate of C is a continuous function of the coordinates of the others.) Some authors denote this spatial distance by $d\sigma$ [33, pp. 237–238] and others by dl [32, pp. 272–273];

(iii) the noninvariant measure $ds_1 = G\,dt$ of the one-dimensional time interval between two events.

The crucial difference between the measure ds_4, on the one hand, and the measures ds_3 and ds_1, on the other, has a decisive bearing, as we shall see, on whether there are in the GTR any noninvariant and hence incompatible congruences as countenanced by Riemann. This will become evident from the special case of comparing respectively the time congruences and space congruences of an inertial system with those of an accelerated system of reference constituted by a disk rotating with uniform angular velocity ω with respect to an inertial system I. The consideration of the rotating disk to which we now proceed will serve to show that Putnam's argument above is altogether vitiated by his fallacious invocation of the four-dimensional space-time tensor g_{ik} to deduce conclusions pertaining to the metrical coefficients which are associated with the interval measures ds_1 and ds_3. Moreover, it will turn out that *within* such accelerated reference systems as the rotating disk, the GTR *routinely* countenances alternative time-interval measures ds_1 (based respectively on "proper" time and "coordinate" time) which yield incompatible congruences within the same reference system. And §2.6 will show that if the disk is allowed to rotate with a time-dependent, *variable* angular velocity ω, then the GTR yields time-dependent *incompatible spatial congruences* in one and the same disk as well.

2.3. Let X, Y, Z, and T be the usual quasi-Cartesian coordinates in the system I, and let R and θ be polar coordinates in the XY plane of I, i.e., coordinates such that

(1) $\quad X = R \cos \theta \qquad$ and $Y = R \sin \theta.$

On the disk-system S', we can then introduce a system of polar space coordinates r, θ' by means of the transformation equations

(2) $\quad r = R$ and $\theta' = \theta - \omega T,$

where r is confined to values less than c/ω (c being the standard velocity of light), since points on the disk are confined to linear velocities $R\omega < c$ in their circular motions with respect to I. For

$T = 0$, i.e., at instants for which the I-system clocks all read $T = 0$, the polar space coordinates r, θ' of any disk point are the same as the polar coordinates R, θ of the I-system point with which it coincides instantaneously. Let each of the standard material clocks at the various points of the disk system be set to read $\tau = 0$ at the moment when the I-system clock instantaneously adjacent to it reads $T = 0$. Then the physical assumptions made by Einstein about the behavior of the standard clocks in the I and S' systems [10, §3, p. 116, and 33, p. 225] yield the following transformations between their so-called *proper time* readings T and τ:

$$\tau = T\sqrt{1 - \frac{r^2\omega^2}{c^2}}. \tag{3}$$

And (1), (2), and (3) show that the nonlinear transformations relating the rectangular coordinates X, Y, and Z of system I to the cylindrical space coordinates r, θ', and z and the proper time coordinate τ of system S' are

$$X = r\cos\left(\theta' + \frac{\omega\tau}{\sqrt{1 - \frac{r^2\omega^2}{c^2}}}\right), \quad Y = r\sin\left(\theta' + \frac{\omega\tau}{\sqrt{1 - \frac{r^2\omega^2}{c^2}}}\right), \tag{4}$$

and $Z = z$.

Now consider the following triplet of events A, B, and C: event A is constituted by the emission of light from the permanently coinciding origins O and O' of the I and S' systems at the instant $T = \tau = 0$; event B is the arrival of the light ray at a point $P'(r, \theta')$ on the disk which coincides with a point $P(R, \theta)$ in the XY plane of I; and event C is the return of the light ray to the common origin of I and S' after its reflection at PP'. It is clear from equation (3) for $r = 0$ and from the fact that the round-trip velocity in I is c that upon the occurrence of event C, the clocks at O and O' read the *same* time numbers $T = \tau = 2r/c$. Thus, both the I clock at O and the S' clock at O' yield the measure $2r/c$ for the time interval between the events A and C. Since events A and C both occur at the same space point O in I and also at the same space point O' in S', the proper time *interval* between them yielded by either of the clocks at O and O' is a measure of the *four*-dimensional interval between the events A and C. And, as required by the GTR, that four-dimensional measure ds_4 is *invariant*. But this invariance does *not* prevent the following fundamental *non*invariance of the time congruence yielded by the *proper* time measures of the time intervals between A and B, on the one hand, and B and C, on the other: the time intervals $\overline{AB}$ and $\overline{BC}$, as measured by the differences between the proper times of occurrence, are congruent in I but incongruent in S'. For the I clock at P reads the time $T = r/c$ upon the occurrence of event B, whereas—by equation (3)—the S' clock at P' reads the lesser time

$$\tau = \frac{r}{c}\sqrt{1 - \frac{r^2\omega^2}{c^2}}.$$

And hence in I, the event B divides the continuum of events between A and C into two *equal* time intervals $\overline{AB}$ and $\overline{BC}$, each of magnitude r/c, but in S', these intervals have the respective magnitudes

$$\frac{r}{c}\sqrt{1 - \frac{r^2\omega^2}{c^2}} \text{ and } \frac{r}{c}\left[2 - \sqrt{1 - \frac{r^2\omega^2}{c^2}}\right].$$

Note that the events A, B, and C pairwise *invariantly* sustain the relation of being connectible by a light ray and hence by a causal chain, here being invariantly ordered with respect to earlier and later. Yet the respective standard material clocks in I and S' metrize time such that $\overline{AB}$ and $\overline{BC}$ receive *equal* proper time measures r/c in I but unequal measures respectively less than and greater than r/c in S'.

We just employed two different time coordinatizations T and τ to illustrate the noninvariance of the time congruences that are associated with the time metrics given respectively by *differences* between the T and τ coordinates. But we can use the very same discrepancies in the behavior of the standard clocks in I and S' to generate different time metrics ds_1 and ds_1' which issue in *incompatible* time-interval congruences *in the very same manifold of co-*

ordinates. To see this, let us employ the *same* time *coordinates* T in both I and S', while being mindful of the fact that this coordinatization still leaves it open whether the time *metric* will be given by mere differences between the time coordinates or not. And then let us go on to specify that the respective incompatible time metrics ds_1 and ds_1' in I and S' are the following:

(a) ds_1 is given by the *difference* between the time coordinates T, i.e., by $ds_1 = T_2 - T_1 = dT$,

but

(b) ds_1' is given by the following different function of T_2 and T_1:

$$ds_1' = T_2\sqrt{1 - \frac{r_2^2\omega^2}{c^2}} - T_1\sqrt{1 - \frac{r_1^2\omega^2}{c^2}} = d\tau.$$

Upon thus basing our time metrics in I and S' respectively on the same coordinatization, we can say that the standard clocks in I and S' respectively have generated alternative time metrizations ds_1 and ds_1' *in the same manifold of coordinates* such that incompatible congruences result among the time intervals in the given manifold.

Neither the one-way outgoing transit time of light from O to P nor from O' to P' yields the measure of the *space-time* interval between events A and B. And neither the one-way return transit time of light from P to O nor that from P' to O' yields the measure of the *space-time* interval between events B and C.

2.4. It is well known that, depending on the requirements of a particular problem, the GTR routinely employs either the proper time furnished by standard material clocks or the so-called "coordinate time" furnished by "coordinate clocks." I noted in the place to which Putnam alludes [1962, pp. 463–465; 1968, pp. 70–73] that in both his 1911 and 1916 papers, Einstein cited examples of numerous cases in which the use of proper time leads to an undesirably complicated description of nature. Thus, I had explained that in the case of our rotating disk, the use of the proper time τ would issue in (1) a dependence of the one-way velocity of light on the time, and (2) the arrival at the origin O' of fewer light waves per unit of O'-clock time than are emitted at a point P' on the disk per unit of P'-clock time. For the sake of *descriptive simplicity*, one can therefore introduce a so-called coordinate time t on the disk which is defined by the relation

$$(5) \qquad \tau = t\sqrt{1 - \frac{r^2\omega^2}{c^2}}.$$

And, in virtue of (3), we have

$$(6) \qquad t = T,$$

i.e., the artificial *coordinate* clocks in S' assign the same numbers to events as the corresponding *standard* material clocks *in system I*.

The use of the coordinate time t in S' avoids the undesirable descriptive complexities which are associated with the *non*artificial proper time τ in that system, while the time coordinates t and τ *each* fully allow the invariance of the four-dimensional interval ds_4, since the metrical coefficients g_{ik} $(i, k = 1, 2, 3, 4)$ associated with the set of S' coordinates that comprise t are simply suitably different from those for the S' coordinates which include τ. Thus, t and τ are factually co-legitimate. Note, however, that if the time metric in S' is given by differences in the coordinate time t, then the time intervals $\overline{AB}$ and $\overline{BC}$ are *congruent*, whereas these intervals are incongruent *in the very same reference system* S' if we use *proper time differences* instead! Moreover, the incompatibility of the time congruences associated with differences between the t coordinates and τ coordinates respectively will become even more striking in §8.2 below. For it will turn out there that there exist two events at O' and P' respectively which are linkable by a light ray, but which are assigned the same time number τ by the standard clocks in S', a fact which has the following consequence: the one-way transit time $\Delta\tau$ for the ray from O' to P' is *zero* in this case, while the return time for the ray from P' to O' is $2r/c$ in the corresponding case. But in t time, these two oppositely directed light rays require equal one-way transit times r/c.

The time congruences associated with the coordinate time variable

t in S' are the *same* as those yielded by the proper time variable T in I. And hence the identical t and T congruences are each generally different from those associated with the proper time variable τ in S'. But in *pre*-GTR physics, the proper time congruences furnished by the standard clocks in any frame were the *customary* ones of that physics. The incompatibility of the τ and t congruences in S' therefore prompted me to speak of the t congruences in S' as "noncustomary" [1962, pp. 418, 419, 453, 463; 1968, pp. 18, 19, 59, 71].

Hence not only as between different systems I and S' but even within the same reference system S' the GTR confronts us with alternative congruences in the time continuum just as is countenanced by Riemann's doctrine.

2.5. If we use the transformation equations $R = r$, $\theta = \theta' + \omega t$, $Z = z$, and $T = t$, then we can express the invariant

$$ds_4^2 = dX^2 + dY^2 + dZ^2 - c^2\,dT^2 = dR^2 + R^2\,d\theta^2 + dZ^2 - c^2\,dT^2$$

in terms of the S' coordinates r, θ', z, and t, obtaining

$$(7)\qquad \begin{aligned} ds_4^2 &= dr^2 + r^2\,d\theta'^2 + dz^2 + 2\omega r^2\,d\theta'\,dt - (c^2 - r^2\omega^2)\,dt^2 \\ &= g_{ik}\,dx^i\,dx^k. \end{aligned}$$

Now let the fourth coordinate x^4 be ct, so that $dt = dx^4/c$. Then the four-dimensional metrical coefficients g_{ik} corresponding to these particular S' coordinates x^i are

$$(8)\qquad g_{11} = 1,\ g_{22} = r^2,\ g_{33} = 1,\ g_{44} = -\left(1 - \frac{r^2\omega^2}{c^2}\right),\ g_{24} = g_{42} = \frac{\omega r^2}{c},$$

all other components of g_{ik} being zero. Thus if we use the time coordinate ct (along with r, θ', and z), then for that set of coordinates the metrical coefficient g_{44} of the four-dimensional interval ds_4 is given by

$$-\left(1 - \frac{r^2\omega^2}{c^2}\right)$$

while

$$g_{24} = \frac{\omega r^2}{c}.$$

But using the *same* time coordinate ct, and letting the one-dimensional time-interval measure ds_1 be given by the difference between the values of this time coordinate, we find that $ds_1 = c\,dt$ so that the one-dimensional metrical coefficient G in the equation $ds_1 = G\,dx^4$ is 1. Clearly, the use of *coordinate clock numbers ct* as time coordinates instead of proper time numbers τ, and the use of a one-dimensional time metric ds_1 which is given by the differences between the coordinate clock times of events, are not at all required to assure or preserve the invariance of the four-dimensional interval measure ds_4: this 4-measure can be expressed in terms of the increments in the *proper* time coordinates τ assigned to events by *standard* clocks (along with other terms) no less than in terms of the increments in the coordinate times t (or ct) assigned to the events by coordinate clocks (along with other terms), the latter expression being given by equation (7). For whatever the time (and space) coordinates, and *whatever the one-dimensional metric* ds_1, the four-dimensional metrical coefficients g_{i4} ($i = 1, 2, 3, 4$) pertaining to terms involving increments in the time coordinate x^4 (along with the remaining space-time coefficients g_{ik}) are so chosen as to yield the invariant 4-measure ds_4. This mathematical point will be illustrated in §2.6.

Hence we see already, in advance of stating Putnam's main errors, how misguided it is for him to speak in the cited passage of "the fact that the [four-dimensional] g_{ik} tensor of the theory [GTR] is to be preserved over the readings of a clock situated on a rotating disk" [p. 242]. For there is and can be no need at all for the 4-tensor "to be preserved" over the readings of a standard clock or any other clock. This is evident upon being mindful of the distinction between the four-dimensional coefficients g_{i4} ($i = 1, 2, 3, 4$) which pertain to increments in the time coordinate and the one-dimensional temporal metrical coefficient G which is ingredient in ds_1.

2.6. A further distinction that must be borne in mind here concerns the three-dimensional *spatial* metric $d\sigma$: in general, the *spatial part* $g_{\iota K}$ ($\iota, K = 1, 2, 3$ only) of the space-time tensor g_{ik} ($i, k = 1, 2, 3$, and 4) is *not* the same as the spatial metric tensor $\gamma_{\iota K}$ ($\iota, K = 1$,

2, 3) which determines the spatial geometry. For the spatial metric tensor turns out [33, p. 238, and 32, p. 273] to be given by

$$(9)\qquad \gamma_{\iota K} = g_{\iota K} - \frac{g_{\iota 4} g_{K4}}{g_{44}} \ (\iota, K = 1, 2, 3 \text{ only}).$$

And it is seen [33, p. 238] that the $\gamma_{\iota K}$ can be the same as the $g_{\iota K}$ only in the very special case in which $g_{\iota 4} g_{K4} = 0$.

The case of our rotating disk provides a good illustration of the fact that, in general, the spatial part $g_{\iota K}$ of the space-time tensor g_{ik} is *not* identical with the spatial metric tensor $\gamma_{\iota K}$. For note by reference to equations (8) and (9) that although $g_{11} = \gamma_{11} = 1$ and $g_{33} = \gamma_{33} = 1$, we have $g_{22} \neq \gamma_{22}$, since $g_{22} = r^2$ whereas

$$\gamma_{22} = \frac{r^2}{1 - \frac{r^2 \omega^2}{c^2}}.$$

The nonidentity of $\gamma_{\iota K}$ and $g_{\iota K}$ involved here can be seen at a glance upon noting that

$$(10)\qquad ds_3^2 = d\sigma^2 = \gamma_{\iota K}\, dx^\iota\, dx^K = dr^2 + \frac{r^2}{1 - \frac{r^2\omega^2}{c^2}} d\theta'^2 + dz^2$$

and comparing this spatial equation (10) with the space-time equations (7) and (8). Furthermore, calculation shows the following: if we use the standard clock readings τ instead of the coordinate-clock readings t as time coordinates on the disk S' while retaining the space coordinates r, θ', and z, then we obtain the following quantities, among others:

$$g_{11} = 1 - \frac{\tau^2 r^2 \omega^4 c^2}{(c^2 - r^2\omega^2)^2},$$

$$g_{14} = -\frac{\tau r \omega^2 c^2}{c^2 - r^2\omega^2},$$

and

$$g_{44} = -c^2,$$

but (9) yields

$$\gamma_{11} = 1.$$

Thus, if τ time is used instead of t time, then the $g_{\iota K}$ ($\iota, K = 1, 2, 3$ *only*) become functions of the proper time τ while the $\gamma_{\iota K}$ ($\iota, K = 1, 2, 3$) remain independent of the time under the given condition of a *constant* angular velocity ω.

The rotating disk also affords a very simple illustration of the *non*invariance of spatial congruence as between the systems S' and I. To see this, suppose that a *unit* rod is first put down on the disk in the tangential direction of increasing θ' at a radial distance $r = r_1$ and then again tangentially at the different radial distance $r = r_2$. Since the rod will have unit length in both of these locations, equation (10) tells us that for this case

$$(11) \qquad d\sigma = 1 = \frac{r_1\, d\theta_1'}{\sqrt{1 - \frac{r_1^2\omega^2}{c^2}}} = \frac{r_2\, d\theta_2'}{\sqrt{1 - \frac{r_2^2\omega^2}{c^2}}}.$$

Now since $r_1 \neq r_2$, the two denominators in these equations are unequal, and hence by equation (11), $r_1\, d\theta_1' \neq r_2\, d\theta_2'$. But $r_1\, d\theta_1'$ and $r_2\, d\theta_2'$ are the respective Lorentz-contracted lengths which are attributed to the rod by the system I in its two locations at $r = r_1$ and $r = r_2$.[5] It follows that whereas the two space intervals with which the rod successively coincides are congruent in S', each being of unit length in S', these space intervals are incongruent in I where they have the respective unequal lengths $r_1\, d\theta_1'$ and $r_2\, d\theta_2'$. Accordingly, in general, there is *non*invariance of the spatial congruences as between S' and I. And hence the GTR confronts us with alternative congruences for physical space in this sense. But we shall see apropos of the *time-dependent* spatial geometries of the GTR later on in our §2.12 that *the GTR also presents us with physical realizations of alternative congruences by means of solid rods in one and the same spatial manifold of a given reference system*! Indeed, it is

[5] For the general principles relevant to the rotating rod lengths in system I, see [10, §3, p. 116, and 33, p. 223].

already demonstrable by letting ω successively take on *different* values, say 0, 1, or 2, in equations (10) and (11) here that the following is true: two entirely tangential intervals of respective coordinate increments $d\theta_1'$ and $d\theta_2'$ at $r = r_1$ and $r = r_2$ respectively which have *equal lengths* $d\sigma$, say $d\sigma = 1$, when the angular velocity of the disk has one of these values will no longer have equal lengths, i.e., will no longer be congruent, when that velocity has a different value.

2.7. The space-time measure ds_4 is an invariant both as between different reference frames and with respect to the use of different space and time coordinates in any given frame. It follows that two *space-time* intervals which are congruent in one reference frame in virtue of having equal measures ds_4 in that frame will likewise have equal measures ds_4 and hence will also be congruent in any other frame. Accordingly, there are no alternative infinitesimal *space-time* congruences in the GTR, and the latter kinds of congruences are thus unique.

2.8. Having stated seven items of technical preliminaries, we are now ready to make explicit the several errors which vitiate the argument by Putnam cited above in §2.0. Of these errors, the cardinal one is his supposition that the uniqueness of the four-dimensional space-time congruences of the GTR can gainsay or detract from the significance of the following claim of mine: the GTR actually employs alternative congruences in *each* of the physically important continua of time and space in precisely the sense countenanced philosophically by Riemann's doctrine. It is now plain that nothing about my claim that the insight contained in Riemann's doctrine has substantial relevance to the GTR requires that there be alternative *space-time* congruences *in addition to* the undeniably present alternative time congruences and alternative space congruences. And, as will become clear in the sequel, Riemann's conception accommodates the GTR's unique space-time congruences as well as the alternative congruences of time and space.

When I wrote about alternative congruences in the GTR in the essay which Putnam criticized [1962, pp. 418, 419, 453, and 463–

465; 1968, pp. 18, 19, 59, and 70–73], I was explicitly discussing the *time* metric and *not* the space-time metric. And in the present chapter, I added the considerations pertaining to the *space* metric and its associated metric tensor $\gamma_{\iota K}$. Putnam myopically clings to ds_4 while blithely disregarding the metric ds_1 and the metric ds_3 by writing that "we are *not* free in general relativity to employ any metric tensor other than g_{ik}" [p. 242]. Indeed he regards attention to these other physically significant metrics as not relevant to the congruence issue, and he charges me with having misleadingly made ds_1 appear relevant by "how Grünbaum employs his terms" [p. 242]. Specifically, he regards my claim that in the system S' the t-time congruences are used as an alternative to the co-legitimate τ-time congruences as objectionable on the following grounds. According to Putnam: (1) when calling attention to the use of the t congruences "What he [Grünbaum] is pointing out is the fact that the customary rule of correspondence (employing a clock at rest in the reference system to determine local time) is not valid under exceptional circumstances [*sic*!] (on a rotating disk). But correspondence rules always break down under some exceptional circumstances." And (2) by calling attention to the coordinate time congruence as an alternative to proper time congruence, "Grünbaum obscures" the fact that "we are *not* allowed to 'choose' a space-time metric" [p. 242]. But it will now become evident that insofar as these statements of Putnam's are true at all, they are utterly unavailing as grounds for invalidating my position.

To begin with, Putnam is simply uninformed when he claims that the use of coordinate time t in place of proper time τ in the GTR is a matter of "exceptional circumstances" and that its use on the rotating disk is atypical. One need only turn the pages of any textbook on the GTR and look at the expressions for the line elements ds_4 to see that it is replete with uses of coordinate time: whenever the metric coefficient g_{44} in ds_4 is a function of the coordinates rather than a constant ($-c^2$ or -1), we know that the time coordinate is t rather than τ, since the use of the latter would give rise to a constant (intrinsically negative) g_{44}. For example, in Schwarzschild's expression for the ds_4 in the gravitational field surrounding the sun, coordinate time is used, as is evident from the fact that g_{44} is given by a function of one of the space coordinates [46, pp. 202–205]. But, more fundamentally, while the use of t time instead of τ time to formulate the theoretical statements of the GTR does obviously call for an appropriately different correspondence rule, supplanting τ by *t also* issues in an *alternative set of congruence classes in the time continuum* as constituted by a light ray or by another process which is represented by a world-line! Putnam begs the question by seeking to *confine* attention to the fact that the correspondence rule appropriate to theoretical statements couched in t time is different from the one required for their counterparts in τ time. Clearly the use of t time involves such a difference in the rule of correspondence. By the same token, t time calls for a *non*constant g_{44} in the line element ds_4, in place of the constant g_{44} associated with τ time. And, as we saw under §§2.3 and 2.4 above, if differences in t time rather than in τ time furnish the time metric, then the radial components of the to-and-fro velocities of light in the rotating disk become equal rather than unequal as well as independent of time.[6]

But all this (and more) cannot detract from the crucially relevant fact that the *t congruences* and the *τ congruences* in the time continuum are *different*, while the choice between them in any given case is a matter of descriptive rather than inductive simplicity. And it is this fact, as well as its spatial counterpart, which justifies my contention that the GTR has implemented Riemann's philosophy of congruence. Indeed, we shall see below in §2.12 that in precise opposition to Putnam, my statement derives further telling support from the explanatory role of the concept of the metrical field in the GTR, as was emphasized by Hermann Weyl in his definitive classic *Space-Time-Matter* [47, pp. 97–98]. It is now evident that there is nothing misleading here at all in "how Grünbaum uses his terms";

[6] We shall have occasion to discuss t-time light velocities on the rotating disk more fully in §8.2.

instead, Putnam's allegation to this effect coupled with the manner in which he appeals to the change in the rule of correspondence constitutes an *ignoratio elenchi*.

Moreover, nothing in my account provides any basis whatever for Putnam's belief that I consider the use of t time in place of τ time as jeopardizing the invariance of the ds_4, i.e., as impugning the integrity of the four-dimensional tensor g_{ik}. And he erects a straw man on a foundation of sheer nonsense when he exclaims that "no special insight [of the kind which Grünbaum attributed to Riemann's doctrine of congruence] is needed to allow for the fact [*sic*!] that the [four-dimensional] g_{ik} tensor of the theory is to be preserved over the readings of a clock situated on a rotating disk!" [p. 242]. For, as I indicated preliminarily under §2.5, the purported imperative that the 4-tensor g_{ik} "is to be preserved over the readings of a [standard] clock" arises from Putnam's unawareness that the bearing of the time coordinatization on the metric ds_4 is quite *different* from its bearing on the metric ds_1: since the g_{ik} $(i, k = 1, 2, 3, 4)$ can be made appropriate to either τ time or t time, the desirability of using t time in place of τ time does *not* arise from the necessity to safeguard the invariance of ds_4 and the integrity of its 4-tensor, any more than the invariance of ds_3 in Euclidean 3-space could possibly require the use of spherical space coordinates r, θ, and ϕ in place of the cylindrical coordinates r, θ, and z!

What the use of t time can meaningfully and correctly be said to have "*preserved*" in system S' is *descriptive* simplicity which inheres, for example, in *precluding* the *time dependence* of the one-way velocity of light. And the preservation of the descriptive simplicity exemplified by the absence of such a time dependence is effected by *the alternative time congruences associated with t time*. But it is precisely this virtue which I had claimed for these alternative congruences in my 1962 essay while calling attention to the fact that their *metrical legitimacy* is vouchsafed by the Riemannian insight that Putnam decries. And hence it is a travesty on Putnam's part to saddle me with the false claim that the GTR exercises the choice of an alternative *space-time* metric "*at the cost of complicating the laws of nature*" [p. 242, his italics].

If one uses the *broader* of the two definitions of what constitutes the "special theory of relativity" (STR), then our results can be used to claim that alternative congruences of time and of space are each encountered not only in the GTR but even in the STR. By the STR in its original, narrower sense, we understand the theory of *inertial* frames in "flat" space-time as linearly related by the Lorentz transformations. In this narrower sense, accelerated (noninertial) frames like our rotating disk are excluded from the purview of the STR. But on the broader, more recent construal of the scope of the STR, the latter comprises noninertial frames and extends to all reference systems which are covariantly characterized by the fact that the Riemann curvature-tensor is zero.[7] This is the necessary and sufficient condition for the feasibility of choosing space-time coordinates for which the components of the metric 4-tensor g_{ik} are *constants*. But since the space-time metric of our system I is characterized by this constancy of the g_{ik} with respect to the quasi-rectangular coordinates X, Y, Z, and T, the Riemann tensor is zero in I, and hence

[7] Thus P. Bergmann writes: "We can formulate the special theory of relativity in terms of curvilinear coordinate systems and general coordinate transformations in a four dimensional world" [3, p. 158] and "A non-inertial frame of reference in the special theory of relativity . . . will include both rectilinear and curvilinear coordinate systems engaged in arbitrary motion" [2, p. 207]. As A. Janis has noted, this introduction of noninertial frames into the STR is entirely analogous to their introduction into Newtonian mechanics: when Newton's second law of motion is referred to noninertial frames, it ceases to have the form $F = ma$, and the statement of the law then includes Coriolis and centrifugal terms among others [36, p. 104].

The Riemann curvature-tensor is constructed solely from the components of the 4-tensor g_{ik} and from their first and second derivatives with respect to the coordinates. And this tensor enables us to speak of the gravitational field in an *absolute* sense. Thus J. L. Synge writes: "In Einstein's theory, either there is a gravitational field or there is none, according as the Riemann tensor does not or does vanish. This is an absolute property; it has nothing to do with any observer's world-line. Space-time is either flat or curved, and [one must] . . . separate truly gravitational effects due to curvature of space-time from those due to curvature of the observer's world-line (in most ordinary cases the latter predominate)" [45, p. ix].

also in the noninertial system S' of the rotating disk. Accordingly, on the broader construal of the STR as characterizing the laws in any system for which the Riemann tensor vanishes, our accelerated system S' falls within the purview of the STR. And in this sense of STR, the system S' illustrates that even the STR confronts us with alternative congruences of time and also of space. Indeed, as we saw in §2.1 by reference to rotating a rod from the y' into the x' direction in a moving *inertial* system and noting its corresponding *unequal* lengths in the rest inertial system, even the STR in the *narrow* sense presents us with cases of noninvariance of spatial congruence.

Thus it is seen to be an undeniable fact that the GTR does employ alternative *physical* congruences. Hence we can proceed to a fuller articulation than I have given heretofore of my thesis that Riemann's view of congruence provides a philosophical underpinning for these different congruence classes of time intervals and space intervals respectively. Putnam believes that though pertinent to a purely abstract mathematical continuum, Riemann's view is unsound as applied to mathematically continuous *physical* manifolds whose congruences are specified by clusters of laws of nature. By contrast, the thesis which I have been defending is that the GTR and even the STR show Riemann's conception to be *physically* illuminating precisely because pertinent relativistic laws tell us that material clocks and rods respectively yield physical realizations of the alternative temporal and spatial congruences which he had countenanced philosophically. To be sure, spatial and temporal congruence relations are each open multiple-criteria relations in the sense that either kind of congruence class can be specified by an open-ended set of compatible physical criteria rather than by a single criterion [26, pp. 14–15]. But this cluster character of congruence must *not* be allowed to confer plausibility on Putnam's supposition that the members of the cluster of physical laws which *involve* space and time congruences all yield one and the same congruence class in each of these two manifolds! For in the light of the relativistic results that I have adduced, this supposition is patently false.

2.9. A brief mention of the ideas of some of Riemann's philosophical precursors will be useful as a historical background for my articulation of his congruence doctrine regarding the continua of space and of time.

During the Middle Ages, Robert Grosseteste and other members of the Oxford school of natural philosophers considered the bearing of the Pythagorean theorem on the hypothesis that physical space might be discrete or—in modern parlance—quantized. And their view was that the *incommensurable* spatial intervals whose existence is entailed by that theorem militated strongly against the quantization of space. Incommensurability suggested that (i) line intervals are infinite aggregates of extensionless physical points rather than finite aggregates of minimal spatial elements of positive extension (space atoms), and (ii) since all physical space intervals are infinite sets of points, their measures (lengths) cannot be given by the cardinal number of their point elements and hence cannot be determined by counting these elements. On the other hand, if physical space were granular (discrete, atomic, quantized), the measure of any given interval could be furnished by the cardinal number of quanta constituting it, and thus an extensional measure would be *built into the spatial intervals themselves.* Hence Walter Burley concluded that "in a continuum [of extensionless points] there is no primary and unique measure [i.e., no "*built-in*" measure of spatial extension] according to Nature, but only according to the institution of men." [8] A similar conclusion was reached by David Hume [31, Part II, Section IV].

It is clear from the considerations offered by these thinkers that whereas one atom of space or any *integral* multiple thereof constitutes a unit of measurement which is *built into* each interval of a discrete space, no unit of measurement is built into the intervals of a continuous space of physical points. Thus the continuity of

[8] Quoted in [48, p. 170].

physical space theoretically allows an unrestricted conventional choice of a unit of length. By contrast, as we noted in Chapter II, in an atomic space, no such *unrestricted* conventional choice obtains: for example, a proposed unit of *one-half* of a space atom would have no physical realization. Accordingly, the reflections of Riemann's philosophical precursors already suggest the following: the latitude for conventional choice in the specification of the spatial metric depends on facts which are not themselves matters of convention.

2.10. Now consider an interval AB in the mathematically continuous physical space of, say, a given blackboard, and also an interval T_0T_1 in the continuum of instants constituted by, say, the movement of a classical particle. By contrast to the situation in an atomic space, neither the cardinality of AB nor any other property *built into* the interval provides a measure of its particular spatial extension, and similarly for the temporal extension of T_0T_1. For AB has the same cardinality as any of its *proper* subintervals and also as any other nondegenerate interval CD. Corresponding remarks apply to the time interval T_0T_1. If the intervals of physical space or time *did* possess a built-in measure or "*intrinsic metric*," then relations of *congruence* (and also of incongruence) would obtain among disjoint space intervals AB and CD on the strength of that intrinsic metric. And in that hypothetical case, neither the existence of congruence relations among disjoint intervals nor their epistemic ascertainment would logically involve the iterative application and transport of any length standard. But the intervals of mathematically continuous physical space and time are devoid of a built-in metric. And in the absence of such an intrinsic metric, the basis for the extensional measure of an interval of physical space or time must be furnished by a relation of the interval to a body or process which is applied to it from outside itself and is thereby "*extrinsic*" to the interval. Hence *the very existence* and not merely the epistemic ascertainment of relations of congruence (and of incongruence) among disjoint space intervals AB and CD of continuous physical space will depend on the respective relations sustained by such intervals to an extrinsic metric standard which is applied to them. Thus, *whether two disjoint intervals are congruent at all or not will depend on the particular coincidence behavior of the extrinsic metric standard under transport* and not only on the particular intervals AB and CD. Similarly for the case of disjoint time intervals T_0T_1 and T_2T_3, and the role of clocks.

Moreover, the failure of the intervals of physical space to possess an intrinsic metric—a failure which compels recourse to an extrinsic transported metric standard to begin with—has the consequence that the continuous structure of physical space *cannot* certify the self-congruence (rigidity) of any extrinsic standard *under transport*. And similarly for physical time and the uniformity (isochronism) of clocks. Thus, for precisely this reason, the following two claims, made at least implicitly in the first Scholium of Newton's *Principia* [34, pp. 17–22], are unsound: (1) The criterion for the adequacy of an extrinsic length standard is that it is self-congruent (rigid) under transport as a matter of spatial fact. (2) Of two extrinsic length standards which yield *incompatible* congruence findings for disjoint intervals, there might be one which is *truly* spatially self-congruent (rigid) under transport. It should therefore not be misleading to render our verdict of unsoundness in regard to these two claims as follows: an extrinsic metric standard is self-congruent under transport *as a matter of convention* and not as a matter of spatial fact, although any *concordance* between its congruence findings and those of another such standard is indeed a matter of *fact*. If the term "transport" is suitably generalized so as to pertain as well to the extrinsic metric standards applied to intervals of time and of space-time respectively, then this claim of conventionality holds not only for time but also *mutatis mutandis* for the *space-time continuum of punctal events*!

Let us now combine this conclusion with our earlier one that the very *existence* of congruences among disjoint intervals depends on their empirically ascertainable relations to the conventionally self-congruent transported standard. Then it becomes evident that the existence of congruence relations among disjoint intervals is (1) a matter of convention precisely to the extent that the self-congruence of the extrinsic metric standard under transport is conventional, and (2) a matter of fact precisely to the extent to which the respective relations of the intervals to the applied extrinsic metric standard are matters of fact. It is therefore an error to think with Putnam that if there is a cluster of physical laws (e.g., Newton's laws of motion, Hooke's law, *et al.*) which specify that all the members of a certain class C of spatial congruence standards must yield the same congruence findings under transport, then all the members of C are alike self-congruent under transport *as a matter of spatial fact.* Thus the presence of a very important conventional ingredient in the congruence of disjoint intervals is not gainsaid at all if this congruence obtains alike with respect to each of an entire cluster of extrinsic standards rather than only with respect to the single standard constituted by the unperturbed solid rod. For what matters is the extrinsicality of every congruence standard and the resulting conventionality of the self-congruence of any and all of them under transport. Indeed, Putnam overlooked that I had called attention to this point [1962, pp. 477–478, 483; 1968, pp. 87–88, 94].

Relations of congruence among intervals of space, time, and space-time respectively are specified by equal measures ds_3, ds_1, and ds_4 respectively. And since these respective intervals do not possess intrinsic metrics ds_3, ds_1, and ds_4, the respective *congruences* among them are extrinsic. Thus, metrics and congruences are extrinsic to the *intervals* of the continuous manifolds of space, time, and space-time, but not to these manifolds themselves. Nonetheless, *for the sake of brevity,* I shall speak of this state of affairs by saying that these continuous manifolds (continua) are devoid of intrinsic metrics.

The conception of congruence set forth here countenances alternative metrizations of continua of like punctal elements which issue in incompatible congruence relations; this is illustrated by my examples from the GTR and by the case of the ND-conjecture of

Chapter II. But nothing in this conception precludes the use of the criterion of descriptive simplicity and convenience to employ a particular kind of metrization and thereby to select a unique class of classes of congruent intervals to the exclusion of others in certain theoretical contexts. Thus, nothing in this conception enjoins us *not* to appeal to the congruences from the physics of daily life as a basis for the geometry of a blackboard or table top. By the same token, my view of congruence fully allows that there are excellent reasons of descriptive simplicity—as explained in Chapter I, §4(i)—for formulating the empirical content of Newtonian mechanics by reference to the standard astronomical time congruence rather than on the basis of the time congruence furnished by the uncorrected rotational motion of the earth. Again nothing in this view of congruence calls upon Einstein to complicate the equations of the GTR enormously by using a *space-time* congruence different from the one he actually does use. At the same time, however, my view legitimates philosophically the cases in which alternative congruences of one kind or another are actually used in science as explained above!

2.11. My criticism of Newton's view of the status of congruence in the continua of physical space and time appeals only to their continuity, which is assumed by him, and *not* to the supplanting of the content of his laws of physics by subsequent theories. And the account of congruence which I have offered in opposition to Newton's is an articulation of what I take Riemann to have espoused cryptically in the following sentences of his Inaugural Lecture in regard to space and time:

> Determinate parts of a manifold, distinguished by a mark or by a boundary, are called quanta. Their comparison as to quantity comes in discrete magnitudes by counting, in continuous magnitude by measurement. Measuring consists in superposition of the magnitudes to be compared; for measurement there is requisite some means of carrying forward one magnitude as a measure for the other. In default of this, one can compare two magnitudes only when the one is a part of the other, and even then one can only decide upon the question of more and less, not upon the question of how many.

> . . . in the question concerning the ultimate basis of relations of size in space . . . the above remark is applicable, namely that while in a discrete manifold the principle of metric relations is implicit in [i.e., intrinsic to] the notion of this manifold, it must come from somewhere else [i.e., be extrinsic] in the case of a continuous manifold [42, pp. 413, 424–425].

This much of the Riemannian conception was also advocated by Poincaré, Reichenbach, Clifford, and me among others. And though denied by Newton himself, it applies alike to Newtonian physics, the STR and the GTR, since each of these theories alike affirms the continuity of the manifolds of space, time, and space-time. Thus, Clifford, who died in 1879, twenty-six years before the advent of the STR, wrote as follows:

> The reader will probably have observed that we have defined length or distance by means of a measure which can be carried about *without changing its length.* But how then is this property of the measure to be tested? We may carry about a yard measure in the form of a stick, to test our tape with; but all we can prove in that way is that the two things are always of the same length when they are in the same place; not that this length is unaltered. . . .
>
> Is it possible, however, that lengths do really change by mere moving about, without our knowing it?
>
> Whoever likes to meditate seriously upon this question will find that it is wholly devoid of meaning [4, pp. 49–50].

Contrast this Riemannian conception of (spatial) congruence set forth by Clifford with the view of (temporal) congruence which Newton espoused by speaking of "the true, or equable, progress of absolute time" [34, p. 19]. And note that Clifford was careful to point out that "We may carry about a yard measure in the form of a stick, to test our tape with; *but all we can prove in that way is that the two things are always of the same length when they are in the same place; not that this length is unaltered*" (my italics). In the same vein, I stated above that the concordance between the congruence findings of two or more extrinsic metric standards is a matter of *fact* while these standards are alike conventionally self-congruent under transport. Indeed, in agreement with Schlick [43, pp. 183–184, and 44, p. 40] and Reichenbach [40, pp. 16–17 and 116], I asserted in my earlier writings [e.g., 26, p. 222, n. 2] that the *concordance* in the behavior of different free Newtonian particles is a matter of *fact* constituting the empirical content of the law of inertia, while the congruence of the successive time intervals required alike by all free particles to traverse Newton's congruent space intervals is a matter of convention in the previously specified sense. Clearly there is complete compatibility between the recognition that the behavioral concordance is a matter of fact which is rendered by physical law, on the one hand, and the thesis of the conventionality of the self-congruence of clocks under temporal "transport," on the other. For the factual concordance rendered by the law of inertia does *not* entail that inertial clocks are each self-congruent under temporal "transport" *as a matter of temporal fact.*

Given this compatibility, it is quite beside the mark and unavailing to try to impugn my conception of congruence by adducing, as Putnam does, (1) the factual concordances asserted by Newton's laws, and (2) the invariance of spatial separation and of temporal separation as between Newtonian inertial systems. Thus, Putnam writes:

> In a Newtonian world, space and time do nonetheless have physical significance, however, inasmuch as physical processes depend upon true temporal and spatial separation in a simple way. Moreover, this is not a matter of mere descriptive simplicity. If we transform to a different metric, or to a different chronometry, so that time, for example, [in a new usage] becomes what we would ordinarily call $(\text{time})^3$, then it will still be true that the behavior of clocks in a Newtonian world depends in a simple way upon the time elapsed, only this invariant fact will now be explained by saying that clocks measure $(\text{time})^3$. Whether we express this fact by the words "clocks measure time" or by the words "clocks measure $(\text{time})^3$," the fact is the same. And it is the same magnitude that enjoys the same significance [p. 212].

(Presumably by $(\text{time})^3$ Putnam means to say $\sqrt[3]{\text{“time”}}$.) And speaking of the Newtonian treatment of the relations between two events, Putnam says: "the temporal separation of the two events and their spatial separation . . . are invariant under the transformations which obtain between 'privileged' coordinate systems in a Newtonian world" [p. 211].

My reply to these statements is fourfold:

i. The fact adduced by Putnam in the first quotation cannot serve to impugn my conception of congruence, which fully allows for the fact in question, as I already noted when pointing out their complete compatibility.

ii. Putnam tells us that the dependence of physical processes "upon true temporal and spatial separation in a simple way . . . is not a matter of mere descriptive simplicity." Here Putnam has a quite mistaken idea of what Schlick, Reichenbach, and I mean by saying that two descriptions differ only in "descriptive simplicity" as contrasted with differing in factual content (or in inductive simplicity): when saying that the difference between the descriptions associated with various alternative metrizations of the Newtonian time continuum is one of mere descriptive simplicity, we are surely *not* saying that there are no facts to be described; for the very concept of descriptive simplicity presupposes that there *are* facts to be described more or less simply. I should have thought that this state of affairs was made completely clear in my detailed discussion of the two forms of Newtonian mechanics which correspond to the astronomical and "diurnal" time congruences respectively, a discussion which appeared in §4(i) of the 1962 essay of mine on which Putnam rests his critique. In short, I have maintained that the particular *form* of a physical law which is associated with a particular metric (of space or time) is more or less descriptively simple, but *not* that there is no factual content to be rendered by the law!

iii. Putnam's specific example of an alternative metric is to the following effect: instead of assigning the numbers t read by standard clocks to events as time numbers, we shall alternatively metrize the given one-dimensional continuum of Newton's theory by calling "time" numbers those different numbers T which are specified by the transformation $t^3 = T$. Thus physical clocks will yield values of $\sqrt[3]{T}$, and the increments on the two different time scales are related by the equation $3t^2\,dt = dT$. Hence the t and T scales yield different congruences in the time continuum, since two intervals of equal dt would not, in general, be intervals of equal increments dT. Putnam believes that this example of the physically uninteresting T congruences constitutes an argument against my position. But what is the force of this example of his? I say that it can serve neither to invalidate *nor to trivialize* the Riemannian thesis of the alternative metrizability of the time continuum (or of a world-line). For, in the first place, the claim that the self-congruence of clocks under temporal "transport" is conventional certainly does *not* assert that in *every* theory which asserts the continuity of time a useful descriptive or other purpose would be served in all cases by abandoning the t metrization in favor of an alternative one such as T. And, in the second place, although Putnam's example of replacing the t scale by the rival T in Newton's physics cannot serve to illustrate the significance and scientific relevance of Riemann's thesis, my earlier examples from the GTR clearly do demonstrate that Riemann's view of congruence is eminently pertinent as a vindicator. Thus Putnam overlooks the telling cases which exhibit the power and relevance of Riemann's thesis and considers only those cases in Newton's physics and in the GTR to which that thesis is not pertinent in any interesting way.

iv. In his second quotation, Putnam appeals to the invariance of spatial and temporal separations under the Galilean transformations of Newton's physics, just as he had invoked the invariance of (infinitesimal) *space-time* separations in the case of the GTR. But none of these invariances militate in the slightest against the conventionality of the self-congruence of extrinsic standards under spatial and/or temporal transport, since these invariances do *not* make for an intrinsic metric. That Einstein himself adhered to Riemann's conception of congruence in the GTR seems to me to emerge clearly from the following statement of his in the fundamental paper of 1916, a statement which would otherwise have been quite pointless: "Thus Euclidean geometry does not hold even to a first approximation in the gravitational field, if we wish to take one and the same rod, independently of its place and orientation, as a realization of the same interval" [10, p. 161]. How would Putnam explain the presence of the if-clause in this statement except as an expression of the *legitimacy* of an alternative congruence? Surely Einstein is not suddenly treating us here to a reminder of the need for rules of correspondence in physical geometry or to the platitude that the standard rod is customarily used to measure length in physics!

Putnam simply glosses over my explicit pinpointed recognition of the ingredience of factual elements in metrical claims pertaining to space, time, or space-time. He does so by appealing irrelevantly to "the objectivity and nonconventionality" [p. 212] of these factual ingredients in the mistaken belief that he can thereby invalidate the Riemannian conception of congruence espoused by me. He writes:

> The argument I shall use to defend the objectivity and nonconventionality of spatio-temporal separation may thus be easily carried over to defend the objectivity and non-conventionality of spatial separation and temporal separation in a *Newtonian* world. Thus this paper is in part a defense of Newton against those who have sought to criticize him, as Riemann and Grünbaum have done, on philosophical grounds. . . . Newton's three-dimensional "ether" has been replaced by a four-dimensional "field," but Newton has not thereby lost standing as a philosopher but has only been advanced upon as a physicist [p. 212].

We saw that the coincidences of an extrinsic standard with various intervals and the concordances among the congruence findings of diverse extrinsic standards are factual (objective and nonconventional). But this objectivity cannot detract from the conventionality of the self-congruence of the extrinsic standards under transport. Moreover, it will be recalled that my criticism of Newton's view of temporal and spatial congruence appeals only to the continuity of space and time assumed by him and *not* to the supplanting of the specifics of his physics by those of Einstein's. If Riemann's denial of an intrinsic metric in continuous physical space and time is to be regarded as a philosophical doctrine, then Newton has indeed "lost standing as a philosopher" in the light of Riemann's view of congruence. *For the latter exhibits Newton's belief in a truly uniform elapsing of time to be an erroneous philosophical addition to his physical theory*, as will be explained further in §3.3.

It now becomes apparent why I believe that Putnam does not score against me in the least when he writes:

Grünbaum is tremendously impressed by the fact that there is nothing in the nature of space-time that requires us to use the term "congruent" to mean congruent according to one g_{ik} tensor rather than another. This is true. Similarly, it is true that there is nothing in the nature of space-time or in the nature of the bodies occupying space-time that requires us to use the word "mass" to mean mass rather than charge. But there is an objective property of bodies which is their mass whether *we call it mass* or call it $(mass)^3$, and there is an objective relation between two spatial locations which we ordinarily call the distance between those spatial locations and which enters into many physical laws (e.g., consider the dependence of *force* on *length* in Hooke's law), whether we refer to that magnitude as *distance* or as *length* or as $(distance)^3$ or as $(length)^3$ or as any other function of distance. If we decide by "distance" to mean distance according to some other metric, then in stating Hooke's law we shall have to say that force depends not on length but on some quite complicated function of length; but that quite complicated function of length would be just what we ordinarily mean by "length" [pp. 219–220].

I claim that Putnam does not score against me here for the following reasons:

1. His appeal to distance being "an objective relation between two spatial locations . . . which enters into many physical laws . . . whether we refer to that magnitude as *distance* . . . or as $(distance)^3$" is to the same effect as his earlier abortive example of referring to the *cube* of a clock reading as "time." Thus, his argument here concerning distance is partly a straw man and partly an *ignoratio elenchi*.

It is a straw man because I explicitly noted the factual ingredients of metrical statements about space no less than in the case of time. And it is an *ignoratio elenchi* because it cannot impugn the conventionality of self-congruence under transport: it overlooks the telling point made by Clifford when he said: "We may carry about a yard measure in the form of a stick, to test our tape with; but all we can prove in that way is that the two things are always of the same length when they are in the same place; not that this length is unaltered" [4, p. 49].

2. Having overlooked Clifford's point concerning self-congruence under transport, Putnam is unable to discern any conventionality in the *existence* of the relation of spatial equality (congruence) among disjoint intervals. Thus he is led to conclude that conventionality is present here *solely* in the exclusively *linguistic* manner in which it is conventional that the intension of the unpreempted noise "mass" is the physical property mass rather than the charge. And then Putnam proceeds to caricature my spatiotemporal thesis by saying: "Grünbaum is tremendously impressed by the fact that there is nothing in the nature of space-time that requires us to use the term 'congruent' to mean congruent according to one g_{ik} tensor rather than another." But the options pertaining to the potential uses of "congruent" *qua merely initially unpreempted noise* do not impress me at all as indicative of the nature of space or of space-time; instead, what does impress me in concert with Riemann and Clifford is that *after* the term "congruent" is already preempted to mean extensional equality, there is nothing in the nature of these continuous physical manifolds which would require us to *ascribe congruence*, i.e., *spatial or spatiotemporal equality*, to certain disjoint intervals as opposed to others. But in a *granular* space, an interval of three space atoms could *not* be held to be spatially congruent to one of only two space atoms.

There is also an important disanalogy between those alternative uses of the term "congruent" which are associated with suitably different metric tensors, and Putnam's alternative uses of the term "mass" here. The intensions of the alternative uses of "mass" in his example are plainly different. But since the alternative uses of "congruent" associated with suitably different metric tensors each pertain to the relation of equality among space-time intervals, they differ only in regard to the *extension* of the term "congruent" and are *conceptually the same*. This conceptual sameness is not matched by Putnam's intensionally alternative uses of the term "mass."[9] Indeed, the conceptual sameness of the congruence relation amid the differing extensions of the term "congruent" associated with suitably different metrics is a consequence of the following prior result of ours: the obtaining of the relation of self-congruence under transport is conventional rather than a matter of spatiotemporal fact.

It is apparent now that the warrant for asserting spatial self-congruence under transport is very much more strongly conventional instead of factual than the warrant for letting the unpreempted word "mass" have one intension rather than another: in the former case, it is a matter of space-time not having intrinsic congruences, whereas in the latter case it is just a matter of the failure of mere noises to have unique intensions automatically. I therefore cannot see that Putnam has adequately come to grips here with my earlier statements along these lines [1962, pp. 420–421; 1968, pp. 20–21], let alone refuted them. And I shall return to Putnam's analogy between "congruent" and "mass" in §3.2.

2.12. Immediately following Riemann's statement "while in a discrete manifold the principle of metric relations is implicit in the notion of this manifold, it must come from somewhere else in the case of a continuous manifold," he wrote in his Inaugural Lecture: "Either then the actual things forming the groundwork of a space must constitute a discrete manifold, or else the basis of metric relations must be sought for outside that actuality, in colligating forces that operate upon it" [42, p. 425]. It is well known that Riemann's

[9] Theories in which the mechanical mass of, say, an electron turns out to be entirely electromagnetic (see [36, p. 528]) are, of course, excluded here as incompatible with Putnam's purpose to use the term "mass" in intensionally alternative ways.

latter sentence, which I cited as well in my 1962 essay [p. 412], was prophetic concerning the GTR's program to provide a *physical explanation* of the spatial geometry by seeking to implement what Einstein called "Mach's Principle" [13]. I speak here of the GTR's 'program'' to explain the spatial geometry by "seeking" to implement Mach's Principle, because the GTR succeeded in explaining only the *deviation* of the spatial geometry from Euclideanism (where it does so deviate) and failed to implement Mach's Principle [26, Chapter 14]. We shall see presently that Riemann's vision of explaining the spatial geometry is an organic logical outgrowth of his account of the status of congruence and is (partly) justified by it. It was therefore hardly cogent for Putnam to have indicted my Riemannian view of congruence and the conception of the spatial geometry associated with it as prejudicial to the GTR's program of explaining the spatial geometry physically. Putnam did so by characterizing his disagreement with me as follows: "The fundamental issue is whether the laws of modern *dynamical geometry* (i.e., laws describing a geometry of variable curvature changing with time) explain or only describe the behavior of solid bodies, clocks, etc." [p. 207]. As evidence for his mistaken claim that *in the dynamical context of the GTR* I opt for mere description, he cites my statement that by the physical geometry "we understand the articulation of the system of relations obtaining between bodies and transported solid rods quite apart from their substance-specific distortions" [1962, p. 510; 1968, p. 126]. But, as Einstein makes explicit apropos of Poincaré in the discussion between himself and Reichenbach to which my statement here pertains, the context of this discussion is "classical physics" [14, p. 677] and *not* the GTR. Here as elsewhere, Putnam unfairly saddles me with having taken the GTR for granted throughout as a basis for my account of spatial geometry and chronometry [pp. 205–209]. But I had been careful *not* to make such a restrictive assumption [see 1962, p. 521; 1968, p. 140] and instead to state explicitly in what specific contexts I was dealing with the GTR. And by carelessly equivocating upon the three- and four-dimensional meanings of the term "physical geometry," Putnam takes my explicitly spatial statements to have referred to four-dimensional *space-time* geometry. Thus he himself clearly refers to the *spatial* geometry *without* necessarily presupposing the GTR when he says at the outset: "I do agree [with Grünbaum] that the ordinary standard of congruence in physical geometry is the solid rod" [p. 205]. But almost immediately [p. 205n] there is a meaning switch such that the word "geometry" refers to the GTR's specification of "the metrical properties of *space-time*" [p. 208, my italics]. Similar meaning switches occur on pages 219–220 and 220–221. By this sleight-of-hand, Putnam succeeds in accusing me of having disregarded the GTR's (attempted) explanation of the spatial geometry and is able to chide me unfoundedly and irrelevantly. But I do not think that at this stage it can at all be reasonably said with Putnam that "general relativity is well confirmed," a statement which would surprise the most ardent adherents and well-wishers of the GTR [p. 254; cf. 6].

I must now justify my earlier assertion that Riemann's envisionment of explaining the spatial geometry physically, which was a precursor of Einstein's original quest to implement Mach's Principle in the GTR, is an organic logical outgrowth of Riemann's conception of congruence, a conception which I showed to have been endorsed by Einstein in the GTR both implicitly and explicitly. The twofold organic connection between Riemann's conception of congruence and his explanatory program is the following:

1. By leaving the coincidence behavior of a solid rod under transport completely *indeterminate*, the continuous structure of physical space allows an infinitude of alternative patterns of coincidence. Therefore the existence of a coincidence behavior *common* to all kinds of unperturbed solid rods strongly suggests an important possibility, viz., that this common transport behavior and hence the spatial geometry is *not* an irreducible fact but depends causally on the matter distribution and can change with that distribution. Thus, for any one such distribution, every kind of meterstick "knows"

alike with which intervals it must coincide under transport, because each such meterstick is guided alike by the *metrical field* arising from the matter distribution. It is strange that Putnam did not take *any* note of my explicit statement [1962, p. 478; 1968, p. 88] that "we find a striking concordance between the time congruence" defined by various clocks "for which the GTR has sought to provide an explanation through its conception of the metrical field, just as it has endeavored to account for the corresponding concordance in the coincidence behavior of various kinds of solid rods."

Recalling Putnam's derogation of Riemann's philosophical criticism of Newton's view of congruence, I confront this derogation with (a) the logical potentiality of Riemann's congruence doctrine for suggesting the general direction of a physical advance beyond Newton's theory *without* the benefit of the physical considerations which prompted Einstein to propound the GTR and (b) the basis on which Riemann actually put forward this prophetic conjecture.

And I call attention to the fact that immediately after quoting the very passages from Riemann's Inaugural Lecture on which I rest my claims here, H. Weyl wrote:

> If we discard the first possibility, "that the reality which underlies space forms a discrete manifold"—although we do not by this in any way mean to deny finally, particularly nowadays in view of the results of the quantum-theory, that the ultimate solution of the problem of space may after all be found in just this possibility—we see that Riemann rejects the opinion that had prevailed up to his own time, namely, that the metrical structure of space is fixed and inherently independent of the physical phenomena for which it serves as a background, and that the real content takes possession of it as of residential flats. *He asserts, on the contrary, that space in itself is nothing more than a three-dimensional* [continuous] *manifold devoid of all* [metrical] *form; it acquires a definite form only through the advent of the material content filling it and determining its metric relations.* There remains the problem of ascertaining the laws in accordance with which this is brought about. In any case, however, the metrical groundform will alter in the course of time just as the disposition of matter in the world changes [47, pp. 97–98].

Thus, Riemann's view of congruence is seen to suggest an explanatory metrical field by suggesting a question on how the solid rod—and other physical agencies yielding the same congruences as the rod—"knows" (a) what particular disjoint intervals of physical space it is to *render congruent* by its coincidence behavior under transport, and hence (b) what particular metrical geometry (as specified by Riemann's general curvatures at the various points) it is to *impart* to the continuous physical space in the absence of "residential flats."

2. With a suitable *change* in the matter distribution and hence in the metrical field, metersticks will coincide with different disjoint intervals under transport in the *same* region of space of a given reference system. But *since there is no intrinsic metric, the stick can be considered self-congruent under transport in the given region in* EACH *of infinitely many* DIFFERENT *patterns of coincidence behavior under transport. And hence one can justifiably regard each one of these alternative patterns as* GENERATING *the metric geometry prevailing at the time in opposition to the conception of "residential flats" espoused by Newton in the first Scholium of his Principia.* Thus with a suitable resulting change in the metric tensor of space, *the same sticks will alike confer a different metric geometry than before upon the same region of physical space.* And this is clearly a case of alternatively metrizing *the same spatial manifold*, the alternative metrizations being physically realized by alternative coincidence patterns of metersticks at different times! Can Putnam tell us which of these incompatible time-dependent coincidence patterns of the metersticks correspond to his "objective," true and unique spatial congruences, and *why*? He cannot.

I submit that the illumination which Riemann's congruence doctrine provides of the very conception of time-dependent spatial geometries and of the explanatory role of the metrical field contrasts strikingly with the otioseness here of Putnam's opposing emphasis on the "objectivity and nonconventionality" of space-time distance in the GTR! And clearly the germaneness of Riemann's congruence doctrine to the quest for a dynamical explanation of the metric geometry is enhanced rather than lessened by the multiple-criterion or cluster character of the congruences with respect to which the metric geometry obtains. For the concordance of the congruences furnished by criteria other than metersticks with those yielded by metersticks is even more impressive than the concordance among the various kinds of metersticks alone.

It is quite apparent that my defense of Riemann's conception of congruence in no way suggests that we be simply content to "only describe the behavior of solid bodies" codified by the spatial geometry to the neglect of also seeking to *explain* that geometry. But since much of what I wrote avowedly pertained to pre-GTR physics, I treated some of the basic laws of geometric behavior of rigid solid bodies as themselves unexplained regularities. A glance at the axioms pertaining to solid bodies in Reichenbach's axiomatization of the special theory of relativity [39, §19] indicates the appropriateness of such a conception. And Einstein's Preface to his fundamental 1905 paper on the STR makes clear that this conception of the role of the rigid body underlies that theory, for he wrote there: "The theory to be developed is based—like all electrodynamics—on the kinematics of the rigid body, since the assertions of any such theory have to do with the relationships between rigid bodies (systems of coordinates), clocks, and electromagnetic processes" [12, p. 38].

2.13. I wish to conclude this clarification and defense of the Riemann-Clifford conception of congruence by dealing with further criticisms offered by Putnam of the contention that space, time, and space-time are devoid of intrinsic metrics, i.e., are intrinsically metrically amorphous. Two of these criticisms seem to be inconsistent with one another, since the first is to the effect that the contention is senseless, whereas the second is to the effect that the contention is tautologically true.

Thus, Putnam declares first:

What Grünbaum means when he says that the metric is not intrinsic is that whereas there *really is* such a thing as color, whether we call it "color" or not, there is not really any such thing as distance except a complicated relational property among solid bodies. To put it positively, he believes that distance really is a relation to a solid body which has been properly corrected for differential forces. Unfortunately, this has no clear sense apart from the contention, which Grünbaum rejects, that statements about distance are translatable into statements about the behavior of solid bodies. Grünbaum rejects this for the quite correct reason that statements about distance are translatable only into statements about the behavior of properly corrected solid bodies, and since the laws used in correcting for the action of differential forces must themselves employ the notion of "distance," to offer such a "translation" as an *analysis* of the notion of "distance" would be circular.

Grünbaum attempts to break the circularity by suggesting, in effect, that we choose a metric arbitrarily, then discover the set of true physical laws (assuming we can do this), expressing them in terms of that metric, and then finally determine that remetricization which agrees with what he takes to be the congruence definition we actually employ, namely, that the length of a solid body should remain constant when transported after we have properly corrected for the action of differential forces [p. 220].

And, still speaking of the procedure which I had outlined as *possibly* capable of determining the spatial geometry empirically [1962, pp. 519–521; 1968, pp. 138–140], he comments:

Even if we could discover the geometry of our world by faithfully following some such prescription, this would not at all tend to show that distance is *really* a relation to a solid standard "apart from substance-specific distortions" or even to bestow a sense upon that proposition [pp. 221–222].

But on the heels of this indictment of my contention as senseless, Putnam likewise charges my claim with being tautological, writing:

According to Grünbaum . . . a nondenumerable dense space has no intrinsic metric. To say of a space that it has no intrinsic metric is in fact exactly to say that it is neither finite nor denumerable but nondenumerable and dense. *On Grünbaum's usage of the terms,* the statement "space S is intrinsically metrically amorphous" . . . has . . . the meaning "space S is nondenumerable and dense" [pp. 222–223].

Amplifying the charge that when making the statement of intrinsic metrical amorphousness, I "*forced*" the latter meaning upon it, Putnam proceeds to construct the following kind of argument: (i) The conventionality of congruence is said to inhere in the intrinsic metrical amorphousness of physical space, but (ii) to say that dense, nondenumerable physical space is intrinsically metrically amorphous is merely to state the tautology that this space is dense and non-denumerable; hence (iii) to assert the conventionality of congruence is to discern a structural feature of space only in the trivial sense of the latter tautology, and therefore (iv) the epistemological conception of metric geometry associated with the conventionality of congruence does not render any nontrivial information concerning the structure of dense, nondenumerable space or space-time [pp. 222–223]. Putnam writes:

> What Grünbaum means by saying that the conventionality of geometry [i.e., congruence] reveals important structural features of space-time is that that conventionality, *when restricted in the way that Grünbaum arbitrarily restricts it*, reveals the fact that space-time is a nondenumerable, dense continuum. It reveals that fact because in the case of a denumerable continuum Grünbaum does not allow us to choose any metric except the one based on a measure which assigns zero to point sets [p. 222].

Putnam then elaborates in a footnote: "More precisely he [Grünbaum] insists on a *countably additive measure*. . . . The 'intrinsic measure' of every distance, according to Grünbaum, is *zero* in any denumerable space" [pp. 222–223n].[10]

I shall now comment, in turn, on Putnam's inconsistent charges that my thesis of the intrinsic metrical amorphousness of physical space, time, and space-time is (1) senseless and (2) tautological.

1. In my Riemannian comparison of mathematically continuous physical space with granular space, I noted that the *former* has no intrinsic metric in the plain sense that its intervals have no *built-in*

[10] In the text that I omitted between the two sentences cited here, Putnam claims to correct "a major error" in my writings on Zeno's paradoxes. I shall deal with this allegation in §5 below.

metric. And I stated the consequences of this intrinsic metric amorphousness for the existence of congruence relations among disjoint intervals by exhibiting the important conventional ingredient of such congruences: the conventionality of the self-congruence of the extrinsic standard under transport. Hence I see no gain in clarity whatever in employing the term "really" to say with Putnam: "What Grünbaum means when he says that the metric is not intrinsic is that . . . there is not really any such thing as distance except a complicated relational property among solid bodies . . . properly corrected for differential forces" [p. 220]. But even if I *were* to formulate my account of the metrical attributes of intervals in this potentially misleading way, I would not wish to speak of "a complicated relational property among solid bodies"; instead, I would speak of a relational property among intervals and *extrinsic metric standards.* The latter formulation would capture the fact that rigid solid bodies are not alone in providing a standard of congruence in physics: for example, in inertial systems, the round-trip times of light do likewise. In particular, the latter formulation would allow for the fact, rightly noted by Putnam, that "even solid bodies do not provide a universal standard of congruence; to define congruence in dimensions which get close to 10^{-13} centimeters, it is necessary to introduce radiative processes, as Reichenbach does" [p. 207n].

But suppose merely for the sake of argument that Putnam did correctly state here "What Grünbaum means when he says that the metric is not intrinsic." Even in that event, Putnam would still be flatly wrong in presuming me to have done the following: to have offered my suggested g_{ik} determination procedure as *grounds* for either the intelligibility or the truth of my denial that physical space has an intrinsic metric. He is therefore shooting down a mere straw man by telling us that "Even if we could discover the geometry of our world by faithfully following some such prescription, this would not at all tend to show that distance is *really* a relation to a solid standard 'apart from substance-specific distortions' or even to bestow a sense upon that proposition" [pp. 221–222].

Of my suggested procedure for determining the spatial g_{ik} empirically, Putnam also says that it "is not successful" as such [p. 221]. He rests this evaluation on the Appendix to his essay whose refutation I must defer until §9 below. Though it will turn out in §9 that Putnam's basis for indicting my suggested procedure is false, I remind the reader that in §2(ii) of Chapter II, I weakened the original claim that I made for it, but for the quite different reason given by Fine [15].

2. His second charge against my thesis of the intrinsic metrical amorphousness of physical space is one of tautology. Referring to my usage of the relevant terms in the statement "space *S* is intrinsically metrically amorphous," Putnam says altogether incorrectly: "he [Grünbaum] has *forced* upon these words the meaning 'space *S* is nondenumerable and dense' " [p. 223]. But the evidence adduced by Putnam in support of this allegation shows *at best* that, on my usage, the phrases "being intrinsically metrically amorphous" and "being mathematically continuous (nondenumerable and dense)" have the same *extension* when applied to manifolds of punctal elements, *not* that they have the same *intension*. Even if I had actually said *without qualification*, i.e., without supported contextual assumptions on the additivity of the measure, etc., that a mathematically continuous space happens to be the *only* intrinsically metrically amorphous space, Putnam's allegation of sameness of *intension* could not be sustained. Yet, to be valid, his charge of tautology requires that my usage did impart the same *intension* to the two phrases in question: If the man who was President of the United States in 1940 is the same as the man who then owned the world's foremost private stamp collection, the statement asserting this fact of *extensional* identity would hardly be tautological (uninformative) and could surely not be said to have forced the same *intensions* upon the phrases "being President of the United States in 1940" and "being the owner of the finest private stamp collection in 1940."

Moreover, note that I used the term "intrinsic" synonymously with "built-in" and that I used its negate "extrinsic" correspondingly.

Hence my Riemannian assertion of the inevitable *extrinsicality* of the metric in continuous physical space fully allows the following simultaneous affirmation: The "intervals" of a *denumerable* dense space, just like those of a continuous space, are devoid of an intrinsic metric, and self-congruence under transport is conventional in both kinds of space; thus, in conjunction with suitably different rules of additivity of the measure, etc., a denumerable space can likewise consistently have various *extrinsically based* positive measures. To say that the conventionality of the interval congruences of a space betokens a structural feature of that space is to say that this conventionality betokens the intrinsic metrical amorphousness of the space in the sense which I have spelled out; it is *not* to say that only one kind of space can possess that structural feature. Hence I had not claimed at all that the structural feature of physical space which is betokened by the conventionality of its interval-congruences *must necessarily* be its full-blown continuity! It happens that Riemann himself was unaware of the distinction between a dense space which is denumerable and one which is nondenumerable when claiming that a continuum has no intrinsic metric.

Indeed, when I wrote that "the continuity we postulate for physical space and time furnishes a sufficient condition for their intrinsic metrical amorphousness" [1962, p. 413; 1968, p. 13], I was careful to say at once: "Clearly, this does not preclude the existence of sufficient conditions *other than continuity* for the intrinsic metrical amorphousness of sets" [*ibid.*, n. 5]. And this remark shows that I was careful *not* to claim *unqualifiedly* even that continuous space, time, and space-time *happen to be* the *only* structures lacking an intrinsic metric. Putnam may have felt justified in disregarding this remark, because he overlooked that when I went on to discuss *denumerable* dense point sets, I avowedly did so in the context of assuming such principles of the *standard* mathematical theory as the countable additivity of the measure. But contrary to Putnam, it was not *my* alleged insistence on the countable additivity of the measure which led me to claim that the measure of every denumerable point set is zero; instead, when I did so, I was setting forth metrical results of the standard mathematical theory used in physics. And when I said more specifically that the standard mathematical theory allows the deduction that the measure of a denumerably infinite point set is *intrinsically* zero [1962, p. 413n5; 1968, p. 13n5], I was asserting the following: *in the standard* theory, this result is deducible *without* any reference to the *congruences* and *unit of length* furnished by an *extrinsic* metric standard. I shall elaborate on this point in §5.

I conclude that Putnam's verdict of tautology against me is not cogent. This verdict is his basis for a further judgment as follows: to say as I do, that the conventionality of congruence betokens an important structural feature of space (or of time or of space-time) "is just to say that Grünbaum finds it an *interesting* fact that a non-denumerable and dense continuum possesses an infinitude of Riemannian metrics" [p. 223]. I trust that my examples from the GTR show why not only I but also Putnam *ought* to find it "an *interesting* fact" that continuous *physical* time and space are legitimately metrizable in significantly alternative ways belonging to the infinite class of Riemannian metrics. And I leave it to the reader to decide whether Putnam has succeeded in showing that my "notion of structural features of the domain in question [i.e., space, time, or space-time] gives the entire show away" [p. 222].

It should be noted that not only the concepts of intrinsic metric, intrinsic metrical amorphousness, and conventionality of congruence have been employed by other authors in essentially my sense as noted above, but also the *terms* representing these concepts. For example, they are used in a well-known work by A. d'Abro [5, Chapters III and IV], and by J. D. North, who writes in his recent history of modern physical cosmology:

> . . . we find Bertrand Russell answering Poincaré's arguments at the time of their publication with the claim that space and time have intrinsic metrics. Whitehead, in fact, agreed, offering psychological considerations as support. . . . Einstein's announcement that the simultaneity of spatially separated events is conventional is an ex-

cellent example of the introduction of a non-trivial change in terminology. The innovation was made in the self-same spirit as Poincaré's statement that to ascribe spatial and temporal congruence is a matter of convention. . . . All will agree that there are conventions, and of many sorts, within all branches of human activity, not least within science. If writers disagree, it is over the question of what is conventional and what is not. We find, therefore, Poincaré arguing that congruence is a matter for definition and Russell denying it. We find Milne (mistakenly) accusing Einstein of not appreciating that there is an element of convention in the notion of a rigid body as one whose rest length is invariant under transport [35, pp. 282, 283, 284].

2.14. In his Inaugural Lecture, Riemann discusses not only the status of spatial congruence, i.e., the basis for assigning equal measures ds to intervals, but also the question of the mathematical dependence of the measure ds on the coordinate differentials and on the coordinates [see Part II, §1]. To arrive at the expression for that functional dependence, he lays down a number of requirements which include the following, among others: (1) If all of the coordinate increments (differentials) are changed in sign, the value of ds is to remain unchanged. This requirement makes only functions of *even* degree in the coordinate differentials admissible. (2) If all of the coordinate differentials are changed in the same ratio, then ds will likewise be changed in that ratio. This second condition demands that the degree of ds be the same as the degree of the homogeneous function in the coordinate differentials. Given these restrictions, we are still confronted with a choice among metrics, among which the quadratic and quartic metrics $ds^2 = g_{ik}\, dx^i\, dx^k$ $(i, k = 1, 2, 3)$ and $ds^4 = g_{iklm}\, dx^i\, dx^k\, dx^l\, dx^m$ $(i, k, l, m = 1, 2, 3)$ are the two simplest candidates. It is well known that if we metrize the space in the standard, customary way by considering an unperturbed solid rod as self-congruent under transport, then it is a matter of empirical fact that the coincidence behavior of such a rod is infinitesimally Pythagorean, i.e., the nine metrical coefficients g_{ik} of the quadratic metric (second-order line element) suffice to relate the coordinate differentials and the coordinates uniquely to the length ds. And in the standard physics this empirical result obviates resorting to the quartic metric ds^4 with its eighty-one metrical coefficients [cf. 40, §§39 and 40]. What may not be so well known is that the nine metrical coefficients of the second-order line element also suffice to yield the ds of the rod even if we do *not* consider the unperturbed rod self-congruent under transport but instead do the following: assign a length ds to it which is a *non*constant function of the independent variables of spatial position and orientation satisfying Riemann's several requirements! Thus, a quadratic metric furnishes a sufficient number of metrical coefficients to cover the coincidence behavior of a rod whether we metrize the space by means of the rod in the customary way *or not*. For example, if we have a rectangular coordinatization x, y, the rod can still be assigned suitably different lengths ds in various positions so as to yield the metrical coefficients of the hyperbolic Poincaré metric

$$ds^2 = \frac{dx^2 + dy^2}{y^2}$$

on a table top instead of those of the standard Euclidean metric: the varying rod lengths ds in question will be given by Poincaré's function upon substituting in it the coordinate differentials dx, dy of those intervals with which the rod in fact coincides at various places y. Thus, in this case, the rod must be assigned the position-dependent length $1/y$. As a second example, consider a table top coordinatized by the polar coordinates ρ and θ. The rod coincides everywhere with intervals for which $\sqrt{d\rho^2 + \rho^2\, d\theta^2} = 1$. But let us *not* assign the length 1 to the rod in all orientations and hence *not* adopt the metric $ds = \sqrt{d\rho^2 + \rho^2\, d\theta^2}$, which would assign the length 1 to *any* such interval. Instead, let us (1) assign the length 1 to the rod in its *radial* orientation, i.e., when it coincides with intervals for which $d\theta = 0$, (2) assign the length *2* to the rod in its *tangential* orientation, i.e., when it coincides with intervals for which $d\rho = 0$, and (3) when it is oriented such that *neither* $d\rho$ nor $d\theta$ is zero, assign to the rod the *non*-unit length $\sqrt{d\rho^2 + 4\rho^2\, d\theta^2}$ which *varies* between 1 and 2 for different orientations. Thus, when

$$d\rho = \rho\, d\theta = \frac{1}{\sqrt{2}},$$

the rod length is $\sqrt{2.5}$. And we see that in this case, the coincidence behavior of the rod yields the *non*standard quadratic metric $ds = \sqrt{d\rho^2 + 4\rho^2\, d\theta^2}$. Hence there is no need to resort to a quartic metric even if we employ a *non*standard metrization so long as we use the rod at all as a basis for the metrization.

Putnam had upbraided me for supposing that physical geometry is "about" the coincidence behavior of unperturbed solid rods [for example, on pp. 207, 217, and 221]. It is therefore very strange indeed that he also indicts me even for having used the solid rod à la Riemann [cf. 42, p. 417] to rule out the fourth-order line element and others of still higher even degree. Having assumed familiarity with Riemann's rationale for the second-order line element, I felt entitled to restrict myself to quadratic metrics [1962, p. 414; 1968, p. 13], and I made it clear that the solid rod was to play either a customary or a noncustomary role in the metrization [1962, p. 415; 1968, p. 15]. Therefore, Putnam does not have a leg to stand on when he criticizes me as follows:

Grünbaum moreover seems to assume that we are not free to assume such differential forms as

$$ds^4 = \sum_{u,o,v,d} g_{uovd}\, du\, do\, dv\, dd$$

We assign a Riemannian metric, i.e., one of the form

$$ds^2 = \sum_{ik} g_{ik}\, di\, dk$$

The choice of any such metric is a matter of pure convention; a continuum cannot intrinsically have one metric as opposed to another; but the choice of a non-Riemannian metric would represent a change in the meaning of "congruent," and would not be the sort of thing that Grünbaum has in mind. . . . Moreover, although space is said to be intrinsically metrically amorphous, it is admissible only to choose a Riemannian metric. This position seems to me to verge on downright contradiction [pp. 217–218].

2.15. Finally, I never said that the justification for postulating one topology of space, time, or space-time respectively as against another is one of *mere* descriptive simplicity. In fact, it seems to me that there is ample good reason for thinking, for example, that our world would be objectively different physically if its time had the closed topology of a circle instead of the open topology of a Euclidean (straight) line, and similarly for such topological properties as the three-dimensionality of space as against a different dimensionality [cf. 26, pp. 197–203 and 330–337]. By contrast, I have put forward positive structural reasons à la Riemann for holding that the alternative between different metrics and hence between their associated metric geometries is one of mere descriptive convenience within the confines of the same topology. Thus the burden of proof is on anyone maintaining that as in the case of the metric, the justification for postulating a particular topological property as against another must be, in principle, a matter of mere descriptive simplicity. And it is surely not established that, in general, there is an inconsistency in according a different philosophical status to the alternative between physical topologies than to the alternative between metrics, as I have done [26, Chapter 11]. I therefore cannot see any merit at all in the following complaint of Putnam's:

Grünbaum supposes that there is an objective topology of physical space determined by the consideration that causal chains shall not be described as discontinuous. . . . It is to [be] emphasized that the choice of what Grünbaum regards as the true or intrinsic topology of space is justified by the fact that it leads to simpler laws, e.g., causal chains are never said to be discontinuous. It is also to be emphasized that Grünbaum is not claiming that all of the admissible metrics lead to equally simple physical laws. On the contrary, Grünbaum admits that if we choose the customary metric in the general theory of relativity given by the general-relativistic g_{ik} tensor, then we obtain much simpler laws than if we choose any other g_{ik} tensor. But even the choice of a g_{ik} tensor in which the description of physical facts becomes incredibly complex represents a permissible convention for Grünbaum. In short, space has an intrinsic topology; but it does not have an intrinsic metric *even though the consequences of changing the metric would be essentially the same as the consequences of changing the topology*, namely, that physical laws would become incredibly complex [pp. 217–218].

But it is not clear at all that the hypothesis of mathematically continuous causal chains is *descriptively simpler* than each one of the hypotheses which would assert them to be discontinuous in some way or another. And, in any case, I had most certainly *not* offered descriptive simplicity as the justification for postulating the collection of topological properties comprised in the topology of space as against an alternative set of them; in fact, I did just the opposite. I conclude that Putnam's charge of inconsistency against me here is not cogent at all, since it rests on his altogether unfounded assertion that the *sole* consequence of changing the topology is to increase the *descriptive complexity* of physical laws.

3. Newton's Conception of Congruence vis-à-vis Geochronometric Conventionalism ("GC") and Trivial Semantical Conventionalism ("TSC")

3.0. Let us recall my statement of the difference between the conventionality of the spatial (or temporal) congruence relation and the much weaker conventionality of assigning a particular intension to the as yet unpreempted *verbal noise* "congruent." In my 1962 essay I referred to the thesis asserting the former, stronger conventionality —or equivalently the nonexistence of intrinsic congruences—as "geochronometric conventionalism" ("GC")[11] [1962, pp. 419–421; 1968, pp. 20–22]. And I referred to the much weaker commonplace that any as yet uncommitted verbal noise (e.g., the *word* "congruent") can be assigned whatever intension we please as "trivial semantical

[11] I disregard as irrelevant for now the broader usage of "GC" in which the chronometric part of the thesis asserts the conventionality of metrical simultaneity as well as of temporal congruence, since simultaneity will be treated in §8 below.

conventionalism" ("TSC") [1962, pp. 420–421; 1968, pp. 20–21]. By affirming "the true, or equable, progress of absolute time" [34, p. 19], Newton ascribed intrinsic congruences to physical time. If we now refer to this claim of his (and/or to its spatial counterpart) as "geochronometric intrinsicalism" (abbreviated to "GI"), then the following logical relations clearly obtain: (i) GC and GI are logically incompatible (indeed, contextually they are contradictories). (ii) Newton's (and the early Russell's) GI are compatible with TSC and hence, in particular, with asserting TSC of the *word* "congruent." (iii) GC and TSC are compatible, but GC is *not* a subthesis of TSC, since GC asserts more than the affirmation of TSC with respect to the word "congruent."

As part of my 1962 exposition, I discussed two other distinct though overlapping theses (besides GC, GI, and TSC). These are as follows:

a. A thesis to which I referred [1962, p. 472; 1968, p. 80] as the "model-theoretic trivialization" of the congruence issue between Newton and Riemann (hereafter referred to as "MTT") and which asserts the following: GC *and* GI are both *meaningless* as they stand, i.e., as statements about space or time, but when they are *both reinterpreted* to become statements about *language*, then GC becomes an uninteresting though true subthesis of TSC, while GI becomes the obviously false denial of TSC.

b. A compound thesis of Eddington's [8, pp. 10–12; 9, pp. 9–10; and 7, p. 230] which states the following: Newton's GI is *true* as a statement about space and time; hence GC is literally *false* as it stands; and GC *becomes* true, though trivially so, *only* if reinterpreted to be a subthesis of TSC. This claim of Eddington's is a conjunction of Newton's assertion of GI (i.e., denial of GC) with an addendum stating that GC is true only if reinterpreted as a subthesis of TSC. Let us now coin the abbreviation "EAN" to refer to that *addendum* of Eddington's to Newton. Note that MTT and Eddington's compound thesis overlap only to the extent that they both affirm EAN, but that otherwise they differ fundamentally. In particular, they differ concerning the status of GI as a claim about space and time: For Eddington, there can be no question of reinterpreting GI to become a statement about language as there is for the proponent of MTT. I thought that amid noting the overlap between MTT and Eddington's compound thesis, I had conveyed this difference as to the meaning and truth status attributed to GI. For I spoke of "Eddington's conclusion that only the use of the *word* 'congruent' but *not* the ascription of the congruence *relation* can be held to be a matter of convention" [1962, p. 406; 1968, p. 5]. This conclusion states that GC is *false*, not—as MTT claims—that GC is meaningless.

3.1. Preparatory to dealing with Putnam's critique of my views on Newton, let me point out that I defended Newton and the early Russell against MTT's superficial dismissal of GI as follows:

> What would be the verdict of the Newtonian proponent of the intrinsicality of the metric on the examples of alternative metrizability which we gave both for space (Poincaré's hyperbolic metrization of the half plane) and also for time (general theory of relativity and Milne's cosmology)? He would first note correctly that once it is understood that the term "congruent," as applied to intervals, is to denote a reflexive, symmetrical, and transitive relation in this class of geometrical configurations, then the use of this term is restricted to designating a spatial equality relation. But then the Newtonian would proceed to claim *unjustifiably* that the spatial equality obtaining between congruent line segments of physical space (or between regions of surfaces and of 3-space respectively) consists in their each containing *the same intrinsic amount of space*. And having introduced this false premise, he would feel entitled to contend that (1) it is *never* legitimate to choose arbitrarily what specific intervals are to be regarded as congruent, and (2) as a corollary of this lack of choice, there is no room for selecting the lines which are to be regarded as straight and hence no choice among alternative geometric descriptions of actual physical space, since the geodesic requirement $\delta \int ds = 0$ which must be satisfied by the straight lines is subject to the restriction that only the members of the unique class of *intrinsically equal* line segments may ever be assigned the same length *ds*. . . .

It is of the utmost importance to realize clearly that the thesis of the conventionality of congruence is, in the first instance, a claim concerning *structural properties of physical space and time;* only the semantical *corollary* of that thesis concerns the *language* of the geochronometric description of the physical world. Having failed to appreciate this fact, some philosophers were led to give a shallow caricature of the debate between the Newtonian, who affirms the factuality of congruence on the strength of the alleged intrinsicality of the metric, and his Riemannian conventionalistic critic. According to the burlesqued version of this controversy, the Riemannian is offering no more than a semantical truism, and the Newtonian assertion of metric absolutism can be dismissed as an evident absurdity on purely semantical grounds, since it is the denial of that mere truism. . . .

We have argued that the Newtonian position espoused by Russell is untenable. But our critique of the model-theoretic trivialization of the conventionality of congruence shows that we must reject as inadequate the following kind of criticism of Russell's position, which he would have regarded as a *petitio principii:* "Russell's claim is an absurdity, because it is the denial of the truism that we are at liberty to give whatever physical interpretations we like to such abstract signs as 'congruent line segments' and 'straight line' and then to inquire whether the system of objects and relations thus arbitrarily named is a model of one or another abstract geometric axiom system. Hence, these linguistic considerations suffice to show that there can be no question, as Russell would have it, whether two non-coinciding segments are truly equal or not and whether measurement is being carried out with a standard yielding results that are true in that sense. Accordingly, awareness of the model-theoretic conception of geometry would have shown Russell that alternative metrizability of spatial and temporal continua should never have been either startling or a matter for dispute. And, by the same token, Poincaré could have spared himself a polemic against Russell in which he spoke misleadingly of the conventionality of congruence as a philosophical doctrine pertaining to the structure of space."

Since this model-theoretic argument fails to come to grips with Russell's root assumption of an intrinsic metric, he would have dismissed it as a *petitio* by raising exactly the same objections that the Newtonian would adduce . . . [1962, pp. 419, 419–420, 472; 1968, pp. 19, 19–20, 80].

Here I was *not* rebutting EAN by invoking Newton and Russell as certainly affirming rather than denying TSC; instead, I was objecting here to MTT's use of question-begging to dismiss both Newton's GI *and* GC incomprehendingly. Indeed, the long statement concerning Newton's GI which I just cited from my 1962 essay was *not* included, as Putnam should have observed, in the *separate part* of that essay which I devoted to my critique of Eddington's EAN [§5(iii)]. As shown by its title, it was in that separate part that I presented my arguments against Eddington's EAN and against its elaboration by Putnam and Feyerabend.

I must emphasize that it is *not* EAN but MTT which I regard as rejecting GI. Not realizing that I distinguish between EAN and MTT, Putnam charges me with regarding EAN and GI "as contrary to each other" [p. 213]. And he thinks that I do not distinguish between EAN and MTT, because he misreads my passage above as offering a charge of *petitio* against EAN rather than against MTT, a charge in which I had indeed stated that MTT rejects GI. Furthermore, he overlooks that I made clear the difference between the grounds on which EAN is espoused by defenders of GI like Eddington or Putnam, on the one hand, and by the proponents of MTT who reject GI, on the other: the former espouse EAN on the grounds that Newton's GI is true and GC false, whereas the latter do so because they fail to understand GI as well as GC and see no issue between them. Thus, I did attribute to the proponents of *MTT* what Putnam rightly calls the "ridiculous . . . interpretation that Newton was denying even semantic conventionality" [p. 214]. But I did not make this attribution to Putnam, as he mistakenly believes [pp. 214–215], or to Eddington, both of whom I identified with EAN.

3.2. These oversights of Putnam's enter into the criticisms offered by him in this connection, as we shall see shortly.

First, let us note that he presents his version of EAN as follows:

. . . the choice of a metric is a matter of convention *only in the trivial sense* that an "uncommitted noise" may be assigned any meaning we wish. Thus, if we do not require that the customary meaning of the term "congruent" should be preserved, then we are free to press upon the *noise* "congruent" any new meaning we wish. If we take "*X* and *Y* are congruent" to mean that *X* and *Y* have the same length according to *M*, where *M* is an unconventional metric that we have invented, then we will simply have changed the meaning of the English word "congruent." Only in the trivial sense that speakers of English have explicitly, or implicitly, to decide what meaning to give to the noise "congruent" if they are to employ it as a meaningful word is the choice of a metric a matter of convention at all [p. 213].

Here I am struck, alas, by the oversimplification inherent in Putnam's claim that if, for example, we characterize two disjoint intervals on a sheet of graph paper as congruent according to Poincaré's non-standard metric

$$ds^2 = \frac{dx^2 + dy^2}{y^2},$$

then "we will simply have changed the meaning of the English word 'congruent.'" Here Putnam is insensitive to the fact that whereas the meaning of "congruent" has been changed with respect to the *extension* of the term, its meaning has not been changed at all with respect to designating a spatial equality relation! Indeed as the logician N. Belnap has pointed out to me, the case of congruence calls for a generalization of the classical account of the relation between the intensional and extensional components of the meaning of a term. According to that account, the intension of a term determines its extension uniquely. But the fact that "being spatially congruent" means sustaining the relation of spatial equality does not suffice at all to determine its extension uniquely in the class of spatial intervals. In the face of the classical account, this nonuniqueness prompted me to refrain from saying that the relation of spatial equality is the "intension" of "spatially congruent"; by the same token, I refrained from saying that the latter term has the same intension in the context of a nonstandard metric as when used with a standard metric. But since the use of "spatially congruent" in conjunction with *any one* of the metrics $ds^2 = g_{ik}\,dx^i\,dx^k$ does mean sustaining the spatial equality relation, I shall refer to this fact by saying that "congruent" has the same "nonclassical intension" in any of these uses.

I am struck all the more by Putnam's insensitivity to this sameness of nonclassical intension because of his very helpful comparison elsewhere [38] between the meanings of the word "true" in the context of standard two-valued and nonstandard three-valued logic respectively. In the latter case, he very properly did *not* say that when we use the word "true" in nonstandard logic, we "simply have changed the meaning" it had in standard logic; instead, Putnam spoke usefully of the "core meaning" shared by the two usages of the term.

Incidentally, even as a statement of Putnam's own position, it is too strong to say, as he does, that the choice of a metric is a matter of convention *only* in the sense that we must decide what meaning to give to the noise "congruent": the choice of a *metric* also involves a convention about the *unit* which is to be used. This emendation does not, of course, pertain to the congruence issue posed by EAN. But I mention it, because in my view the continuous structure of space makes not only for the conventionality of self-congruence under transport but also for the enormous latitude in the conventional choice of a unit on which the metric is based. For we recall that in a granular space, there is no conventionality of congruence and far less latitude for the conventional choice of a *physically realized* unit than in a continuous space: in a granular space, this choice is confined to an *integral* multiple of one granule. And this state of affairs shows, as we anticipated in §2.9, that the *latitude* for conventional choice in the specification of the spatial metric depends on objective facts which are not themselves matters of convention.

This cardinal point of mine concerning the role of objective facts is to be borne in mind along with my endorsement of Reichenbach's recognition that the congruence convention based on the rigid rod is made possible by the following: the rod's coincidences with disjoint intervals are independent from its path of transport. As Reichenbach puts it, "*It is again a matter of fact that our world admits of a simple definition* [i.e., convention] *of congruence because of the factual relations holding for the behavior of rigid rods; but this fact does not deprive the simple definition* [i.e., convention] *of its definitional* [i.e., conventional] *character*" [40, p. 17; italics in the original]. Thus, my critique of GI is to be taken in the context of my twofold emphasis on the role of the following presumed *facts:* facts pertaining to structure (spatial continuity) and objective facts regarding the behavior of extrinsic congruence standards under transport. These facts, I claim, make for the conventionality of self-congruence under transport and for the enormous latitude in the choice of a unit on which the metric is based. Therefore if my critique of GI is sound, there is no justification for Putnam's complaint that "the extent to which and the way in which the choice of a metric is a matter of 'convention' is vastly exaggerated by Grünbaum" [p. 210].

In my charge of *petitio* against MTT, I had explicitly made the point that GI is compatible with asserting TSC of the word "congruent" [1962, pp. 419–420, 472; 1968, pp. 19–20, 80]. Putnam is therefore being redundant *with me* when he writes:

> No one who wished to maintain that space possessed an intrinsic metric would wish to *deny* the thesis of "trivial semantic conventionalism" (TSC). An example should make this point clear. Grünbaum regards phenomenal qualities as intrinsic. Suppose that I ascribe a headache to you. Then I may say that I regard it as an objective fact, and in no sense a matter of convention, that you have a headache, and at the same time recognize that that objective fact can be expressed in the way it is expressed, namely by using the English word "headache," only because certain semantic conventions have been laid down [pp. 213–214].

In the mistaken belief to be addressing himself to my argument *against EAN*, he goes on to say:

> This rather obvious remark permits us to dispose of one argument used by Grünbaum. If, Grünbaum argues, the thesis that the choice of a metric is a matter of convention is nothing but a tautology, then Newton was denying a tautology; but Newton clearly did not mean to deny a tautology! It is easy to see what has gone wrong. Newton meant to affirm, not deny, TSC, and to affirm further that, given the received meaning of "congruent," questions of congruence are matters of objective physical fact [p. 214].

To this I reply the following:

1. It is surely not appropriate here to use the word "tautology," as Putnam does, and I never used it either in stating MTT or EAN. To affirm TSC of the occurrences of "congruent" in an uninterpreted formal calculus is to state a *truism* about linguistic symbols but not a tautology; let me therefore assume that we have replaced "tautology" by "truism."

2. In the 1962 passage cited in §3.1 above, I had charged MTT with a *petitio principii* in its treatment of *both* GI and GC. Putnam simply misstates this *petitio* charge against MTT as a *reductio ad absurdum* argument against EAN. But I had *not* offered the charge of *petitio* as my defense of GC against EAN!

3. The statement which Putnam gives here of Newton's position is given more clearly in my 1962 passage cited above, since I am more specific there about how we are to construe Putnam's phrase "the received meaning of 'congruent.'"

Putnam continues:

> GC similarly affirms that, given the received *definition* of congruence, questions of congruence are matters of objective physical fact; but that there are many alternative admissible *definitions*, leading to different metrics, and that to adopt a noncustomary one is *not* to change the *meaning* of "congruent" (thus denying *both* TSC *and* Newton) [p. 214].

In regard to this, several comments are likewise in order:

1. If it is clear at the outset, as Putnam rightly says [p. 213] it is, that no proponent of GI would wish to deny TSC, then I submit that it is equally clear that no champion of GC would be guilty of "deny-

ing *both* TSC *and* Newton"; surely what is involved is that along with denying Newton's GI, the exponent of GC denies *EAN* but not TSC! I am assuming that Putnam did not altogether disregard the meaning which I stipulated for "TSC" and that he is therefore *not* using this term to refer to the thesis which I have now called "EAN."

2. When Putnam (both here and on his page 225) represents GC as asserting that to adopt a noncustomary congruence convention "is *not* to change the *meaning* of 'congruent,'" he is being seriously misleading: as I noted above in the present §3.2 apropos of his version of EAN, GC asserts this statement only with respect to the nonclassical intension of "congruent" but denies it, of course, with respect to the extension of this term.

3.3. Putnam endeavors to obtain added leverage for his indictment of my rejection of Newton's GI as follows: although in 1957 I had published a paper entitled "The Philosophical Retention of Absolute Space in Einstein's General Theory of Relativity" [27], Putnam alleges recklessly that I rejected Newton's doctrine of absolute motion *and* claims that I did so "*a priori*" on the same Riemannian grounds which led me to reject Newton's GI. Putnam is led to this double imputation because he conflates, as I do not, the following logically different components of Newton's theory of absolute space and time: (a) absoluteness in the sense of possessing intrinsic congruences of space and of time as claimed by GI, (b) absoluteness of rotational or other accelerated motion in the sense of the interpretation which Newton placed upon his rotating bucket experiment; this dynamical absoluteness is entirely compatible with the relativity of uniform translational motion (i.e., with "Newtonian relativity") as shown by the invariance of Newton's dynamical laws under the Galilean transformations which relate his inertial systems; this absoluteness of accelerated motion is therefore *not* tantamount to absoluteness of *position*, and (c) absoluteness of *uniform translatory motion* (no less than of accelerated motion) and hence absoluteness of position in the sense of the original nineteenth-century *ether theory* of light-wave propagation, as embedded in Newtonian mechanics.

Let us suppose that, contrary to actual fact, all the empirical facts of our world did conform to both (b) *and* (c), so that any Michelson-Morley experiment, for example, would have the positive outcome originally expected by the ether theory. My Riemannian analysis has shown that even in that case, the kind of absoluteness asserted by GI under (a) would *not* obtain. In fact, in my GTR paper of 1957 [27], I noted that while the GTR rejects absoluteness in the sense of (a), its failure to implement Mach's Principle constitutes the retention of dynamical absoluteness in the sense of (b) to a significant extent!

Putnam conflates absoluteness in sense (a), on the one hand, with absoluteness in the quite different sense of (c), on the other. And he is not reluctant to foist upon me the grotesque claim that Newton was mistaken in regard to (c) and (b) *merely* because Newton was mistaken in regard to (a). Indulging once again in undocumented imputation, Putnam writes:

> What Newton did maintain is that it is an objective property of a body to be at the same (absolute) place at two different times. The reasons he gave for this were, as is well known, physical and not metaphysical ones: Within the framework of Newtonian physics we obtain different predictions in two universes all of whose *relative* positions and velocities are the same at one instant if one of the two universes consists of bodies at rest relative to the ether frame and the other is in motion relative to the ether frame. In short, the ether frame, or absolute space, represents a theoretical notion which Newton introduces in order to explain observable phenomena. Grünbaum's position is that Newton was *a priori* mistaken. No continuous space [here Putnam says in a footnote "Here Grünbaum means a dense nondenumerable space"] can have an intrinsic metric. Of course, Grünbaum is not saying that the predictions of Newtonian physics are *a priori* wrong: The world might have been Newtonian. But in that case Grünbaum would still not agree with Newton that bodies possess a non-relational property which is their sameness of (absolute) location at different times. Possibly what he would say

is that this property ought to be analyzed as the property of being at rest relative to a solid body which is such that if that solid body is taken as the reference system, then the laws of nature assume a simplest form. If no such solid body in fact exists (because no solid body is actually at rest in absolute space), then counterfactuals would have to be used [pp. 215–216].

In addition to the central error already pointed out, this passage contains the following further defects:

1. It designates Riemann's grounds for rejecting GI as "*a priori*" and "metaphysical" in a sense invidiously contrasted with Newton's "physical" grounds for absoluteness in sense (c). But I should have thought that the grounds for postulating the continuity rather than atomicity of physical space and time cannot be sharply distinguished *epistemically* from those which Newton invoked in support of (b) and (c).

If Putnam had studied more patiently the 1962 essay by me on which he rests his critique, he would not have said here that "Grünbaum's position is that . . . No continuous space [i.e., manifold] can have an intrinsic metric." For he could then not have failed to note that all of part (iii) of §2 of that essay is devoted to emphasizing "that continuity cannot be held with Riemann to furnish a sufficient condition for the intrinsic metric amorphousness of any manifold *independently of the character of its elements*" [1962, pp. 430–431; 1968, pp. 33–35]: I maintain that Riemann's claim does not hold for cases other than space and time such as the real number continuum *itself* in which the elements differ individually in magnitude.

Having erroneously conjectured that, even if the world were empirically Newtonian as stated in (c), I would reject absoluteness of position and would resort to "'analyzing away' the notion of absolute space [*sense* (*c*)!]," Putnam proceeds to explain irrelevantly but illuminatingly what is wrong with such "analyzing away." And then he concludes: "What has been said about sameness of absolute position applies equally well to the relation of congruence" [p. 217].

I have endeavored to show that none of the arguments offered by Putnam are capable of sustaining this conclusion.

3.4. Putnam flies in the face of my writings, is unmindful of an important caveat of mine, and saddles me with an error of Eddington's, writing:

One mistake that Grünbaum frequently makes must be guarded against from the beginning. This is the mistake of supposing that because it is a tautology that we are free to use an uncommitted word to mean whatever we like, all of the *consequences* or *entailments* of any particular decision must likewise be tautological. That such and such will result as a matter of physical or mathematical fact if we make one choice rather than another may be far from tautological. In particular, then, given the topology of space-time it is a fact that certain definitions of "congruent" will lead to a Riemannian metric.

Grünbaum argues that this fact, and indeed the fact that any continuous space can be metricized in a great number of different ways, would all be tautologies if TSC were true. This is just a mistake. TSC does not assert, and it is not a tautology, that there exists even one way of using the word "congruent" that leads to a Riemannian metric in the technical sense; that is a matter of mathematical fact. That any space which possesses a Riemannian metric possesses a great many of these is also a matter of mathematical fact. That of these infinitely numerous possible Riemannian metrics for space-time there appears to be exactly one (apart from the choice of a unit, which is not at issue) which leads to simple and manageable physical laws is a matter of objective physical fact [pp. 218–219].

And in amplification of the first sentence of this passage, Putnam says:

Grünbaum *explicitly* commits the mistake of supposing that according to TSC our *entailed* semantical decisions must be an instance of trivial semantic conventionality on p. 491 of his article, when he writes: "Sophus Lie then showed that, in the context of this group-theoretical characterization of metric geometry, the conventionality of congruence issues in the following results: (i) the set of all the continuous groups in space having the property of displacements in a bounded region fall into three types which respectively characterize the geometries of Euclid, Lobachevski-Bolyai, and Riemann, and (ii) for *each* of these metrical geometries, there is *not* one but an *infinitude* of difference [I had said "different"] congruence classes . . . On the Eddington-Putnam thesis, Lie's profound and justly celebrated results no less than the relativity of simultaneity and the conventionality of temporal congruence must be consigned absurdly to the limbo of trivial semantical conventionality . . ." [pp. 218–219n].

I now confront these statements by Putnam in which he should surely have used the words "truism on a par with TSC" in place of "tautology" with the following.

1. In my 1962 essay, I had written:

. . . consider a surface on which some set of generalized curvilinear (or "Gaussian") coordinates has been introduced and onto which a metric $ds^2 = g_{ik}\,dx^i\,dx^k$ is then put quite arbitrarily by a capricious choice of a set of functions g_{ik} of the given coordinates. . . . Is it then correct to say that since this metrization provides no information at all about the coincidence behavior of a rod under transport on the surface, it conveys no factual information whatever about the surface or physical reality? That such an inference is mistaken can be seen from the following: depending upon whether the Gaussian curvature K associated with the stipulated g_{ik} is positive (spherical geometry), zero (Euclidean geometry), or negative (hyperbolic geometry), *it is an objective fact* [italics new] about the surface that through a point outside a given geodesic of the chosen metric, there will be respectively 0, 1, or infinitely many other such geodesics which will not *intersect* the given geodesic [1962, pp. 444–445; 1968, pp. 49–50].

2. In another 1962 paper and elsewhere [21, pp. 21–22; 26, p. 113], I had occasion to issue a caveat apropos of precisely the point which Putnam claims I have failed to understand. For there I considered the case of two isomers of trinitrobenzene one of which is not toxic while the other is highly toxic. I then pointed out that it is not at all a linguistic triviality but rather "*a fact about the world*" of chemistry that the statement "trinitrobenzene is highly toxic" is true if we use the noise "trinitrobenzene" to refer to the second of the two isomers, whereas that statement is false if the same noise refers

instead to the first isomer. More particularly, I noted that the possibility of making respectively a true and a false statement here upon using the word "trinitrobenzene" so as to have the same core meaning while undergoing the specified partial change of intension reflects the following chemical *fact:* "the existence of isomeric substances of radically different degrees of toxicity (allergenicity)."

3. Eddington, whose views on this issue I had originally understood Putnam to be endorsing and elaborating (at a 1958 Princeton meeting sponsored by the Minnesota Center for Philosophy of Science), wrote as follows:

> Kinematical cosmology seems to me perverted from the start by this mistaken impression that the customary reckoning of length and time becomes ambiguous in applications on so vast a scale; so that we need to introduce a new system of metrical ideas designed specially for the cosmological problem. It contemplates physicists, who . . . contrive to set up (in very specialised conditions) a system of space and time reckoning by interchanging light signals and communicating to one another the results. . . .
>
> *The actual results reached in Milne's kinematical cosmology are only surprising to those who have not realised that if you alter the meaning of words you can make any statement true.* As Milne does not define length and time in the customary way, his conclusions require translating [7, p. 230; italics mine].

Clearly, it is Eddington who is guilty here of the error of supposing that a specified alteration in the meanings of certain words can *by itself* assure—as a matter of *mere linguistic triviality*—that particular statements in which these words function with the given altered meanings will be factually true. And in a statement by Eddington which I cited in my original essay [1962, p. 486; 1968, p. 97], he had declared: "I admit that space [i.e., the particular metric geometry found to prevail upon assigning a particular extension to the equality predicate 'congruent'] is conventional—for that matter, the meaning of every word in the language is conventional" [9, p. 9].

4. Again in my original essay I endorsed the position attributed to Poincaré by L. Rougier, whom I cited as follows: "The conven-

tions fix the language of science which can be indefinitely varied: once these conventions are accepted, the facts expressed by science necessarily are either true or false" [1962, p. 506; 1968, p. 122].

To this documentation, I add the following:

1. Nowhere had I overlooked that "given the topology of space-time it is a fact that certain definitions of 'congruent' will lead to a Riemannian metric." Indeed, I found it necessary to point out to Putnam in §2.14 of this chapter that awareness of precisely this fact was the basis for Riemann's rationale in ruling out quartic metrics in favor of quadratic ones.

2. Absolutely nowhere do *I* argue that the fact just mentioned (under 1) "and indeed the fact that any continuous space can be metricized in a great number of different ways, would all be tautologies [truisms on a par with TSC] if TSC [*sic*!] were true." For along with other even semiliterate people, I devoutly espouse TSC—i.e., that all unpreempted noises can be used to mean what we wish—and never offer any kind of *reductio ad absurdum* argument against it! What I did say originally and have now documented anew is that in his elaboration of *EAN*, Eddington alleges that the alternative metrizability of physical space *and* the geometries associated with metrics other than the standard one obtain *in virtue of TSC*. And I submit that this contention of Eddington's plainly does trivialize Lie's important result. Instead of saddling me with Eddington's error, let Putnam disavow Eddington's elaboration of EAN.

It emerges therefore that there is agreement between Putnam and me in the following limited sense: even if the conventionality of congruence held in the sense of EAN—as he maintains—rather than holding in the sense of GC—as I maintain it does—the geometry predicated on any one *non*customary use of the equality predicate "congruent" is *not* a truism holding on the strength of TSC.

4. The Metrization of Pressure Phenomena

4.0. In my original essay, I examined a corollary of EAN, which I stated as follows: "GC must be a subthesis of TSC because GC has bona fide analogues in every branch of human inquiry [each obtaining in virtue of TSC], such that GC cannot be construed as an insight into the structure of space or time" [1962, p. 487; 1968, p. 98]. And then I stated an illustration given by Eddington of a purported analogue to GC, saying

instead of revising Boyle's law $pv = RT$ in favor of van der Waals' law

$$\left(p + \frac{a}{v^2}\right) \cdot (v - b) = RT,$$

we could preserve the statement of Boyle's law by merely redefining "pressure"—now to be symbolized by "P" in its new usage—putting

$$P =_{\text{Def}} \left(p + \frac{a}{v^2}\right)\left(1 - \frac{b}{v}\right)$$

[1962, p. 487; 1968, pp. 98–99].

In criticism of his pressure example, I then wrote:

The customary concept of pressure has geochronometric ingredients (force, area), and any alterations made in the geochronometric congruence definitions will, of course, issue in changes as to what pressures will be held to be equal. But the conventionality of the geochronometric ingredients is *not* of course at issue, and we ask: Of what *structural feature* of the domain of pressure phenomena does the possibility of Eddington's above linguistic transcription render testimony? The answer clearly is *of none*. Unlike GC, the thesis of the "conventionality of pressure," if put forward on the basis of Eddington's example . . . is thus merely a special case of TSC. We observe, incidentally, that two pressures which are equal on the customary definition will also be equal (congruent) on the suggested redefinition of that term: *apart* from the distinctly geochronometric ingredients *not* here at issue, the domain of pressure phenomena does not present us with any structural property as the counterpart of the lack of an intrinsic metric of space which would be reflected by the alternative definitions of "pressure" [1962, pp. 490–491; 1968, p. 103].

And I concluded by saying:

These objections against the Eddington-Putnam claim that GC has bona fide analogues in every empirical domain are not intended to deny the existence of one or another genuine analogue but to deny only that GC may be deemed to be trivial on the strength of such relatively few bona fide analogues as may obtain [1962, p. 492; 1968, p. 104].

As against this, Putnam does not rest his case on Eddington's example but offers a new justification for EAN in opposition to GC, writing:

> Suppose now that I stipulate that in remetricizing pressure (P) one may only mean by the word "pressure" $f(P)$, where f is an arbitrarily chosen bicontinuous mapping. There will then be infinitely many permissible "pressure metrics." That there are infinitely many is a mathematical theorem, not an instance of trivial semantical conventionality. The theorem indeed is a rather trivial one, namely, the theorem that there exist infinitely many one-one bicontinuous functions. But this is nonetheless a theorem. Deliberately mocking Grünbaum, I may now go on to say that the important thesis of the conventionality of pressure is not an instance of trivial semantic conventionality but reveals an important structural property of the pressure domain, namely, its intrinsic metrical amorphousness [pp. 223–224].

My original statement and Putnam's reply prompt several observations:

1. It is not at issue, of course, that the choice of a unit exists in the case of pressure no less than in the case of space and time, and that there is a choice of the zero of pressure no less than of a zero of time. For example, pressure can be reckoned from atmospheric pressure as a zero or from vacuum as zero. It is to be understood that I shall ignore here as not germane to the issue qualifications that would be needed to take cognizance of the matter of units and of the zero.

2. Let us take the elementary case in which the force F is perpendicular to the area A, so that the pressure is just the scalar dF/dA, where

$$F = ma = m\frac{d^2x}{dt^2}$$

but a is only the acceleration that would be produced if F were the only force acting on the particle of mass m; furthermore $dA = \sqrt{g}\, dx^1\, dx^2$, g being the determinant of the metric coefficients g_{ik}. In conformity to Putnam's notation, we shall use the capital letter P hereafter to denote the ordinary pressure of standard physics, thereby abandoning the notation of Eddington's example.

In order to seek a *bona fide* analogue to the case of congruences among space intervals, we shall wish to consider *intervals* in the continuum of pressures, i.e., pressure *changes*. But in seeking such an analogue, one is struck at the outset by an important disanalogy. In the case of space and time, intervals as such are constituted merely by the points and instants *without* as yet involving any metric; but the *metrical* attribute of pressure is indispensable *ab initio* to confer identity on the elements of the continuum of pressures, and—unlike points and instants—the elements of the pressure continuum each have distinctive magnitudes of their own. Hence the difference between the individual magnitudes of the elements of the pressure continuum furnishes the *intervals* of the latter continuum with an intrinsic metric, in contradistinction to the intervals of space and time! This disanalogy must be borne in mind in assessing the analogical significance of the existence of alternative congruences among pressure-changes (intervals) which may arise as follows.

Suppose that with Putnam we introduce a new metrization $P' = f(P)$, where f is a biunique and bicontinuous but otherwise arbitrary mapping. If f is nonlinear, equal (congruent) pressure changes dP will have unequal (incongruent) magnitudes dP'. The disanalogy just mentioned makes itself felt here in the fact that the alternative pressure-interval congruences arise from the alternative metrizations which first *generate* and are *constitutive* of the respective *elements* of the functionally correlated pressure continua P and P', whereas in the case of space and time, alternative congruences among the already constituted intervals of points (instants) first arise from *extrinsically* metrizing these already nonmetrically given intervals in alternative ways. As I mentioned in §3.3, I had called attention to a similar point as a basis for qualifying Riemann's views on congruence [1962, p. 431; 1968, p. 33].

I shall refer to the important disanalogy just discussed as "the initial disanalogy."

3. As already mentioned in the quotations from my 1962 essay, new pressure metrics P' will be generated upon adhering to the equation $P = dF/dA$ while introducing nonstandard space and time metrics, which determine the physical attributes of area and acceleration ingredient in the pressure. It is an open question in physics whether mass will turn out to be quantized in the sense that all (rest) masses are integral multiples of one mass-"granule." If that should be so, the mass metric could affect only the *unit* in the pressure metric. And in that case, there would be a large class K of *new* pressure metrics P' which are assured—to within a constant factor—*solely* as a consequence of the alternative congruences of space and time. Moreover if this class K includes, as it presumably does, metrics P' which are *non*linear functions of the original standard pressure metric P, then in their case two pressure changes (intervals) $P_2 - P_1$ and $P_3 - P_2$ which are equal (congruent) in the standard metric will have unequal (incongruent) magnitudes $P'_2 - P'_1$ and $P'_3 - P'_2$ in the P' scale. Under these conditions, the domain of pressure-changes has alternative congruences which are "inherited" *solely* from those of space and time. It would assuredly be most wrongheaded to regard such inherited alternative metrizability on the part of the dependent physical magnitude of pressure-change as trivializing or detracting from the significance of the alternative congruences of space and time! And we must recognize that because of the pervasive ingression of the metrics of space and time into other physical magnitudes, there will be an enormous range of inherited alternative metrizability of the latter which redounds to the scientific relevance of the intrinsic metrical amorphousness of space and time rather than serving to support EAN!

Suppose on the other hand that mass is continuous, it being understood that this is not the only possible alternative to its being quantized in the above sense. In that case, it would be possible—though not physically interesting—to countenance an infinitude of alternative mass metrics m' which are biunique and bicontinuous functions of the standard mass m. As we shall see, it is innocuous for my position that the introducibility of these new mass metrics is not vouchsafed by TSC alone but also requires the mathematical existence of infinitely many functions of the required kind. In the present case of continuous mass, the new mass metrics m' would join the alternative metrics of space and time to generate a class K' of alternative pressure metrics P', which may be wider than K. Thus there might be a class K'' of just those new pressure metrics P' which are generated by the nonstandard mass metrics m' *without* any *non*standard metrics of space or time. And this class K'' may have members yielding pressure interval congruences incompatible with those of the P metric.

4. Eddington's particular example of a new concept symbolized by the word "pressure" is not an instance of a remetrization $P' = f(P)$. Among other things, Eddington's "pressure" is a function of the volume of a gas under pressure and not only of the ordinary pressure. I had charged that Eddington's example is not an analogue of the spatial case of alternative interval congruences but merely an instance of TSC. I had not claimed, however, that every alternative pressure metric P' of the form $P' = f(P)$ is guaranteed by TSC alone, and my defense of GC does *not* depend on making this kind of claim implicitly regarding domains other than space and time. Putnam is wrong, I maintain, in thinking that he can base a *reductio ad absurdum* of my position on the fact that the existence of infinitely many $P' = f(P)$ depends on a mathematical theorem and is not assured by TSC alone. That he cannot justify EAN against GC on this basis becomes clear the moment we recall the following: my reason for espousing GC and rejecting EAN was not at all the purely negative one that alternative space and time congruences are not vouchsafed by TSC alone; instead I articulated what it is positively that gives substance to GC independently of TSC: the intrinsic

metrical amorphousness of space and time with respect to the interval congruences. Thus, for example, I would *not* say that GC is an interesting thesis concerning the structure of space just because the real number coordinates presupposed by the metrical coefficients g_{ik} cannot be introduced and assigned on the strength of TSC alone.

5. Suppose now that there are pressure metrics P' in class K'' (i.e., P' metrics generated by new mass metrics m' without any nonstandard metrics of space or time) such that their pressure-interval congruences are incompatible with those of the P metric. Let us call these particular P' metrics "M metrics." What then would be the significance of the M metrics for the issue between GC and EAN? I say: (a) the M metrics cannot trivialize GC as a thesis concerning space and time in favor of EAN: the M metrics cannot detract from the fact that the alternative space and time metrics countenanced by GC have physically interesting relevance to the theory of relativity as I showed in §2, whereas I know of no corresponding relevance of the M pressure metrics or of the mass metrics m' (for rest-mass) on which they depend, (b) the M metrics fall within the scope of what I called the "initial disanalogy" above, and (c) even if the M metrics were not actually contrasted with those of space and time in these important respects, they would illustrate the following earlier claim of mine: "These objections against the Eddington-Putnam claim that GC has bona fide analogues in every empirical domain are not intended to deny the existence of one or another genuine analogue but to deny only that GC may be deemed to be trivial on the strength of such relatively few bona fide analogues as may obtain" [1962, p. 492; 1968, p. 104].

4.1. The various facets of the analysis given in §4.0 exhibit the extent of the distortion of my position in Putnam's declaration that "Reichenbach does not insist, as Grünbaum does, that the dependence of geometrical statements upon 'arbitrary definitions' (in the sense of 'arbitrary' relevant to TSC) applies only to geometry" [pp. 242–243]. Putnam goes on to say: "Unlike Grünbaum, Reichenbach does not derive from the possibility of remetricizing space-time any epistemological moral which does not apply to every physical state" [p. 247]. What I have written here from §2 onward, especially in the present §4, will enable the reader to determine whether such divergence as may exist on this score between Reichenbach's views and mine is justified or not.

5. Zeno's Paradoxes

In my earlier writings on Zeno's paradox of extension [17, especially pp. 297–298 and 302–304, and 24], I had argued that within the framework of the additivity rules for length laid down by the standard mathematics used in physics, the assumption that a line segment consists of the *real* points cannot consistently be replaced by the postulate that it consists instead entirely of a merely denumerable set of points, such as the rational or even the real algebraic points alone. Thus, in one of these two prior publications, I wrote explicitly: "unless substantial modifications are made simultaneously throughout the body of analytic geometry, the proposal of some writers that we replace the Cantorean conception of the line as continuous by the postulate that it consists of only a *denumerable*, discontinuous infinity of points must be rejected on logical grounds of inconsistency alone" [24, p. 236]. And in another of these two earlier publications, I had explicitly emphasized that countable additivity is integral to the standard mathematical theory within whose framework I was going to treat the resolution of Zeno's metrical paradox [17, pp. 297–298].

It is therefore very puzzling indeed that these prior publications prompted the following altogether undocumented and mistaken criticism by Putnam:

> . . . in the case of a denumerable continuum Grünbaum does not allow us to choose any metric except the one based on a measure which assigns zero to point sets.
>
> More precisely he insists on a *countably additive measure*. I wish to call attention here to a major error in Grünbaum's writings. Grünbaum has asserted that Zeno's paradox *depends for its solution* on

countable additivity, and hence on the non-denumerability of space-time. I submit that this is something that the mathematical community simply knows to be false. Even if we assume there are only space-time points with rational coordinates "in reality," world lines, and hence *motion*, are perfectly possible. Moreover, Achilles can catch the tortoise in the conventional way: by being at $x = 1/2$ at $t = 1/2$, at $x = 3/4$ at $t = 3/4$, at $x = 7/8$ at $t = 7/8$ . . . (note that only rational space-time points are involved). Grünbaum's mistake appears to be a simple quantifier confusion: from the fact that the countable sum $1/2 + 1/4 + 1/8 + \ldots = 1$ it follows that the measure is *sometimes* countably additive, not that it must *always* be countably additive.

Let us assume that only points with rational coordinates exist (using some specified coordinate system), and that the physically significant notion of distance can be taken to be given by the square root of the sum of the squares of the coordinate differences (Pythagorean theorem). Then this measure is only finitely additive in this denumerable space; but no serious reason exists for supposing that we would not use this measure for that reason. The "intrinsic measure" of every distance, according to Grünbaum is *zero* in any denumerable space, but to accept this would be to abandon the notion of distance altogether.

Grünbaum also suggests that to use finitely additive measures would require going over to mathematical intuitionism. This is an incorrect remark [pp. 222 and 222–223n].

Putnam's criticism contains the following misunderstandings of my earlier writings:

1. When I wrote that the length (i.e., measure) of a denumerable point set is zero, I was setting forth metrical results of the standard mathematical theory, as should have been evident from my documentation from standard treatises. And I called attention to the fact that the penalty incurred by the postulation of a denumerable space is *either* the deducibility of Zeno's metrical paradox of extension (as *distinct* from the paradoxes of motion!) *or* the truncation of the standard mathematics consequent upon sacrificing the countable additivity of length (and measure). It is a travesty for Putnam to

have characterized this claim of mine by saying that "Grünbaum . . . insists on a *countably additive* measure."

2. Putnam unconscionably distorts my thesis by the completely undocumented contention that I have "asserted that Zeno's paradox *depends for its solution* on countable additivity, and hence on the non-denumerability of space-time." Since I never argued for super-denumerability in this manner, this statement is *not* a "major error in Grünbaum's writings" but only a baffling piece of free association on Putnam's part. What I did say instead, of course, is that *given* the countable additivity present in the standard mathematics, *Zeno's metrical paradox of extension* depends for its solution on the super-denumerability of the space or time interval.

In other words, I said that if space is postulated to be denumerable, then the simultaneous affirmation of countable additivity in the theory issues in Zeno's paradox. This is very different from saying that Zeno's paradox is deducible in every theory that affirms denumerability independently of what it asserts about additivity. And only free association can identify my assertion with the inanity that Zeno's paradox depends for its solution on countable additivity. But such additivity would surely have been affirmed by Zeno in the context of the axioms that scholars such as Luria have attributed to him [25, p. 117].

Putnam failed to heed my explicit caveat that "Zeno's mathematical (metrical) paradoxes of plurality [extension] . . . be distinguished from his paradoxes of motion" [17, p. 288], which raise distinctive issues of their own. This caveat of mine and even a cursory reading of what I wrote should thereafter have made it abundantly clear that I had adduced super-denumerability to deal with the paradox of extension and *not* to refute Zeno's arguments on *motion*. Having carelessly attributed to me the distortion that I believe the solution of Zeno's paradox of motion to depend on countable additivity, Putnam declaims sonorously that "the mathematical community simply knows [it] to be false." And then he proceeds to obfuscate the issue to which I had addressed myself by pointing out irrelevantly that *motion* is possible in a denumerable space of only rational points for which countable additivity is *not* affirmed. This exercise in irrelevancy does indeed show that motion is possible without countable additivity. But it does not contribute one iota to a responsible criticism of what I wrote about the relevance of super-denumerability to the avoidance of Zeno's paradox of *extension*.

3. Having demolished his straw man, Putnam goes on to offer a conjecture about the reasoning by which he presumes me to have arrived at my alleged *insistence* on the countable additivity of the measure. His conjecture is a mere invention. For he overlooks that instead of fallaciously *deducing* what he calls "always" being countably additive from what he calls "*sometimes*" being countably additive, I was not *deducing* countable additivity at all. Instead, I was being mindful of the incontestable fact that the standard mathematical theory does affirm countable additivity.

Incidentally, it is misleading for Putnam to speak of a theoretical situation in which the measure is "*sometimes*" countably additive. For the measure is either countably additive or it is not. And the case of the decomposition of Achilles' unit path into subintervals of lengths ½, ¼, ⅛, . . . , which Putnam considers under the rubric of "*sometimes* countably additive," is more properly characterized as follows: The measure is *not* countably additive, since we do not possess in this restricted theory a rule enabling us to *infer* the length of the total interval by countably adding the denumerable sequence of length-numbers of the decreasing subintervals into which the total interval has been decomposed; instead of being inferable via an additivity rule, the length of the total interval is known from a direct application of the metric to its end points.

4. Putnam's example of a denumerable space in which the measure is *only finitely additive* presumably rests on the following rules:[12]

[12] I am indebted to my mathematical colleague Albert Wilansky for clarifying comments on the presumed meaning of some of Putnam's remarks.

(1) The measure of every unit point-set is zero. (2) The measure of an interval is given by its length as specified by the Cartesian form of the Pythagorean metric. And (3) the measure of any *finite* disjoint union of point sets is equal to the arithmetic sum of the measures of the individual sets. Putnam comments on this only finitely additive measure that "no serious reason exists for supposing that we would not use this measure for that reason." In so doing, he argues completely at cross purposes with those of my earlier writings on which his criticism must rest. I refer the reader to my very recent book *Modern Science and Zeno's Paradoxes* [25, Chapter II, §2, A, and Chapter III], where I have given physical arguments against the claim Putnam makes here.

5. Putnam misconstrues my Riemannian account of "intrinsic measure" and carelessly omits the explicit qualifying reference to "ordinary analytic geometry" and to its countable additivity which I made in that connection:[13] he declares (again without documentation) that "the 'intrinsic measure' of every distance, according to Grünbaum, is zero in any denumerable space, but to accept this would be to abandon the notion of distance altogether." As will be recalled from §2.13, I had explained what I meant by maintaining that the "intrinsic length" of a denumerable "interval" of only rational points is zero within the standard mathematical theory: I meant that in that theory this result is deducible *without* any reference to the congruences and unit of length furnished by a transported length standard, which is *extrinsic* or external to the "interval."

Putnam does not tell us on what basis he asserts that the acceptance of the claim which he incorrectly imputes to me here would require us "to abandon the notion of distance altogether." Why would it do so within the framework of the standard theory? Perhaps he confuses here the following two claims: (1) the *true* claim that

[13] [1962, p. 413n5; 1968, p. 13n5.] The two references which I give in that footnote also make it quite clear that my statements in that footnote are predicated on the countable additivity of the standard mathematical theory.

if we consider a set *S* of points *containing only* the totality of rational points between two fixed rational points, then the intrinsic measure of *S* within the standard mathematical theory is zero, and (2) the *false* claim that in the standard theory, the length of a full-blown *interval bounded* by two fixed rational points but containing *all* the *real* points between them is zero.

Putnam objects quite justifiably to the following inessential remark of mine: "The measure of a denumerable point set is always zero [cf. 30, p. 166], unless one succeeds in developing a very restrictive intuitionistic measure theory of some sort" [1962, p. 413n5; 1968, p. 13n5]. As is evident from my reference to Hobson's treatment and from my reference to "ordinary analytic geometry" in the same footnote, the first part of my assertion pertains to the standard mathematical theory, which is my universe of discourse throughout. Hence the first part of my assertion is true. But Putnam notes usefully that I stated incorrectly that the only way to avoid countable additivity is to erect measure theory on the foundations of some sort of intuitionistic substitute for classical mathematics.

6. A-Conventionality and B-Conventionality

6.0. In order to present my critique of Eddington's pressure example in my 1962 essay, I wished to provide a mere illustration of the clash between GC and EAN and to introduce a convenient abbreviatory terminology for use in subsequently discussing purported analogues of GC. I wrote:

> To state my objections to the Eddington-Putnam thesis [i.e., to EAN], I call attention to the following two sentences:
>
> (A) Person X does not have a gall bladder.
>
> (B) The platinum-iridium bar in the custody of the Bureau of Weights and Measures in Paris (Sèvres) is 1 meter long everywhere rather than some other number of meters (after allowance for "differential forces").
>
> I maintain that there is a *fundamental difference* between the senses in which each of these statements can possibly be held to be con-

ventional, and I shall refer to these respective senses as "A-conventional" and "B-conventional" . . . [1962, p. 489; 1968, p. 101].

Putnam quotes this much, but does not quote the immediately following text of mine which is linked to what he quotes by a colon and reads:

. . . in the case of statement (A), what is conventional is *only* the use of the given *sentence* to render the *proposition* of X's not having a gall bladder, *not* the factual *proposition* expressed by the sentence. This A-conventionality is of the trivial weak kind affirmed by TSC. On the other hand, (B) is conventional not merely in the *trivial* sense that the English sentence used could have been replaced by one in French or some other language but in the much *stronger* and *deeper* sense that it is *not* a factual proposition [i.e., not a proposition stating a spatial fact] that the Paris bar has everywhere a length *unity* in the meter scale even *after* we have specified what sentence or string of noises will express this proposition [1962, *ibid.;* 1968, *ibid.*].

Putnam saw fit to give the following outrageous paraphrase of the sense in which (B) is conventional: "(B) is conventional in (roughly) the sense of being an analytic statement" [p. 224]. This, even though I had *never* employed the term "analytic" to characterize the congruence convention *and* had given examples to explain explicitly that "for an axiomatized physical theory containing a geochronometry, it is *gratuitous* to single out the postulates of the theory as having been prompted by *empirical* findings in contradistinction to deeming the *definitions of congruence* to be wholly a priori, or vice versa" [1962, pp. 498–499; 1968, pp. 112–113]. Clearly, I had claimed (B) to be conventional as an example of an *initial* congruence specification, *not* as an addition to a prior specification of the extension of the spatial equality relation. And qua initial congruence specification, (B) is not analytic but is a *stipulative* specification of part of the extension of the spatial equality term "congruent," just like the following statement: "Disjoint space intervals which can each be brought into coincidence with the same unperturbed transported rod are hereby stipulated to be spatially equal."

Nor can the congruence claim contained in (B) be regarded as analytic even if (B) were regarded as a consistent addition to a prior congruence specification, telling us that the Paris meter is to be our unit. To see this, let us first recall from §3.2 that Reichenbach and I had emphasized the following: The consistent interchangeability everywhere and always of all initially coinciding unperturbed solid bodies to specify congruence depends on an important concordance in their behavior. In 1921, Einstein stated the relevant empirical law of concordance, writing:

> All practical geometry is based upon a principle which is accessible to experience, and which we will now try to realise. We will call that which is enclosed between two boundaries, marked upon a practically-rigid body, a tract. We imagine two practically-rigid bodies, each with a tract marked out on it. These two tracts are said to be "equal to one another" if the boundaries of the one tract can be brought to coincide permanently with the boundaries of the other. We now assume that:
>
> If two tracts are found to be equal once and anywhere, they are equal always and everywhere.
>
> Not only the practical geometry of Euclid, but also its nearest generalisation, the practical geometry of Riemann and therewith the general theory of relativity, rest upon this assumption [11, p. 192].

I shall refer to the empirical assumption just formulated by Einstein as "Riemann's concordance assumption." Given Riemann's concordance assumption, we can use the class of all other solid bodies initially coinciding in Sèvres with the meterstick there to specify congruence interchangeably everywhere and at all times. And then the conjunction of this prior specification with Riemann's concordance assumption will entail the congruence claim contained in my statement (B) as a logical consequence. But this entailment hardly makes my statement (B) analytic, since the congruence claim contained in (B) would be an entailed statement of *fact* under the posited circumstances. Indeed, Putnam himself had recognized in his account of my GC that relatively to a prior specification of an extrinsic congruence standard, the *objective facts* of spatial coincidence determine whether any other body which is to serve interchangeably with the prior initial standard meets the requirement of self-congruence under transport or not [p. 214].

Having misparaphrased my account of the conventionality (B) as stating that (B) is "(roughly) . . . analytic," Putnam carried coals to Newcastle by noting that asserting (B) to be analytic "without qualification" is contrary to GC [p. 224]; for if (B) were held to be analytic, alternative congruence specifications, which are incompatible with it, would be self-contradictory. And in that case there would be no option of choosing one of the alternative specifications instead of (B), whereas GC asserts that option.

Furthermore, once his straw man that I construe (B) to be "(roughly) . . . analytic" is erected, he argues that in my endeavor "to distinguish GC from the trivial semantical conventionality allowed by TSC . . . the distinction between A-conventionality and B-conventionality cannot possibly help Grünbaum" [p. 224] for the following reasons:

The thesis of GC is not that the "customary congruence definition" is *uniquely* analytic in the way in which the usual definition of "bachelor" is analytic; but the thesis of GC is rather that the use of geochronometrical language presupposes that we render either the sentence (B) *or some other sentence tying geometrical statements to statements about the behavior of rods and clocks* true by stipulation. It further asserts that any convention consistent with the alleged intrinsic topology of space-time is admissible if it leads to a Riemannian metric. And it finally asserts that the choice of an admissible convention according to which the rod is said to change its length when transported is *not* a change of *meaning.* On this one page the reader is surprised to find that Grünbaum suddenly restricts his claim to the claim that the customary congruence definition is a stipulation. But this claim, although it happens to be false, is not at all incompatible, as we have already seen, with the Newtonian view that the metric of space is something perfectly objective or with the analogous view of Wheeler in connection with the metric of space-time. If (B) were an analytic statement, it would be an intra-theoretical analytic statement and would not imply any translatability of statements about distance into statements about the actual behavior of solid rods. In fact, as we already noted, it is compatible with the adoption of any metric whatsoever, although one has to change the physical laws appropriately (in a manner explained in the Appendix) if one changes the metric [p. 225].

But in §3.2, I pointed out that GC is misstated by the assertion that the choice of a nonstandard metric "is *not* a change of *meaning*." And having just disposed of the alleged "analyticity" of (B), I am now prompted to offer the following comments:

1. In offering my conception of (B) as an illustration of the clash between GC and EAN, I was not presenting a full statement of GC, but an example of a choice of a congruence specification. And as I argued in §2.11, the existence of alternatives from which to make such a choice is emphasized in the sense of GC by the if-clause in Einstein's GTR statement: "Thus Euclidean geometry does not hold even to a first approximation in the gravitational field, if we wish to take one and the same rod, independently of its place and orientation, as a realization of the same interval." Hence my statement of the sense in which GC conceives (B) to be stipulational as an initial congruence specification does not show at all that "Grünbaum suddenly restricts his claim [i.e., his account of GC] to the claim that the customary congruence definition is a stipulation."

2. It follows from my account of Newton's GI in §3 and of the stipulational content of (B) in the present subsection that Newton's GI regards the extension of the spatial equality predicate to be fixed, whereas GC's claim of B-conventionality regards that extension as open to stipulation. Putnam glosses over the fact that what is being stipulated by (B), apart from the unit, is only the *extension* of spatial equality but *not*—as EAN would have it—the (nonclassical) intension of the mere *noise* "the same length everywhere." Hence, contrary to Putnam, my claim that, qua statement of the customary congruence definition, (B) is a stipulation *is incompatible* "with the Newtonian [GI] view that the metric of space is something perfectly objective."

3. The remainder of what Putnam says here depends upon his claim that (B) "is compatible with the adoption of any metric whatsoever," a claim which rests on an alleged theorem for which he offers a proof in the Appendix to the main body of his paper. I shall endeavor to show in §9 below, however, that this purported theorem is not only a *non sequitur* but false. I have therefore not taken issue before and will not take issue until §9 below with those of Putnam's claims in the main body of his paper which presuppose his alleged theorem.

4. He not only asserts incorrectly here that my characterization of (B) as a stipulation is compatible with Newton's GI, but also that this characterization "happens to be false." I shall therefore consider his argument in support of this claim of falsity in the following subsection.

6.1. Putnam writes:

> Grünbaum does maintain that (B) is conventional, and this would be interesting if true. . . . I do not regard the definition (B) as analytic, or true by convention (or stipulation) in *any* sense. Suppose that the standard meter stick in Paris doubled in length although not subject to any unusual differential forces at all. It would then be twice its previous length relative to everything else in the universe. I am not supposing that this (a stick doubling its length in the absence of forces) could in fact happen (I wish to ignore quantum mechanics, in which it in fact could happen); our question is whether I have described a *self-contradictory state of affairs*. If now (B) were true by convention, we would be required to do one of two things:
>
> (1) *Invent* a differential force [in Reichenbach's sense of a perturbing force] to explain the expansion of the meter stick in Paris notwithstanding the non-existence of any physical source for that force . . . [In this case (B) becomes completely empty, of course.]
>
> (2) *Say* that the rod did *not* change length, and that everything else in the universe underwent a shrinking. It seems to me that neither of these descriptions of the hypothetical case is required by our language, or our stipulations, or by anything else. And this shows that (B) is a *framework principle of physics*, and not a stipulation.

On Grünbaum's view, refusing to give either of the two descriptions mentioned would be *changing the definition* of "congruent." But it seems to me that this is a *reductio ad absurdum* of the view in question [pp. 224, 226].

Putnam's example here is predicated on a hypothetical violation of Riemann's concordance assumption by the meterstick kept at the Bureau of Weights and Measures in Paris (Sèvres) *and* on the empirical correctness of the following restrictively *modified* form of that concordance assumption: If two unperturbed tracts *other than the Paris meterstick* are found to be equal once and anywhere, they are equal always and everywhere, at least for the period of time during which the Paris meterstick is "twice its previous length relative to everything else in the universe." Clearly, the congruence specified interchangeably through time by the members of the class *K* of concordant bodies *other than* the Paris stick is invoked to claim that the Paris meterstick did *not* remain self-congruent at its spatial location but doubled at a certain instant of time. Now, *even though* the original Riemannian concordance assumption, which is coupled with (B) in actual physics, *no longer holds*, Putnam's thesis is that if I regard the congruence claim contained in (B) as a stipulation, then I am logically precluded from doing the following: *alternatively* stipulating congruence through time by means of the latter class *K* of bodies.

Despite emphatic statements by both Reichenbach and me to the contrary (see §3.2 above), Putnam assumes that when we offer what we consider to be a stipulational specification of congruence à la (B), we do so utterly unmindful of the physical validity of such principles as Riemann's concordance assumption, i.e., blind to the theoretical context of the physics in which the congruence specification is to function. Once this fabrication is recognized as such, Putnam's criticism is seen to be yet another straw man, as I shall now show. Nay, it will become apparent that the *framework principle of physics* involved here is *not*, as Putnam has it, the congruence claim itself which is contained in (B); instead, the pertinent framework principle of physics is Riemann's concordance assumption, which is coupled with (B) in actual physics, as Einstein noted (see §6.0).

It is clear that in circumstances in which Riemann's concordance assumption ("RCA") holds for a time period *T*, the unperturbed Paris meterstick will agree during *T* with all other unperturbed solid rods, i.e., with the bodies belonging to *K*, in the following twofold way: (1) there will be concordant findings on the self-congruence (rigidity) of any one space interval at different times, and (2) the unperturbed Paris meterstick will single out one and the same congruence class of disjoint space intervals as the members of *K*. In that case, the Paris stick and the members of *K* can serve interchangeably to specify congruence. But now suppose with Putnam that at time t_0, this concordance gives way to a discordance between the Paris stick and the members of *K* in regard to continuing self-congruence of any one space interval as follows: after t_0, intervals which, according to the congruence furnished by the members of *K*, remain spatially congruent to those states of themselves that antedated t_0 will have shrunk to half their sizes, as judged by the congruence based on the unperturbed Paris stick, while this stick will have doubled in relation to the members of *K*. In that case, logical consistency precludes our continuing to use the Paris stick interchangeably with the members of *K* after t_0 to certify continuing self-congruence in time. And it is perfectly clear from §2 above that in the hypothetical eventuality of a sudden discordance at t_0 between previously concordant criteria of self-congruence, the intrinsic metric amorphousness of space sanctions our choosing from then on *either one* of the suddenly incompatible congruences as follows: (a) we discard the Paris stick from the set of bodies which served interchangeably before to specify self-congruence of space intervals in time, and henceforth use the members of *K or* (b) we retain the Paris stick to the exclusion of the members of *K*, which are no longer interchangeable with it in regard to congruence.

When I assert that (B) and the congruence claim contained in it hold as a matter of stipulation or convention, I plainly do *not* thereby assert that one *must* therefore introduce this particular stipulation under the hypothetical circumstances in which RCA fails or even in circumstances under which RCA is true! In fact, I assert just the opposite. It is true, of course, that if (B) holds by convention, then the assumed facts of discordance could not logically *compel* you to give it up. But it is obviously a mistake to infer that (B) could not be given up in favor of an alternative *K*-congruence convention, just because *the facts of discordance* could not compel you to give it up!

Curiously, this elementary point is not understood by Putnam when he declares quite wrongly that "If now (B) were true by convention, *we would be required* [my italics] to do one of two things: (1) *Invent* a [mythical] differential force to explain the expansion of the meter stick in Paris . . . (2) *Say* that the rod did *not* change length and that everything else in the universe underwent a shrinking" [p. 226]. What is true instead is the following crucially amended statement which provides no basis whatever for impugning the conventionality of (B): *If we adopt the particular congruence convention ingredient in (B)*—which the conventionality of congruence certainly does *not* compel us to do—then we would be required to describe the hypothetical state of affairs imagined by Putnam in one of the two ways outlined by Putnam, or at least in a way that excludes the description that the Paris stick has doubled, since that description contradicts (B) under the assumption of vanishing perturbing forces. Hence when Putnam goes on to say that "neither of these descriptions [(1) and (2)] of the hypothetical case is required by our language, or our stipulations, or by anything else" [p. 226] my comment is: certainly neither of these descriptions (1) and (2) is *required* by the *conventionality of congruence*, since that conventionality countenances the alternative use of the members of *K* to stipulate congruence, a stipulation which issues in the description that the Paris meterstick has doubled. But it is too strong to say that the exclusion of the description that the Paris stick has doubled is not required "by anything else." For the exclusion of this doubling description *is* required by the allowable (though, under the circumstances, highly inconvenient!) use of the Paris stick to specify congruence.

Putnam then offers a glaring *non sequitur* in the following inference: the fact that neither of the descriptions which he calls (1) and (2) is required "shows that (B) is a *framework principle of physics*, and not a stipulation" [p. 226]. I say that the dispensability of both (1) and (2) does not show anything of the kind: all that it does show is that after t_0 we can stipulate congruence by means of the behavior of the members of K rather than à la (B) by means of the Paris stick and that we can *therefore* describe the hypothetical discordance as a doubling of the Paris stick rather than à la Putnam's (1) or (2). This much, however, is not only allowed by the conventionality of congruence but is a direct logical consequence of it under the assumed circumstances. As in the case of the different hypothetical situation discussed in Chapter II and elsewhere [20, pp. 301ff], it is precisely my conception of congruence which co-legitimates alternative descriptions. Thus, it is not (B) qua initial congruence specification but RCA (which is coupled with (B) under actual circumstances) that is the framework principle of physics relevant here, if we understand by such a principle one that has both empirical content and pervasive theoretical relevance. And it is perfectly clear from my account that if we avail ourselves of the congruence specified by the assumedly concordant members of K as our initial congruence specification, then (B) becomes *falsifiable* by the occurrence of the hypothetical discordance.

Putnam is therefore wrongly fancying himself to be correcting my view when he writes: "We can now see the status of (B) more clearly: Within the context of the rest of physics, (B) is falsifiable!" [p. 229]. And this contextual falsifiability does *not* preclude the *alternative* use of the Paris stick instead of the members of K. Indeed, though Putnam denies that (B) is conventional, he, no less than I, allows that it be retained in the face of the hypothetical discordance as a metrical basis for describing the discordance in conjunction with suitably formulated laws of physics [p. 229]. Thus, he allows, as I do, that (B) can serve as a basis for a description of the discordance which is equivalent to one based on the alternative congruence specified by the members of *K*.

Note how I characterized the change made in the congruence specification at t_0 so as to obtain the *K* congruence: I spoke of discarding the Paris meterstick from the set of bodies which served *interchangeably* before t_0 to specify self-congruence of space intervals in time. *Thus understood,* the discarding of (B) at t_0 in favor of the *K* congruences may be said to be a change in the definition (i.e., specification) of congruence. But, of course, this particular kind of change is such as to allow all unperturbed bodies in the universe except the Paris meterstick to retain their sizes after t_0. It is therefore only by crudely trading on there having been a change in the congruence definition without attention to the particular kind of change that Putnam can create a semblance of absurdity (in the last paragraph of the quoted passage) for the following statement: the refusal to give one of the descriptions (1) and (2) which are generated by adherence to (B), and hence the refusal to retain the Paris meterstick after t_0, constitutes a change in the definition of "congruent." And only the oversimplification inherent in Putnam's inveterate concern with the "meaning" of "congruent" (as distinct from the extension of the spatial equality relation) can lend substance to his belief that he can improve on my position when he writes:

> Let us return for a moment to our hypothetical case of the doubling stick. The reader may feel inclined to say that this much of GC is correct: that there are *two* perfectly equivalent admissible descriptions, one according to which the stick doubled its length, and one according to which everything else in the universe underwent a shrinking. Grünbaum may be wrong in maintaining that adopting the former description is *changing* the *definition* of "congruent"; but it would be equally wrong (one might suppose) to say that adopting the latter description is "changing the meaning."
>
> . . . If we accept the view of Professor Feigl, that the meaning of theoretical terms in science is partly determined by the whole "net-

work" of laws, then we can say: The present *meaning* (not "definition") of "congruent" is such that under certain circumstances (B) might turn out to be false. It is adopting the "crazy" description (*everything* shrinking) and infinitely complicating the laws to do it, that *is* "changing the meaning" [pp. 228, 229].

What my conception prompts me to say instead is the following: when the Paris stick assumedly becomes discordant with the members of K at time t_0, the cessation of their interchangeable use for congruence purposes will involve some change in the extension of "congruent" after t_0. But, of course, the *scope* of the change in the extension will be radically greater or smaller according as we adhere to (B) after t_0 or reject (B) in favor of the K congruence.

6.2. It remains in this §6 to deal with Putnam's attempt to discredit my conception of the customary congruence specification on the alleged ground that it entails the following epistemological dichotomy: "In Newtonian physics the second law [of motion] was 'conventional' whereas Euclidean geometry was synthetic" [p. 227]. To establish this gross distortion, Putnam starts out reasonably enough by saying:

Suppose that a solid object does change its shape or size. Let us ask why we are required to say that the body was acted upon by any forces at all. The answer is that such changes presuppose accelerations and that, according to Newton's second law, accelerations require forces to produce them [p. 226].

But then Putnam goes on to resuscitate a hoary error which Schlick and Reichenbach had exposed as such forty years ago. Putnam writes:

Reichenbach and Grünbaum are tacitly assuming that Newton's second law is analytic, or at least that one of its consequences, the Law of Inertia, is built into the customary definition of "congruent." On this view, we are changing the definition of "congruent" (or possibly of "force" and hence *indirectly* of "congruent") if we ever say for any reason whatsoever that an acceleration occurred although no appropriate force was present. Thus, the Newtonian physicist would have been held to be revising the customary "congruence definition" if he gave up . . . the second law . . . but the physicist is said not to have changed the "definitions" of geometrical terms when he abandons the principles of Euclidean geometry. In Newtonian physics the second law was "conventional" whereas Euclidean geometry was synthetic, on this account.

. . . I think it is clear to common sense that this account is a distortion. Newton's second law and the principles of Euclidean geometry functioned on a par in Newtonian physics, and no physicist in his right mind ever considered giving up either one. . . . It is only because he holds the erroneous view that the principles of Euclidean geometry enjoyed a different status from the second law, that the second law was "conventional," and that the second law together with the decision to exclude universal forces uniquely determines the metric, that Grünbaum is able to maintain that the shattering discovery that Euclidean geometry was false was merely an "empirical" discovery. In fact, as I have emphasized elsewhere, it was *not* "empirical" in the narrow sense in which Grünbaum uses the term (although it was empirical in the sense of being about the world), but represented rather an unprecedented revision of *framework principles* in science [pp. 226–228].

This argument is vitiated by the following blunders:

1. Let us recall Einstein's statement (quoted in §6.0) of the conjunction of RCA with the standard congruence specification by means of unperturbed rods. A mere glance at that statement shows that it would be a gross error to do the following: to infer that RCA must be a *stipulational* rather than empirical truth, just because RCA is assumed by or *built into* the customary congruence convention as given by the class of *interchangeable* unperturbed rods. This point was noted by Reichenbach [40, pp. 16–17]. And Schlick made a similar point when he considered the definition of temporal congruence on the basis of the law of inertia in his "Are Natural Laws Conventions?" [43, pp. 183–184]. Schlick explained that the empirical status of the law of inertia derives from the concordance in the behavior of different free Newtonian particles, which allows them to be used interchangeably to stipulate temporal congruence. And Schlick made clear that the empirical content of the law is not

impugned at all by its being assumed in the stipulation of temporal congruence. Reichenbach also emphasized this à propos of the law of inertia [40, p. 116], as I noted above (see §2.11).

Hence Putnam is stating an obvious *non sequitur* when he declares that "Reichenbach and Grünbaum are tacitly assuming that Newton's second law is analytic" [p. 226]. For this conclusion is not deducible from the fact that Newton's second law is presupposed along with RCA in our standard congruence convention in the following sense: we attribute the destruction of the initial congruence of two solids and hence the destruction of the self-congruence at different times of at least one of them to the action of a perturbing force on at least one of them. By the same token, it is innocuous that we assume that "the Law of Inertia is built into the customary definition of 'congruent' " [pp. 226–227]. For the fact that a certain principle (law) is presupposed by a convention simply does not turn the principle itself into a convention!

2. Putnam tells us that according to the Reichenbachian view of congruence espoused by me, the presupposition of the law of inertia by the customary congruence convention entails the following: "we are changing the definition [i.e., extension] of 'congruent' . . . if we ever say for any reason whatsoever that an acceleration occurred although no appropriate force was present" [p. 227]. But if some particular single mass particle at rest were to exhibit an acceleration at some time in the absence of any net forces on it, this would not, of itself, guarantee that there is a single solid rod that once sustained an uncaused acceleration by behaving discordantly in the hypothetical manner of the Paris meterstick in Putnam's example, discussed in §6.1. And since that particle at rest did not enter into the extension of either spatial or temporal congruence, that particle's failure to conform to a law presupposed by the specification of the extension of either kind of congruence would not involve any alteration in the extension of the congruence relation.

Suppose, however, that we concentrate on a case like that of the spontaneously discordant Paris meterstick in Putnam's example above. We saw (at the end of §6.1) that in a case of that kind, the violation of Newton's law makes for a slight reduction in the extension of spatial congruence. But even if the resulting change in the extension of congruence were broad in scope, it would be utterly unavailing to prove Putnam's claim that, according to my view of congruence, "In Newtonian physics the second law was 'conventional.'" For we saw that the original, unmodified congruence convention does not detract from the empirical character of Newton's second law (or of the law of inertia) by being stipulational and by assuming that this law holds unexceptionally; by the same token, neither is the empirical character of the law gainsaid in the least by the fact that certain kinds of exceptions to the law would "change the definition of 'congruent'" in the sense of altering the extension of that term! More generally, the mere fact that a congruence specification presupposes a certain law and that this specification is conventional cannot detract at all from the empirical character of the law, if it is otherwise empirical. Thus if Newton's second law of motion is presupposed by the standard congruence convention while some other physical principle or set of principles is not so presupposed, this fact would *not* show at all that an exponent of the conventionality of congruence is committed to assigning a different *epistemological* status to Newton's law, i.e., a conventional status, as against regarding that other set of principles to be empirical.

3. It follows from the latter considerations that no epistemological difference between Euclidean physical geometry and Newton's second law would be deducible from my conception of congruence even if the following statement *were* true: there is *no* sense in which the customary congruence definition already presupposes the empirical validity of Euclidean geometry in the small, while the empirical correctness of Newton's second law *is* presupposed by that congruence convention. But the first part of this latter statement is not true: the customary congruence is specified by an unperturbed rod, i.e., by a rod suitably corrected for thermal and other distortions, *and*, as I noted in the 1962 essay: "before the corrected rod can be used

to make an empirical determination of the *de facto* geometry, the required corrections must be computed via laws, such as those of elasticity, which involve Euclideanly calculated areas and volumes" [1962, p. 509; 1968, p. 125]. Hence even if Putnam were right in claiming that my conception of congruence gainsays the empirical status of those laws which are presupposed by the customary congruence specification, he would still be mistaken in inferring that my conception commits me to a difference in epistemological status between Newton's second law and Euclidean physical geometry. Indeed, by employing the customary space and time congruences ingredient in its accelerations, the statement of Newton's full-blown empirical second law of motion is predicated on a congruence convention no less than the empirical assertion that the physical geometry of daily life is Euclidean with respect to the customary congruence.

We see that what "is clear to common sense" is that Putnam's account of my position "is a distortion," *not* that I gave the distorted account with which he fallaciously saddles me.

4. I have made no commitments whatever as to what a Newtonian physicist would have said in point of historical fact about the *alleged* claim of a difference in epistemological status between Newton's second law and Euclidean geometry. Hence I indignantly reject as fabrications the claims which Putnam attributes to me here on this score. I have, of course, said things about the status of both Euclidean geometry and Newtonian physics from the point of view of Riemann's philosophy of congruence. Since I recognized that Newton's GI is incompatible with Riemann's GC and have upheld Riemann against Newton's GI, it is hardly surprising that Newton would have rejected some of my actual conclusions. And it is altogether irrelevant that Newton would have rejected the ones that Putnam fallaciously foists on me here.

Thus, I *never* said "that the shattering discovery [by the GTR presumably] that Euclidean geometry was false was merely an 'empirical' discovery" as contrasted with being "an unprecedented revision of *framework principles* in science." On the contrary, I took full cognizance of the original tenacious entrenchment of Euclidean geometry in physics by noting the following [1962, pp. 493–494, 497–498; 1968, pp. 106–107, 111]: Even Poincaré had urged and expected that Euclidean geometry would be preserved in physics in the face of presumably adverse astronomical (parallactic) evidence by abandoning the customary congruence convention, a fact which demonstrates, incidentally, that alternative *spatial* metrics were physically relevant even *historically*, as contrasted with Putnam's alternative mass metric $m' = m^3$, for example. Again, I absolutely never said "that the second law [i.e., $F = ma$] together with the decision to exclude universal forces uniquely determines the [spatial] metric." For I noted that the required knowledge of the perturbing influences acting on the measuring rod pertains to a whole host of correctional physical laws whose content goes well beyond $F = ma$. And it is plain, therefore, that I did not affirm any of the premises which Putnam alleges me to have used to deduce the conclusion that the overthrow of Euclidean geometry was "merely" empirical. Nor did I deduce or otherwise assert the conclusion itself. Alas, Putnam has treated the reader to a tissue of fabrications and fallacious imputations.

Finally, Putnam remarks: "Grünbaum severely berates Nagel for suggesting that the postulating of universal forces (or of deformations not accounted for by differential forces) would be anything other than a change of definition" [p. 227]. And in a footnote, he elaborates:

> Nagel suggested that this would be an *ad hoc* assumption, which is in agreement with the standpoint taken here. Even if Grünbaum were right in supposing that what is *really* happening, in some sense, is that the proponent of universal forces is tacitly changing the meaning of distance, I do not see why he so vigorously disagrees with Nagel here. For surely *ad hoc* hypotheses, in the strong sense in which Grünbaum himself has employed the term (implying *logical* immunity to revision) can always be *regarded* as tacit meaning-changes [p. 227n].

If I understood Nagel correctly in the passage which I had quoted from him [1962, pp. 440–441; 1968, pp. 44–45], he intended his characterization of universal forces as *ad hoc* in a sense that would deny the legitimacy of their introduction on the grounds that they are evidentially unwarranted. And it seems to me that precisely Nagel's attribution of this pejorative sense of *ad hoc* to universal forces precludes Putnam's suggested assimilation of Nagel's conception of their status to the view that their introduction merely involves a tacit change in the meaning (extension) of "congruent." For it seemed to me that by characterizing the introduction of universal forces as *ad hoc*, Nagel meant *not* to countenance them, whereas he hardly would have objected to them, if he had conceived of their introduction as tantamount to a tacit meaning change: meaning changes which are recognized as such are hardly regarded as evidentially unwarranted.

7. The Intrinsicality of Color Attributes

As in the case of pressure discussed in §4, Putnam attempts to adduce examples from the domain of color in an attempt to trivialize my account of the metrical amorphousness and alternative metrizability of space and time.

His original example was that we could adopt a new usage for color words—to be called the "Spenglish" usage—as follows: we take a white piece of chalk, for example, which is moved about in a room, and we lay down the rule that depending (in some specified way) upon the part of the visual field which its appearance occupies, its color will be called "green," "blue," "yellow," etc., rather than "white" under constant conditions of illumination. Putnam had presented this example orally at a 1958 meeting, and I had then construed it as pertaining to *phenomenal* color rather than to physical color, i.e., frequency of electromagnetic vibrations in the optical band. The upshot of the critique I had offered of the example on the basis of a phenomenalist construal [1962, pp. 487–490; 1968, pp. 99–103] had been the following:

> . . . the alternative color descriptions *do not render any structural facts of the color domain* and are therefore purely trivial. . . . while the assertion of the possibility of assigning a space-dependent length to a transported rod reflects linguistically the objective nonexistence of an intrinsic metric, the space-dependent use of color names does *not* reflect a corresponding property of the domain of phenomenal colors in visual space. In short, the phenomenalist color of an appearance is an intrinsic, objective property of that appearance (to within the precision allowed by vagueness), and color congruence is an objective relation. But the length of a body and the congruence of noncoinciding intervals are *not* similarly *non*conventional [1962, p. 490; 1968, p. 102].

In reply, Putnam says first:

> The purpose of the cited example was to show that one could argue (paralleling Grünbaum) that the color properties of things are not "intrinsic" either, and hence that *no* property is, or can be, "intrinsic" in Grünbaum's sense. . . . The reference to "structural properties of the domain" has already been dealt with in the case of *pressure*. But note the retreat to the phenomenal level here! The general point at issue is that the criteria for the application of an "uncommitted" word at one place are *logically* independent of the criteria for the application of that word at any other place (or time). Grünbaum points to the case of phenomenal color properties in phenomenal space (which I wasn't discussing) [pp. 230, 231].

In §4, I argued that Putnam's treatment of the case of pressure was quite unsuccessful. But, furthermore, his 1957 color example fails to support his charge of triviality when construed as pertaining to optical frequencies no less than in the case of the phenomenalist construal which I had given to it. And it fails for the following reasons in the case of the frequency interpretation:

1. If it is understood that the time-metric ingredient in the frequency is not at issue and is fixed, then what I said about a given phenomenal color in my critique of the example applies *mutatis mutandis* to a given optical frequency, i.e., number of vibrations per second.

2. Alternatively, suppose that Putnam intends that the frequency

of any monochromatic component of white light be made to vary with spatial position as specified by an appropriate function, though being spatially constant in the standard physics when a given source of such light is transported from one place to another or when the given monochromatic light beam travels through space. Such a space-dependent frequency is assurable solely as a consequence of the following kind of noncustomary, space-dependent *time* metric: the duration of a time interval at any given space point P is given *not* by the difference between the time numbers which a standard clock at P assigns to the terminal instants of the interval, but rather by a suitable function of both that difference and the coordinates of the spatial position P of the clock. And such alternative metrizability of the physical color spectrum is then merely "inherited" from the alternative metrizability of time. Hence just as in the case of such inherited alternative metrizability of pressure (see §4.0), it plainly boomerangs to adduce the ensuing alternative metrizability of color frequency involved here in an attempt to trivialize the intrinsic metric amorphousness of either time or space!

Indeed the space-dependent kind of time metric involved here is familiar to us from our discussion of coordinate time on the rotating disk of the GTR in §2.1. But the difference between the use of that metric on the rotating disk and in this new version of Putnam's example is the following: whereas its use makes the frequency of radiation *independent* of spatial position in the context of the clock behavior on the disk assumed by the GTR, it does just the opposite in the context of Putnam's example. Hence while the use of a space-dependent time metric achieves descriptive simplicity on the rotating disk of the GTR, it merely issues in pointless descriptive complexity in the case of Putnam's example.

Hence if there is anyone who needs to "retreat to the phenomenal level here" in an attempt to make out his case, it is not I. After this patronizing remark, Putnam endeavors to make his point anew by reference to a color example which is avowedly phenomenalist and writes:

I shall now introduce the word "tred," which is an "uncommitted noise," and which I can, therefore, feel free to use as I like. "Tred" will be used in such a way that an object is called "tred" if it is at one place in my visual field and red, or at a certain other place in my visual field and green, or if it is at a certain third place in my visual field and blue, etc. In addition to introducing the term "tred," I shall introduce quite a number of similar terms, e.g., "grue" (with apologies to Goodman), "breen," etc. Since I am not free to call tred, grue, breen, etc. *colors,* I will call them *grullers.* In addition, two objects in my visual field which possess the same *quale* will be said to possess different "*shmalia.*" The definition of a *shmale* will be left to my readers' imagination; but, roughly, two objects irrespective of their location in my visual field which are both *tred,* or both *grue,* or both *breen,* will be said to possess the same *gruller shmale,* they don't, of course, possess the same *color quale.* Conversely, two objects in my visual field which are both red, or both blue, or both green, will be said to possess the same *color quale,* although they don't possess the same *gruller shmale.*

Now, let us for the moment treat "red" as an "uncommitted noise." Suppose we decide on the criteria to be employed for assigning a truth-value to the assertion that an object at a certain particular place in the visual field is *red.* Then regardless of its color and gruller, it is still open whether any object at any other place in the visual field will or will not be called "red." Note that whether by the words "same color" I decide to mean what we ordinarily mean by "same color" or rather to mean "same gruller," the formal properties of an identity relation—transitivity, symmetry, and congruence [*sic*]—will all be preserved. Moreover, the gruller of a color patch is just as much an objective property of that color patch as its color is. Is it an intrinsic property? Anyone familiar with psychology would certainly venture the conjecture that if our experience were such that patches *moving across our visual field* normally *retained the same gruller* rather than the same color, then the *gruller* of a patch would *quickly seem real and intrinsic* to us. We would begin to *see* grullers. Thus the notion of an intrinsic property seems to be relative to a particular conceptual frame. If we normally describe a patch as e.g., "that red patch moving across my visual field from left to right," then in that conceptual frame *red* is an "intrinsic" property of the patch in question. If, on the other hand, we *normally* employ such descriptions as "that tred patch moving across my visual field from left to right," then *tred* is an "intrinsic" property of the patch in question.

We have seen that insofar as the existence of infinitely many Riemannian metrics is used by Grünbaum as an argument for the "intrinsic metrical amorphousness" of space-time (and insofar as that argument bestows a sense on the assertion in question), we could just as well argue that "phenomenalist colors" are *intrinsically color amorphous*, for there are a large number (though not an infinite number) of possible "color metrics" of the kind discussed. Moreover, this fact *does* reflect "structural properties of the color domain": both psychophysical properties (the existence and number of locations in visual space) and mathematical theorems (the *number* of color metrics is a certain combinatorial function of the number of discriminable locations in the visual field) [pp. 231–232].

I find absolutely nothing in Putnam's reasoning here which can serve to invalidate anything that I said about either congruence of intervals in physical space or sameness of phenomenal color among patches in one's visual field. Note incidentally that when I referred to sameness of phenomenal color as "color congruence," I did *not* say that sameness of color possesses, as Putnam would have it, "the formal properties of an identity relation—transitivity," etc. [p. 232]. For the dyadic relation of sameness of phenomenal color, i.e., visual indistinguishability with respect to color often *fails* to be transitive.

And my reasons for considering Putnam's appeal to his "grullers" to be ill conceived are as follows:

1. Consider the following pair of statements: (A) The same color quale is present at two specified locations of visual space; and (A′) at a fixed location of visual space, the same color quale occurs at two different times. And also consider a second pair of statements as follows: (B) Two disjoint intervals of physical space contain the same amount of space; and (B′) the amount of space in the physical space interval between two points P and P' is the same at different times. I take it to be a matter of objective visual fact whether or not a given color quale inheres in a given color patch at a certain time, and *hence* a matter of objective visual fact whether it is the same as

(or color-congruent to) the quale present in another color patch. By the same token, the color matching or its absence between the states of a fixed location in visual space at two different times is a matter of visual fact, although uncertainty about the accuracy of our memory may sometimes make us hesitate in regard to the reliability of our findings on this fact. Both the color qualia and any congruences between them are built into or intrinsic to the visual patches in which they inhere at any given time. By contrast, my analysis from §2 onward has demonstrated, I believe, that the account which I just gave of the truth criteria pertinent to statements (A) and (A′) does *not* also apply *mutatis mutandis* to the spatial counterparts (B) and (B′) of these statements. It is irrelevant to point out with Putnam [p. 231] that if we decide to call the color of any object at a particular place "red," then it is still open whether we will also use the noise "red" to denote the color of a red object elsewhere. For while the sameness of the *name* of spatially different instances of redness may thus be conventional, the sameness of the color quale named is *not*! Since all true descriptions of the color domain must *agree* on the color *congruences*, there is a basic difference in ontological and epistemic status between sameness of color and spatial congruence. And this difference in status alone shows that permissible alternative descriptions of the color domain cannot be adduced to establish the following claim made by Putnam: "'phenomenalist colors' are *intrinsically color amorphous*" in a sense analogous to the intrinsic metrical amorphousness of space.

2. The fundamental disanalogy just noted also makes itself felt in another way. The term "spatially congruent" has the same (nonclassical) intension and only a different extension in the context of alternative space metrics (see §3.2); but if the words "same color" were used neologistically in Putnam's alternative description to mean same *gruller* rather than same color, then the original and new *intensions* would be avowedly incompatible. For if two visual objects have the same gruller by both being tred, for example, then they must differ with respect to their respective color qualia. I submit that

this fact bespeaks intrinsic color definiteness as opposed to Putnam's color amorphousness. And hence it is unavailing for Putnam to note that color congruence and gruller congruence are each a sameness relation of a certain kind. Thus, his attempt to liken "same color" to "spatially congruent" fails altogether, quite apart from the erroneousness of his claim that each of the relations of having the same color and having the same gruller possesses the formal transitivity property belonging to geometrical congruence.

3. Laws pertaining to the color qualia resulting from combining others, and relations of identity or difference among instances of color qualia, do *not* depend on the locations of these instances in visual space. Indeed the generic truths pertaining to color qualia as such can be stated by reference to suitable instances of the qualia all of which occur at a *single fixed location* in visual space at different times. Although every location in visual space will have some color or other, the number of different locations in visual space is not determined by the number of distinct color qualia: locations in visual space can sometimes be identified by visual criteria other than color such as the visual *shape* of the appearances at these locations. What is pertinent to the domain of color qualia is the number of different color qualia, *not* the variable, time-dependent number of different locations in visual space at which an instance of one or another of these qualia will be found. Putnam apparently thinks that the number of locations existing in visual space at a given time has the status of being a feature of the color domain just because each location in visual space has some color or other. He might as well have argued that the number of collisions occurring in some inertial system of the STR at a given time is a feature of the physical space of that system just because each such collision has a spatial locus in that system. Both the number of distinguishable places in one's visual field and the number of simultaneous collisions in an inertial system are time dependent, whereas the structural features of the phenomenal color domain are not.

Hence it is false to claim with Putnam that "the number of locations in visual space" [p. 232] is a structural property of the color domain which qualifies as a color-counterpart to the continuity of physical space. Yet this false claim is essential to the plausibility of his twin contentions that (a) the invocation of the number of locations in visual space to generate the new gruller description ("color metric") is merely the provision of an alternative description of the color domain, and (b) the co-feasibility of this kind of alternative description bespeaks intrinsic color amorphousness no less than alternative spatial congruences betoken the intrinsic metric amorphousness of space. As against Putnam, I maintain the following: In a description of the actual phenomenal color world we know —as distinct from the hypothetical world in which patches moving across our visual field normally retain the same *gruller*—the coupling of color with spatial specifications in the gruller description irrelevantly imports extraneous spatial information. By contrast, the use of alternative space or time metrics to describe the latter manifolds does *not* involve the intrusion of extraneous information of a respectively nonspatial or nontemporal kind.

4. It is unavailing, though true, that an object which is characterized by the *same* shade of color at two different points of visual space *changes* gruller as between these two points. For this fact cannot be adduced to assimilate the alternative between color congruence and gruller congruence to the alternative between customary and noncustomary space congruences so as to indict my claims about space and about color. Far from detracting from the intrinsicality of color congruences, the co-feasibility of the gruller description merely shows that the color domain can be described both with and without extraneous spatial information.

5. Putnam creates no difficulty whatever for my position by making the following correct point. In the hypothetical world in which patches normally retain the same *gruller* (rather than color!) when moving across our visual field, we would say that grullers are intrinsic properties of patches in the sense of invariantly characterizing them amid their motions. For this fact hardly impugns my claim that in

our actual world colors and color congruences are intrinsic to the patches characterized by them in a way in which the lengths and congruences of intervals in physical space are not. Indeed, Putnam overlooks the fact that I have been at pains to emphasize all along apropos of atomic versus continuous space that the intrinsicality or conventionality of metrical attributes and congruences *depends on the facts of the world rather than being a matter of mere logic or semantics*!

The reader will be able to determine whether my account of the difference between spatial congruence and color congruence is, as Putnam has it, "an interesting example of the way in which Grünbaum deploys his concepts" [p. 231].

8. Simultaneity in the Special and General Theories of Relativity

8.0. The focus of my 1962 essay was on issues pertaining to the congruence relations among intervals of space and time. And hence I made only *illustrative* mention of simultaneity in that essay, when I likened the philosophical status of congruence to that of metrical simultaneity in the STR with respect to involving an important conventional ingredient. It is therefore very difficult to see how Putnam could have gained the impression that my 1962 essay contains my account of the status of simultaneity in the STR, let alone in the GTR. Unfettered by the fact that I had offered an account of the status of simultaneity in other publications, Putnam takes no cognizance of them; instead, he gives free reign to his lively imagination to invent such an account in my name. Indeed, Putnam is so untrammeled by anything that I said as to adduce *against me* a result that I myself had explicitly set forth and emphasized two years before him in the 1961 publication *Current Issues in the Philosophy of Science* [22, pp. 43–53], which was readily available to him when he wrote his critique.

Specifically, in my 1961 paper I discussed the following statement made by Einstein in Section 1 (entitled "Definition of Simultaneity") of his fundamental 1905 paper on the STR: "We have not defined a common 'time' for [the spatially separated points] *A* and *B*; the latter time can now be defined by establishing *by definition* that the 'time' required by light to travel from *A* to *B* equals the 'time' it requires to travel from *B* to *A*" (italics in the original). And the comment I made in that 1961 paper on this statement of Einstein's was as follows:

> The *two* physical assumptions on which this conception of simultaneity rests are the following:
>
> (i) Within the class of physical events, material clocks do *not* define relations of absolute simultaneity under transport: if two clocks U_1 and U_2 are initially synchronized at the *same* place *A* and then transported via paths of *different lengths* to a different place *B* such that their arrivals at *B* coincide, then U_1 and U_2 will no longer be synchronized at *B*. And if U_1 and U_2 were brought to *B* via the *same* path (or via different paths which are of *equal* length) such that their arrivals do *not* coincide, then their initial synchronization would likewise be destroyed.
>
> (ii) Light is the fastest signal *in vacuo* in the following topological sense: no kind of causal chain (moving particles, radiation) emitted *in vacuo* at a given point *A* together with a light pulse can reach any other point *B* earlier—as judged by a local clock at *B* which merely orders events there in a metrically arbitrary fashion—than this light pulse.
>
> . . . The importance of understanding the grounds on which Einstein thought he could safely make assumption (i) can be gauged by the following basic fact: if assumption (i) had been thought to be *false*, then the belief in the truth of (ii) would *not* have warranted the abandonment of the received Newtonian doctrine of absolute simultaneity. And, in that eventuality, the members of the scientific community to whom Einstein addressed his paper of 1905 would have been fully entitled to *reject* his *conventionalist* conception of *one*-way transit times and velocities. . . . In fact, (i) is essential, because a physicist brought up in the Newtonian tradition quite naturally uses *not* signal connectibility but the readings of suitably transported clocks as the fundamental indicators of temporal order. He recognizes, of course, that the truth of (ii) compels such far-reaching revisions in his theoretical edifice as the repudiation of the second

law of motion. But he stoutly and rightly maintains that *if* (i) is *false*, absolute simultaneity remains intact, unencumbered by the truth of (ii) [22, pp. 44–45].

Ironically, I repeated this final conclusion from my 1961 paper in an essay [29, pp. 175–176] published in the very Delaware Seminar volume in which Putnam accuses me of having mistakenly denied it! Putnam refers to a world—to be called "quasi-Newtonian"—in which assumption (i) is false while (ii) is true, and his irresponsible accusation takes the following form:

> . . . I wish to clarify Grünbaum's account to some extent. In order to do this I shall introduce a special kind of fictitious universe, to be called a *quasi-Newtonian* universe. . . .
>
> Now then, what is the situation with respect to simultaneity in the quasi-Newtonian world? In the quasi-Newtonian world, the customary correspondence rule for simultaneity involving the transported clocks does not lead to inconsistencies that cannot be explained as due to the actions of differential forces upon the clocks. There is then, in my view, no reason to regard simultaneity as a notion needing a definition except in the trivial sense in which every notion requires a definition (TSC). . . .
>
> . . . According to Grünbaum, what Einstein realized under the circumstances described above was that (a) in any universe in which there exists a limit to the velocity with which causal signals can travel (in particular, even in a quasi-Newtonian universe) we are free to define simultaneity as we wish subject to one important restriction, (b) the only important restriction is that a cause may not be said to be simultaneous with or later than its effect, and (c) it is *logically* irrelevant that some "definitions" would lead to vastly more complicated physical laws, although it is required that the system of physical laws should not become more complicated in the *one* respect prohibited by (b), i.e., that causes should not be said to be simultaneous with or later than their effects. . . .
>
> I conclude that the alleged insight that we are "free" to employ a noncustomary "definition" of simultaneity even in a quasi-Newtonian universe is correct only in the sense allowed by TSC, and that Grünbaum's interpretation of Einstein therefore *trivializes* Einstein's profound logical-physical insight [pp. 236, 237, 238, 241–242].

It is evident that Putnam has no case whatever against me here but merely belabors a point which I had made two years earlier. Moreover, in my 1962 "Epilogue" for P. W. Bridgman's posthumous *A Sophisticate's Primer of Relativity*, I had discussed a different kind of quasi-Newtonian world and had made the corresponding point concerning that universe [19, p. 185].

It should be noted that Putnam's fabrication here springs from the same imputation to which I called attention apropos of congruence at the end of §7: the false supposition that my claims of conventionality concerning metrical simultaneity are asserted to be true as a matter of mere semantics instead of being grounded on the full range of the known relevant facts of the world.

8.1. I should have thought that the status of metrical simultaneity in the STR as conceived by Einstein would no longer be a matter for philosophical debate. But Putnam's discussion shows that this is not so: in addition to attacking views concerning simultaneity which I have never held, he objects to claims which I regard as integral to a correct account of Einstein's conception of simultaneity in the STR. And in order to come to grips with these objections in a coherent manner, I must ask the reader's indulgence to permit me to give a sketch of what I take such an account to be.

If a physical theory T claims that an attribute or relation of a physical event or object is the same in every reference system in which it is specified, then T can be said to regard the attribute or relation in question as "invariant" or "absolute" in virtue of thus being independent of the reference system. By the same token, attributes or relations of an event or thing are called "covariant" or "relative" in T, if T asserts that they do not obtain alike with respect to all physical reference systems but depend on the particular system.

Einstein held that time relations between noncoincident physical events depend for their very existence on the fact that certain physical relations obtain between these events. He thus espoused a relational conception of time (and space); that is, he regarded time (and space) as systems of relations among physical events and things. Time relations are first constituted by the system of physical relations obtaining among events. Hence, the structure of the temporal order will be determined by the physical attributes which enable events to sustain relations of "simultaneous with," "earlier than," or "later than." Specifically, this structure will depend on whether these attributes define temporal relations unambiguously—i.e., so that *every* pair of events is unambiguously ordered in terms of one of the relations "earlier than," "later than," and "metrically simultaneous with." If the temporal order were thus unambiguous, it would be "absolute," for a time relation between any two events that is an unambiguously obtaining factual relation is wholly independent of any particular reference system and hence is the same in every reference system.

If the transitive relation of metrical simultaneity were thus unambiguously fixed by the physical facts, then *as a matter of physical fact* only one particular event at a given point Q could qualify as simultaneous with a given event occurring at a point P elsewhere in space.

Is there an actual physical basis for relations of absolute simultaneity among spatially separated events? Suppose that the behavior of all kinds of transported clocks whose rates accord with the law of inertia were of the kind assumed by the Newtonian theory—that is, that if two clocks are initially synchronized at essentially the *same* place A, this *contiguous* synchronism will be preserved after they both have been separately transported to some other place B independently of the lengths of their respective paths of transport and of whether or not their arrivals at B coincide. In that case a physical basis for absolute simultaneity would exist in the form of the coincidence of physical events with suitable identical readings of transported clocks of identical constitution. But according to the STR, this physical-clock basis for absolute simultaneity does not exist. Indeed, the STR makes the contrary assumption, which was labeled "(i)" in §8.0 above. Thus transported material clocks fail to define unambiguously obtaining relations of simultaneity within the class of physical events because relations of simultaneity yielded by clock transport depend on the particular clock used: a given pair of spatially separated events E and E^* at space points A and B respectively will or will not be held to be simultaneous depending on which of the discordant clocks U_1 and U_2 serves as the standard. This dependence on the particular clock used prevents transported clocks from defining relations of absolute simultaneity within the class of physical events. It also led Einstein to conclude that even within a single inertial system the simultaneity of two spatially separated events E and E^* cannot be based physically on the criterion that the *spontaneous* numerical readings on two clocks U_1 and U_2 be the same for the event E occurring at the location of U_1 and event E^* occurring at the place of U_2, the clocks having previously been transported to these *separate* places from a common point in space at which they had identical readings at the outset.

The behavior which Einstein attributed to transported clocks thus does not make for relations of absolute simultaneity among spatially separated events. This does not, of course, preclude the existence of some other physical relatedness among events which might make for such relations. Yet spatially separated events can sustain physical relations of one kind or another only because of the presence or absence of their actual or at least physically possible physical connectibility. We must therefore ask under what conditions two such events can be simultaneous if the relations of temporal order between them depend on the obtaining or non-obtaining between them of a physical relation of causal connectibility. Einstein was driven to ask this question because he had postulated that the transport of contiguously synchronized clocks does not furnish a physical basis for the obtaining of relations of simultaneity. He therefore sought to ground the temporal order *to begin with* on physical foundations which are independent of any synchronism of spatially separated clocks.

We are not ready to characterize one of two causally connected (or connectible) events as *the* actual or possible partial or total cause of the other. Hence, the statement I am about to make of the relevance of causal connectibility to simultaneity will use only a symmetric relation of causal connectibility, one that does not involve singling out one of the two causally connected events as *the* (partial or total) cause of the other. Since we are concerned with time relations obtaining independently of any clock synchronism, we can now define "timelike separation" as follows: two events will be said to sustain the relation of timelike separation—the relation of being either earlier or later independently of any clock synchronization—if and only if they sustain the symmetric relations of causal connectibility or connectedness. This usage of the term "timelike separation" is more general than the one familiar from books on relativity, which restrict this term to the case of events linkable by causal chains *slower* than light *in vacuo*, while speaking of events linkable by a light ray as having a "null separation," since $ds_4 = 0$ for the latter pairs of events. Therefore, two noncoincident events will here be said to *lack* a timelike separation, i.e., to be topologically simultaneous and to have a spacelike separation if and only if the specified symmetric causal connectibility does not obtain between them. Two *coinciding* events are topologically simultaneous but do not, of course, have a spacelike separation.

Thus, the physical basis for the relation of topological simultaneity among noncoinciding events is the impossibility of their being the termini of influence chains. We must now ask: are the physically possible causal chains of nature such as to define a relation of topological simultaneity among noncoinciding events which is transitive and thus *unique* in the sense that only *one* event at a point Q could be topologically simultaneous with a given event at another space point P? To answer this question, consider four events E_1, E_2, E_3, and E which satisfy the following conditions represented on the accompanying *events* diagram ("world-line" diagram), from which the "arrow" of time has been omitted: (1) E_1, E_2, and E_3 are causally connectible *by a light ray in vacuo*, and E_2 is temporally between E_1 and E_3. (For a definition of temporal betweenness in terms of causal connectibility, see my [25, p. 59].) (2) E_1, E, and E_3 are causally connectible other than by a light ray, E being temporally between

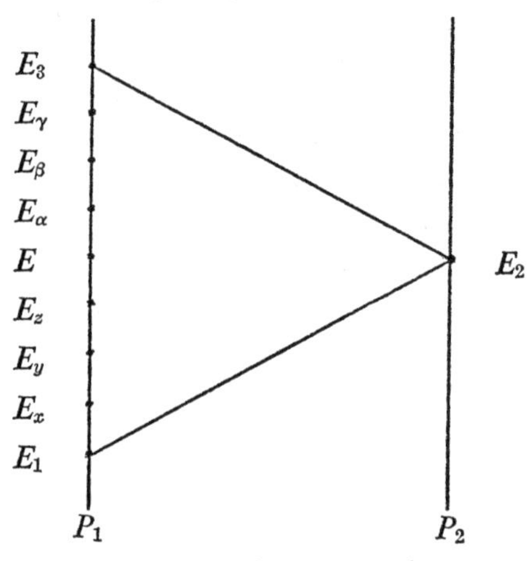

E_1 and E_3; these three events all occur at the *same* space point P_1 of an inertial system, whereas E_2 occurs at a different space point P_2 of that same inertial system. Furthermore, let E_x, E_y, and E_z be events which sustain relations of temporal betweenness to E_1 and E and to one another, as shown on the diagram. Similarly, let E_α, E_β, and E_γ be temporally between E and E_3. The given facts that temporally E_2 is between E_1 and E_3 and that E, E_x, E_y, E_z, E_α, E_β, and E_γ are each between E_1 and E_3 as specified do *not* furnish a basis for any relations of timelike separation or topological simultaneity between E_2 and any one of the events at P_1 lying within the open interval between E_1 and E_3. Whether the latter relations exist will depend, therefore, on whether it is physically possible for there to be causal chains of which E_2 and a particular event in the open interval at P_1 would be the termini.

Newtonian physics can grant that no light ray or other electromagnetic causal chain could provide a link between E_2 and E_z or E_2 and E_α, and it can allow that the *round*-trip time of light is not arbitrarily small but fixed by the path P_1P_2. But it adduces the second law of motion to assert the physical possibility of other causal chains (such as moving particles) that would indeed furnish these links, and the *non*conventionality of its metrical simultaneity rests on this possibility. It is precisely this latter possibility that is denied by Einstein's STR. The STR enunciates the postulate which we labeled "assumption (ii)" in §8.0 and to which we shall refer as the "limiting assumption" when it is coupled with the claim that the round-trip time of light is positive. According to this limiting assumption, it is physically impossible for there to be causal chains that would link E_2 with any event lying *within* the open time interval $\overline{E_1E_3}$. This physical impossibility obtains independently of any inertial system, and its logical consequences therefore have invariant, or absolute, significance.

The STR's affirmation of the limiting role of electromagnetic causal chains *in vacuo* within the class of causal chains has a fundamental consequence: topological simultaneity is not a uniquely obtaining relation, because *each* one of the infinity of events at P_1 *within the open time interval between* E_1 *and* E_3—not just a *single* one of these events—is topologically simultaneous with E_2. Moreover, this nonuniqueness of topological simultaneity is absolute, since it prevails alike in every inertial system, for on Einstein's limiting assumption, *none* of these events at P_1 sustains the relation of timelike separation to E_2—i.e., is either earlier or later than E_2 independently of any clock synchronization. But since *none* of this infinity of events at P_1 is either earlier or later than E_2 in this sense, and since clocks do not furnish consistent relations of simultaneity under transport, an important result follows: no one member of this class of events is *physically* any more entitled than any other to the status of being metrically simultaneous with E_2—that is, to being assigned the same time number by a clock as E_2. In the context of the *special* theory of relativity, this conception of Einstein's does require modification, however, in the light of P. W. Bridgman's particular method of synchronism by *infinitely slow* clock transport, as set forth in the work cited under [19]. But I am concerned here to set forth Einstein's own conception, as stated in his 1905 paper.

Since all members of our class of events at P_1 are equally candidates for being metrically simultaneous with E_2, some one of these events comes to be metrically simultaneous with E_2 to the exclusion of the others *not* as a matter of temporal fact but only by *convention*, or "definition." And this conventional choice of one event E^* in the open interval $\overline{E_1E_3}$ at P_1 *conventionally* renders the remainder of the events *in that interval* either earlier or later than E_2. The choice of a particular event E^* is made by means of a rule for setting the clock at P_2 so as to assign to E_2 the same time number that the clock at P_1 assigns to E^*. And the synchronism of the clocks at P_1 and P_2 is decreed by the fact that the clock at P_2 is set on the basis of such a rule. In Newtonian physics there is no corresponding scope for choice of the event that can warrantedly be metrically simultaneous with E_2 on the basis of the *spontaneous* readings of the members of the class of contiguously synchronized and subsequently transported clocks. In Newton's theory, such clocks yield a consistent, unambiguous time coordinatization, and hence the differences between the readings of any and all such clocks alike furnish the *same* time metric for any pair of spatially separated events, although that time metric is, of course, *not* intrinsic. By contrast, the physical facts postulated by relativity require the introduction, *within a single inertial frame* S, of a convention specifying which pair of topologically simultaneous events at P_1 and P_2 is to be metrically simultaneous. Such a stipulation is necessary in the STR because that theory asserts that the topological simultaneity of noncoincident events or their spacelike separation is not a transitive relation. That is, although E_x is topologically simultaneous with E_2, and E_2 is, in turn, topologically simultaneous with E, the events E_x and E are *not* topologically simultaneous; instead, the relation between E_x and E is one of timelike separation, since they are causally connectible by the motion of the pointer of a dial clock. The objection that the ordinary usage of the term "simultaneous" in common-sense discourse entails the transitivity of *every* kind of simultaneity relation is a *petitio principii*. It is tantamount to a linguistic decree that Newtonian beliefs be retained in every physical theory which makes technical use of homonyms of ordinary temporal terms.

In regard to simultaneity, therefore, Einstein's conceptual innovation can be summarized as follows: Time relations among events are assumed to be first constituted by specific physical relations obtaining between them. These physical relations, in turn, are postulated to be such that the topological simultaneity of events at spatially separated points P_1 and P_2 is not a uniquely obtaining relation. Metrical simultaneity is thus left indeterminate by topological simultaneity and by the behavior which the STR postulates for transported, adjacently synchronized clocks. Therefore, a conventional choice or synchronization *rule* for which there was no scope in Newton's theory must be invoked over and above the relevant physical facts to assert that a given event at P_2 sustains a uniquely obtaining equality relation of metrical simultaneity to an event at P_1. In this sense the relation of metrical simultaneity is *not* a spontaneously obtaining physical relation in the STR but depends on a conventional choice. For any given event E_2 at P_2, this conventional choice consists in the selection of a unique event at P_1 as metrically simultaneous with E_2 from within the infinite class of those events at P_1 which are topologically simultaneous with E_2. And this choice is implemented by the rule for setting the clock at P_2. In brief, Einstein's innovation is that the physical relatedness which makes for the very existence of the temporal order has a structure that precludes the existence of spontaneously and uniquely obtaining relations of metrical simultaneity. Thus, the failure of our measuring operations to disclose relations of absolute simultaneity is only the epistemic consequence of the fact that these relations do not exist.

Einstein's avowal of the conventionality of metrical simultaneity in Section 1 of his STR paper of 1905 explicitly pertains to the situation *within* any given inertial system in the sense of making no reference whatever to the relative motion of different inertial systems. In particular, as is made evident by the account above of the logical status of metrical simultaneity, the conventionality assertion is no commitment to a discordance or relativity of metrical simultaneity as between relatively moving inertial frames. Einstein's repudiation of Newton's belief in the uniqueness of distant simultaneity consists in (1) rejecting clock transport as a basis for absolute metrical simultaneity and (2) asserting the *non*uniqueness of the *absolute* relation of *topological* simultaneity. In addition, this denial of uniqueness leaves open the question whether there will be disagreement among relatively moving inertial systems as to the metrical simultaneity of certain pairs of noncoinciding events pending the specification of the particular synchronization rules to be employed in these systems.

The resulting conventionality of metrical simultaneity does furnish the logical framework within which the *relativity* of simultaneity as between relatively moving inertial systems can first be understood. The nonuniqueness of the invariant relation of topological simultaneity allows each inertial system to adopt its own metrical synchronization rule. Let each system separately adopt for itself the particular, maximally convenient rule selected by Einstein in Section 1 of his fundamental paper, and let the spatial separation of P_1 and P_2 have a component along the line of the relative motion of the inertial frames. Then that relative motion results in the following relativity of simultaneity: each of the relatively moving inertial systems chooses a *different pair of events* as metrically simultaneous from within any given class of topologically simultaneous pairs of events at P_1 and P_2. This result is embodied in the so-called Minkowski light cone diagram. In short, in the STR, the problem of metrical simultaneity is the following: Given the clock behavior formulated in assumption (i) of §8.0 and also the spacelike separations of the events in the left and right portions of the Minkowski diagram, i.e., granted the joint validity of assumptions (i) and (ii), can metrical simultaneity be said to have the same logical status as in Newton's theory? And Einstein's answer in Section 1 of his 1905 paper is: No, metrical simultaneity is established by a convention or "definition" whose philosophical status as a convention is itself the result of anti-Newtonian empirical facts postulated in assumptions (i) and (ii).

Now let the following assignment of time numbers be made to the events at P_1 in our events diagram by means of a local clock stationed there in the given inertial system S: t_1 is the time of E_1, t_3 the time of E_3, and $(t_3 + t_1)/2$ the time of E. The conventionality of metrical simultaneity then expresses itself as follows: the obtaining of metrical simultaneity within S depends on a choice—*not* narrowed by the objective physical facts of the temporal order—of a particular numerical value between t_1 and t_3 as the temporal name to be assigned to E_2 at P_2 by an appropriate setting of a like clock stationed there in S. Thus, depending on the particular event at P_1 that is chosen to be simultaneous with E_2, we set the clock at P_2 upon the occurrence of E_2. To use Reichenbach's notation, the reading t_2 which we impart to this clock as the time of E_2 has the value between t_1 and t_3 given by

$$t_2 = t_1 + \epsilon(t_3 - t_1),$$

where ϵ has the particular value between 0 and 1 appropriate to the choice we have made. For example, suppose that we choose ϵ so that E_y is simultaneous with E_2. Then in system S all the events between E_1 and E_y, by definition (that is, via this rule), will be earlier than E_2, whereas all the events between E_y and E_3, by definition, will be later than E_2. Alternatively, we could choose ϵ so that E_y, instead of being simultaneous with E_2 in S, would be earlier than E_2. Or a different value of ϵ might be chosen, so that E_y is later than E_2 in S.

This freedom to decree time relations numerically by setting the clock at P_2 was restricted by *defining* causally connectible events to be "time-separated." For we thereby *stipulated* that clock settings be made to convey the objective causal connectibility of such events by issuing in the assignment of *different* time numbers to them. We did, moreover, confine choices of ϵ to *within* the open interval between

0 and 1. Once a criterion of metrical simultaneity has been chosen in a given system as indicated, the ensuing assignment of time numbers by synchronized clocks makes it unambiguous which member of any pair of objectively time-separated events is the earlier of the two and which the later. And if the world contains irreversible processes, the lower and higher time numbers thus assigned will have added physical significance [26, Chapter 8].

Einstein selected E to be metrically simultaneous with E_2 in S by choosing the value $\epsilon = 1/2$ in *each* inertial system in order to obtain descriptively simple laws of nature. The conventionality of simultaneity allows but does not entail our choosing the same value $\epsilon = 1/2$ for all directions within every system. In each system this choice assures the *equality* of the one-way velocities of light in opposite directions by yielding *equal* one-way transit times $t_2 - t_1$ and $t_3 - t_2$ for equal distances. The ratio of these one-way transit times is $\epsilon/(1 - \epsilon)$, and therefore, in the case of $\epsilon \neq 1/2$, these one-way times are *unequal.* But on Einstein's conception no fact of nature independent of our descriptive conventions and convenience would be contradicted if we chose values of $\epsilon \neq 1/2$ for each inertial system, thereby making the velocity of light different from c in both senses along each direction in all inertial systems. Of course, for the sake of the resulting descriptive simplicity and convenience, Einstein chose $\epsilon = 1/2$ in the formulation of the empirical content of the STR. And speaking of the particular definition employing $\epsilon = 1/2$, Einstein writes as follows in Section 1 of his fundamental 1905 paper on the STR:

> We assume that this definition of synchronism is free from contradictions, and possible for any number of points; and that the following relations are universally valid:—
>
> 1. If the clock at B synchronizes with the clock at A, the clock at A synchronizes with the clock at B.
>
> 2. If the clock at A synchronizes with the clock at B and also with the clock at C, the clocks at B and C also synchronize with each other.

In Reichenbach's *Axiomatik der relativistischen Raum-Zeit-Lehre* the definition of simultaneity based on Einstein's choice of $\epsilon = \frac{1}{2}$ in the STR is called "Definition 8." And Reichenbach writes there: "As early as in his first paper on the theory of relativity (*Ann. d. Phys.* 1905, *17*, 891), Einstein treated Definition 8 as a *definition* but pointed out that the symmetry and transitivity of the synchronism obtained from it is not a matter of *definition* but of *empirical fact*" [39, p. 34n2]. Likewise the role of empirical law in assuring that Einstein's choice of $\epsilon = \frac{1}{2}$ makes for descriptive simplicity is explicitly recognized in my book of 1963 [26, p. 356].

8.2. Let us now turn to simultaneity in the GTR before evaluating Putnam's contentions.

It must be borne in mind at the outset that the choice of $\epsilon = \frac{1}{2}$ in the inertial systems of the STR can restrict the conventional choice of metrical simultaneity in a *non*inertial system of reference such as that of the rotating disk of our §2. For consider the open interval of events between the optical events A and C of our §2.3 above which occur at the permanently coinciding origins O and O' of the systems I and S', remembering that each of the infinitude of events temporally between A and C at the common origin has a *spacelike* separation from the optical event B which occurs at the instantaneously coinciding points P' and P of S' and I respectively. And suppose that after assigning time coordinates within the inertial system I on the basis of $\epsilon = \frac{1}{2}$, we specify the time (and space) coordinates associated with the noninertial system S' by means of some set of transformation equations relating them to those of the inertial system. Then it is clear that the resulting time coordinatization in S' has already fixed the particular event within the open time interval $\overline{AC}$ at the common origin of I and S' which is to be assigned *the same time number* as the event B occurring elsewhere at the instantaneously coinciding space points P and P'. But thus then to fix simultaneity in S' must not be permitted to detract from the ingredience of a choice in the metrical simultaneity of that noninertial system. For the choice of metrical simultaneity in S' is being made via the transformation equation for the S' time coordinate. Indeed, the metrical simultaneity in S' corresponding to the transformation equation (3) of §2.3, relating the proper time τ in S' to the proper time T in I, is generally incompatible with the metrical simultaneity in S' corresponding to the coordinate time t, which is specified by the transformation equation (6) of §2.4, i.e., by $t = T$. This fact is evident from §2.3. For we saw there that whereas the standard clock at P' in S' assigns the proper time

$$\tau = \frac{r}{c}\sqrt{1 - \frac{r^2\omega^2}{c^2}}$$

to the event B, the coordinate clock there assigns the different time $t = r/c$ to that same event in the same frame S'. And since the standard and coordinate clocks at the common origin agree with each other and with the I-system clock there, the event B will be metrically simultaneous in S' with a *different* event between A and C at the common origin according as we use equal τ numbers or equal t numbers as the basis of metrical simultaneity.

The time interval AC required by light to travel from O' to P' and back is $2r/c$ on the clock at O'. But if we consider light rays emitted at O' at various, successively later times $\tau = k$, we can see that the one-way transit times for the fixed spatial path $O'P'$ have *time-dependent* values $\Delta\tau$ which depend on the emission time k as follows:

$$\Delta\tau = \left(k + \frac{r}{c}\right)\sqrt{1 - \frac{r^2\omega^2}{c^2}} - k = \epsilon_k\left(2\frac{r}{c}\right)$$

or

$$\Delta\tau = k\left[\sqrt{1 - \frac{r^2\omega^2}{c^2}} - 1\right] + \frac{r}{c}\sqrt{1 - \frac{r^2\omega^2}{c^2}}.$$

The first of these two equations becomes obvious upon realization that a light ray emitted at the common origin O and O' at a time $T = t = \tau = k$ will reach P' at a time $T = k + r/c$ on the I-system clock instantaneously adjacent to P' and hence at a time

$$\tau = \left(k + \frac{r}{c}\right)\sqrt{1 - \frac{r^2\omega^2}{c^2}}$$

read by the standard disk clock at P'. More specifically, we see from our two equations here that the following results must be true:

1. The successive one-way transit times from O' to P' corresponding to successively later times of emission $\tau = k$ form a *decreasing* sequence, and indeed this one-way transit time $\Delta\tau$ becomes *zero* for the particular ray emitted at a time $\tau = k_0$ given by

$$k_0 = \frac{\frac{r}{c}\sqrt{1 - \frac{r^2\omega^2}{c^2}}}{1 - \sqrt{1 - \frac{r^2\omega^2}{c^2}}}.$$

Moreover, for emission times k *later than* k_0, the one-way transit time $\Delta\tau$ from O' to P' will be *negative*.

2. For $\Delta\tau = 0$, $\epsilon_k = 0$, and for $\Delta\tau < 0$, $\epsilon_k < 0$. But the use of the coordinate time $t = T$ in S' is tantamount to choosing the alternate synchronism $\epsilon = \frac{1}{2}$ *in the very same reference system* S'.

3. In an inertial system, the choices of ϵ are confined to the range between 0 and 1, but not so in a noninertial system such as our S'. In particular, the case of the light ray whose one-way transit time along $O'P'$ is *zero* shows the following: in the τ-time description, there exists a pair of events which are *metrically* simultaneous but *not* topologically simultaneous. But this state of affairs is indeed countenanced by the GTR in systems other than inertial frames: in *non*inertial frames, there is *no* requirement in the GTR that only topologically simultaneous events may be metrically simultaneous! For the τ numbers constitute an admissible time coordinatization on the disk S'. That this is indeed so emerges from the following considerations.

In the STR, Einstein had maintained that clock transport does not provide a physical basis for time order relations among events and that we are therefore at liberty to decree the time numbers assigned to them by material clocks. On the other hand, the causal connectibility of two events is indeed an objective fact. Therefore, it is a *stipulation* to require that the objective relation sustained by two events which are causally connectible only by a light ray in *vacuo* be rendered in the following particular numerical way: defining events of this kind as "time-separated" via the demand that clocks at their respective spatial loci be set to assign *different* time numbers to them such that ϵ is confined to within the open time interval between 0 and 1. Einstein's choice of $\epsilon = \frac{1}{2}$ in the light-signal method of clock synchronization clearly implements this stipulation within the *special* theory of relativity.

We saw, however, that the enormous descriptive simplicity which is assured in the context of the physical facts pertaining to inertial systems by the synchronization rule $\epsilon = \frac{1}{2}$ is *not* also achieved by a corresponding one-time application of that rule to standard material clocks in the noninertial systems of the GTR. And whereas the stipulation of time-separation for events connectible only by a light ray is built into the Lorentz transformation equations, no corresponding stipulation is built into the laws of the GTR. Hence the ensuing far greater freedom in the assignment of proper times τ to such events allows the assignment of the *same* proper time number to a pair of them.

What better philosophical basis for understanding this entire state of affairs can there be than the conception that metrical simultaneity importantly involves a conventional ingredient over and above any implicit in the use of spontaneous clock readings in Newton's theory?

A light ray emanating from the coinciding origins and reaching the instantaneously coinciding points P and P' of the I and S' systems has a radial path in I and hence does *not* have a radial path in S'. But the radial path in S' is a geodesic path linking O' and P' in the context of the hyperbolic type of non-Euclidean spatial geometry prevailing on the disk [33, pp. 242–243], and, as is clear from equation (10) of §2.6, the length of that path is r. Hence in S' the light ray from O' to P' traverses a distance *greater than* r during the proper τ-time interval

$$\frac{r}{c}\sqrt{1-\frac{r^2\omega^2}{c^2}},$$

which corresponds to a t-time (coordinate-time) interval of magnitude r/c. Hence the τ-time velocity and the t-time velocity of that light ray are each greater than c.

In order to lead up to a further consideration of metrical simultaneity in the rotating disk system S', let us now examine the bearing of the t-time coordinatization on the velocity of light in different directions at a point of that system. It will now turn out that the t-time speed w of a light ray at an arbitrary point P' of the disk is not the same in all directions and generally not the same for light traveling away from a point in *opposite* directions. For whereas the radial components w_r of two differently directed rays are equal, their θ' components $w_{\theta'}$ are not.

To see how this result comes about, let us note first the general expression for the components of the velocity of light in the opposite spatial senses of the increase and decrease of the ιth space coordinate. Let n be a unit vector which points in the direction of the velocity vector w of light, and let n^ι ($\iota = 1, 2,$ and 3) be its components. Then it is shown [33, p. 240] that

$$(1) \qquad w(n^\iota) = \frac{c\sqrt{-g_{44}}}{\gamma_\iota n^\iota + 1},$$

where we have put

$$(2) \qquad \gamma_\iota \equiv \frac{g_{\iota 4}}{\sqrt{-g_{44}}}, \quad \text{and where} \quad (\gamma_\iota n^\iota)^2 < 1.$$

Care must be taken *not* to confuse the γ_ι here with the components $\gamma_{\iota K}$ of the spatial metric tensor associated with the spatial line element $d\sigma$ of our §2.6. It is clear from equation (1) of this subsection that if $\gamma_\iota \neq 0$ ($\iota = 1, 2, 3$) in the coordinate system under consideration, then the magnitude of the one-way t-time light velocity w at a given point will generally depend on the direction of propagation of the light ray, since the n^ι will have opposite signs (plus and minus) for light traveling away from a point in opposite senses.

The application of equation (1) to our rotating disk S' now enables us to see that in the case of the coordinate time $t = T$, the speed of light traveling on the disk will *not* be equal in different directions at a given point unless the light paths are each entirely radial. It will be recalled from §2.5 that for the particular S' coordinates used there $g_{14} = g_{r4} = 0$, while $g_{24} = g_{\theta'4} = wr^2/c$ and

$$g_{44} = -\left(1 - \frac{r^2\omega^2}{c^2}\right).$$

Since $g_{r4} = 0$, it follows from equations (2) and (1) that the radial velocity components w_r of directionally different light rays must be equal at a given point, each having the value $w_r = c\sqrt{-g_{44}} = \sqrt{c^2 - r^2\omega^2}$. But the nonradial components $w_{\theta'}$ will have *unequal* values for $r \neq 0$, since equation (2) becomes

$$\gamma_{\theta'} = \frac{g_{\theta'4}}{\sqrt{-g_{44}}} = \frac{\frac{\omega r^2}{c}}{\sqrt{1 - \frac{r^2\omega^2}{c^2}}} = \frac{\omega r^2}{\sqrt{c^2 - r^2\omega^2}} \neq 0.$$

Since the components $w_{\theta'}$ of the velocities of different light rays traveling away from a given point in S' are unequal while their components w_r are equal, the magnitudes of these t-time velocities are unequal unless the light paths are contrived to be entirely radial by means of constraining mirrors. We saw that the paths of the unconstrained light ray from O' to P' and return are not radial.

The usually different velocities of light traveling away from a given point in opposite directions generally make for an inequality between the transit times required by light to traverse a given closed polygonal path in opposite senses after departing from the same point. Thus, suppose that two light signals are jointly emitted from a point P'' on the periphery of the disk and are each made to traverse that periphery in opposite directions so as to return to P''. Since the spatial metric coefficients are independent of the time in this case, the length $d\sigma$ of any given element of the periphery is not a function of time, and therefore the oppositely directed light pulses traverse equal total distances. But for any given point of the periphery the light velocities in the opposite senses of the periphery are unequal. And the unequal travel times needed by the oppositely directed light pulses to traverse the respective equal elements of the periphery which share the given point combine here to yield unequal total travel times for the entire periphery, as becomes evident upon considering their respective paths in the nonrotating system I. The one light pulse which travels in the direction *opposite* to that of the disk's rotation in I will travel a shorter path in I and hence will return to the disk point P'' earlier than the light pulse traveling in the direction in which the disk rotates.

What are the metrical simultaneity rules implicit in the time coordinatization established on the rotating disk by means of the coordinate time $t = T$? This coordinate time uses $\epsilon = \frac{1}{2}$ to synchronize the clocks at the various disk points *with the central clock at* O'. For t time assures the equality of the to-and-fro transit *times* required by the unconstrained nonradial light rays going from O' to any other disk point P' and then back. But the very clock settings resulting from synchronizing the various disk clocks with the center O' clock via $\epsilon = \frac{1}{2}$ can be shown to issue generally in a synchronism $\epsilon \neq \frac{1}{2}$ as between two disk clocks each of which is stationed at a point *other than* O' [39, pp. 136–137 and 140–141]. That ϵ generally differs from $\frac{1}{2}$ for the latter pairs of disk clocks does not follow from the mere general inequality of the to-and-fro speeds of light linking such clocks. For in the absence of information concerning the length ratio of the respective to-and-fro spatial paths, this inequality of the to-and-fro speeds does not suffice to establish the inequality of the to-and-fro *transit times* of light corresponding to $\epsilon \neq \frac{1}{2}$.

Since $-g_{44} > 0$ [33, p. 235; 32, p. 272], equation (2) of this subsection shows that $\gamma_\iota = 0$, if and only if $g_{\iota 4} = 0$. Applying this result to equation (1), we find that in any coordinate system the condition

(3) $\quad g_{\iota 4} = 0 \; (\iota = 1, 2, 3 \text{ only})$

for every point in 4-space is necessary and sufficient for the directional independence of the speed of light at every space point. Since the vanishing of the metrical coefficients of all the cross-terms $dx^\iota \, dx^K$ $(\iota \neq K)$ in the spatial metric assures the orthogonality of the spatial coordinate lines, a coordinate system characterized by the vanishing of the coefficients $g_{\iota 4}$ of all the space-time cross-terms $dx^{\iota}\,dx^4$ ($\iota = 1, 2, 3$) can be said to have a time axis which is everywhere orthogonal to the spatial coordinate curves. And hence a coordinate system for which $g_{\iota 4} = 0$ ($\iota = 1, 2, 3$) is called "time orthogonal." Incidentally, as noted in §2.6, the time-orthogonal systems, which we just showed to have directionally independent speeds of light at any given point, are also precisely the ones in which the spatial part $g_{\iota K}$ ($\iota, K = 1, 2, 3$ only) of the space-time tensor g_{ik} ($i, k = 1, 2, 3$ *and* 4) is the same as the spatial metric tensor $\gamma_{\iota K}$ ($\iota, K = 1, 2, 3$) which determines the spatial geometry.

The rotating disk system S' does not qualify as a time-orthogonal coordinate system. But the class of time-orthogonal systems is interestingly wider with respect to the propagation properties of light than the class of inertial systems. Specifically, we now note the following: while the to-and-fro coordinate-time transit times of light in time-orthogonal systems are equal for any given pair of fixed space points [39, pp. 129–130], the round-trip times of light are generally not the same for different paths of equal length, and hence a Michelson-Morley experiment would generally not have the null outcome obtained in an inertial system. In addition to $g_{\iota 4} = 0$ ($\iota = 1, 2, 3$), it is here assumed that the g_{ik} ($i, k = 1, 2, 3, 4$) are not time-dependent but only space-dependent. And it is then shown [39, pp. 129–130] that there is a coordinate time such that all clocks are pairwise synchronized with one another on the basis of $\epsilon = \frac{1}{2}$. But the difference between light propagation in an arbitrary time-orthogonal system and that in an inertial system nevertheless makes itself felt, as we shall now see. Since the coordinate-time one-way light velocity w is given by $w = d\sigma/dt$, we have

$$(4) \qquad dt = \frac{d\sigma}{w}.$$

Applying the condition of time orthogonality given by equation (3) to equations (2) and (1), we have

(5) $w = c\sqrt{-g_{44}}$,

so that (4) becomes

(6) $dt = \dfrac{d\sigma}{c\sqrt{-g_{44}}}$.

Now two space paths P_1P_2 and P_1P_3 emanating from a common point such as the light source of the Michelson-Morley experiment will be equal if

$$\int_{P_1}^{P_2} d\sigma = \int_{P_1}^{P_3} d\sigma.$$

But, by equation (6), the corresponding travel times of light $\overline{P_1P_2}$ and $\overline{P_1P_3}$ are respectively given by

$$\int_{P_1}^{P_2} \frac{d\sigma}{c\sqrt{-g_{44}}} \text{ and } \int_{P_1}^{P_3} \frac{d\sigma}{c\sqrt{-g_{44}}},$$

and these one-way travel times in the different space directions P_1P_2 and P_1P_3 will generally be *unequal*, since the space-dependent values of g_{44} are different along these two space paths. But, by integration of equation (6) over each of the space paths P_1P_2 and P_2P_1, the to-and-fro travel times $\overline{P_1P_2}$ and $\overline{P_2P_1}$ are equal to one another in conformity with $\epsilon = 1/2$, and there is likewise equality between the to-and-fro travel times $\overline{P_1P_3}$ and $\overline{P_3P_1}$. Hence $\overline{P_1P_2P_1} = 2\overline{P_1P_2}$ and $\overline{P_1P_3P_1} = 2\overline{P_1P_3}$. But we saw that $\overline{P_1P_2} \neq \overline{P_1P_3}$. Therefore $\overline{P_1P_2P_1} \neq \overline{P_1P_3P_1}$. Thus, there is inequality among the round-trip times of light over equal but differently directed space paths emanating from a common point, unless the g_{44} happen to have the same values along the two space paths. Therefore, if the equal arms of the Michelson-Morley interferometer are large enough for a significant difference in the g_{44}, then, unlike the situation in an inertial system, there is inequality of the round-trip times of light in different spatial directions, although the to-and-fro travel times over any given path are equal.

The GTR no less than the STR makes assumption (ii) and assumes further that the round-trip times of light on any given clock are finite instead of zero. And it incorporates assumption (i) of the STR that standard clocks do not define absolute simultaneity under transport. Thus, for any given event at a fixed space point in a reference frame of the GTR, there is not just one but an infinite class of events at any other space points in the frame which have a spacelike separation from the given event. Hence, for the reasons explained above apropos of the STR, metrical simultaneity in the frames of the GTR importantly involves a nontrivial conventional ingredient in the form of a choice of ϵ. Of course, the time coordinatization for a noninertial frame of the GTR need not be made from scratch but is often effected by means of a coordinate transformation relating the time coordinate of the frame to the time (and space) coordinates of an inertial system I in which $\epsilon = \frac{1}{2}$ has already been chosen. In the latter case, the choice of metrical simultaneity is made in the noninertial frame *implicitly* by the conjunction of the choice of $\epsilon = \frac{1}{2}$ in the inertial system with the coordinate transformation specifying the time coordinate in the noninertial frame in terms of the I-system coordinates. And in that case, of course, the one-way velocities of light in the noninertial frame during a specified time interval, and hence the equality or inequality of the to-and-fro velocities of light in such a frame, are no longer open to stipulation by means of an independent choice of ϵ.

It is therefore evident that both the equality of the to-and-fro velocities of light in noninertial time-orthogonal systems of the GTR, and the general inequality of these velocities in systems such as our disk S' accord entirely with my account of the status of metrical simultaneity. An illustration of the equality of the to-and-fro velocities of light in a *non*inertial time-orthogonal frame of the GTR is given by the situation in the coordinate system employed in Schwarzschild's solution for the field surrounding the sun. This coordinate system is time orthogonal [33, pp. 325–326], and hence the to-and-fro velocities of light are equal in it even though the paths of light rays are *not* spatial geodesics (but only space-time geodesics).

Putnam runs afoul of Schwarzschild's time-orthogonal coordinate system K when claiming that we can circumscribe the class of exactly those "privileged" coordinate systems which are related by the Lorentz transformations on the basis of the following requirement: "we choose any inertial system as our rest body and *define* the to and fro velocities of light to be equal, or in any constant proportion to one another" [p. 239]. Suppose we do choose a local inertial system I_L falling freely toward the sun or alternatively an astronomical inertial system I very far removed from the sun. If we then choose a constant value for the ratio

$$\frac{1 - \epsilon}{\epsilon}$$

of the to-and-fro light velocities in one of these systems I_L or I, Schwarzschild's coordinate system K will surely be a member of the set of all other coordinate systems in which the same choice can also be made consistently. Yet K is not related to the primary system I_L or I by the Lorentz transformations. For as we saw, a Michelson-Morley experiment would generally not have a null outcome in K; by depending on the spatial orientation of the light path, the one-way velocity of light in K would generally not be c, and the light path would *not* be a *spatial* straight line (geodesic) in K. And these results are incompatible with K's being related to one of the I systems by the Lorentz transformations, since the latter leave the light velocity invariantly c. Yet when speaking of the privileged class of coordinate systems related by the Lorentz transformations, Putnam declares: "*As a matter of empirical fact* again, these 'privileged' coordinate systems are exactly the ones we get if we choose any inertial system as our rest body and *define* the to and fro velocities of light to be equal, or in any constant proportion to one another" [p. 239].

Although Putnam does not say so, let us suppose that his claim of the uniqueness of inertial systems with respect to the equality of the to-and-fro velocities of light is predicated on the restriction that we use the *proper* time furnished by standard physical clocks as a basis for the velocities. In that case, he would need to *prove* that under that restriction inertial systems are the *only* systems which are time orthogonal. In order to justify his uniqueness claim here, Putnam would need to furnish such a proof, because the results of the GTR in regard to the velocity of light are generally formulated on the basis of coordinate time rather than proper time. As Einstein explained in Section 3 of his 1916 paper on the GTR: "In the general theory of relativity, space and time cannot be defined in such a way that differences of the spatial co-ordinates can be directly measured by the unit measuring-rod, or differences in the time co-ordinate by a standard clock" [10, p. 117]. The reader will recall, from my §2.4 above, Einstein's illustration of that conclusion by reference to the rotating disk. But let us suppose that under the restriction to *proper* time, inertial systems are the only time-orthogonal frames and that Putnam's uniqueness claim is indeed predicated on proper time and hence true. Would this score against what I said concerning metrical simultaneity? Not at all! For I had *not* said that if $\epsilon = \frac{1}{2}$ is used in inertial systems, then in *arbitrary* coordinate systems employing *standard* clocks, *these* clocks can be synchronized so that the metrical simultaneity furnished by them also yields equal to-and-fro velocities of light in noninertial systems at all places and times. Indeed, I had claimed the *contrary*, when discussing light traveling from the center of a rotating disk to a point on the periphery in my 1962 essay [1962, p. 464; 1968, p. 72], and—as noted in §§2.4 and 2.7 above—Putnam showed full awareness of this fact on page 242 of his essay.

8.3. It will now become apparent that there is no merit at all in the objections raised by Putnam in the following statement of his:

According to Grünbaum, we are free to redefine simultaneity in any way whatsoever as long as we do not conflict with the principle that a cause is never said to be simultaneous with or later than its effect. In particular, then, Einstein is held to have been asserting even in the special theory of relativity that we are free to define the to and fro velocities of light to be equal even when the reference system is not an inertial system. This is contrary to the way in which *everyone*, including Einstein himself, has always interpreted the special theory of relativity. As is well known, the demand of the special theory of relativity is that our reference system *must* be an inertial system. We are free to *define* the to and fro velocities of light to be equal relative to any admissible reference system. We are also free to define the to and fro velocities to be unequal but in a constant proportion to one another: This is equivalent to changing the reference system. We are not free, and Einstein has never suggested that we are free, in the context of the *special* theory of relativity, to stipulate that the to and fro velocities of light are in a variable proportion to one another, or that the to and fro velocities of light are equal relative to a reference system which is not inertial. Only in 1915 did Einstein finally arrive at the Theory of General Relativity according to which the principle of equivalence holds and the distinction between inertial systems and physical systems in accelerated motion loses all meaning. I shall argue that the nature of the logical step from the special theory of relativity to the general theory of relativity is precisely what is obscured and not illuminated if one accepts Grünbaum's account [pp. 235–236].

My account of simultaneity above shows that this statement by Putnam contains the following errors and/or straw men:

1. I ignore as unworthy of documentary rebuttal the puerile straw man that I took the reference frames of the STR of 1905 to be the same as those of the GTR of 1916. But while rejecting this crude imputation, I must call attention to a cardinal fact which I noted above: metrical simultaneity and hence the ratio of the to-and-fro velocities of light have the same nontrivial conventional ingredient in the GTR as in the STR, because the GTR makes assumption (i) and the limiting assumption no less than does the STR. As we saw, it is the conjunction of these two assumptions which constitutes the basis for my account of the status of metrical simultaneity and thereby of the ratio of the to-and-fro velocities of light. This community of assumptions between the STR and GTR clearly goes well beyond the GTR's assertion of the infinitesimal validity of the STR.

Accordingly, what I do claim is the following: in virtue of the specified community of assumptions, Einstein's contention in Section 1 of his 1905 STR paper that the ratio of the to-and-fro velocities of light in an *inertial* system is conventional surely holds for the status of that velocity ratio in the reference systems of the GTR. And Putnam fails altogether to gainsay this community of assumptions by the following attempted *reductio ad absurdum:*

> . . . Einstein did *not* immediately (i.e., with the advent of the special theory of relativity) permit the selection of a body in accelerated motion with respect to the fixed stars as the "body of reference." If, as Grünbaum claims, Einstein's "insight" *from the beginning* was that we are "free" to define the to and fro velocities of light to be equal, even relative to a system in accelerated motion relative to the fixed stars, then an important part of general relativity was already included in the special theory of relativity. In fact, however, Einstein allowed the use of systems in accelerated motion relative to the fixed stars as "fixed bodies of reference" only *after* he became convinced that *as a matter of physical fact* no complication in the form of the laws of nature results thereby [in the sense that the laws of nature are generally covariant; p. 241].

Putnam renders my claim that the STR and GTR share assumption (i) and the limiting assumption as stating that "an important part of general relativity was already included in the special theory of relativity." But this kind of inclusion is altogether *compatible* with the fact that "Einstein allowed the use of systems in accelerated motion relative to the fixed stars as 'fixed bodies of reference' only *after* he became convinced that *as a matter of physical fact* no complication in the form of the laws of nature results thereby" (in the sense that the laws of nature are generally covariant). This compatibility obtains all the more, since Einstein admitted in 1918 the correctness of the following contention by Kretschmann [46, p. 168]: the principle of the general covariance of the laws of nature is a *heuristic regulative principle* of physical inquiry rather than a principle of which one can say with Putnam: "Here we have a purely physical, not a logical insight, namely, the insight that the laws of nature can be written in a simple covariant form, which is then the same no matter what system we choose as the 'fixed body of reference' " [p. 240]. The principle of general covariance as such does not hold, as Putnam has it, "*as a matter of physical fact*" [italics in original; p. 241]: that principle itself does not restrict the actual physical content of the laws of nature [16, p. 395], since "the laws of physics—whatever they might be—could in any case be expressed in invariant form" [46, p. 167].

It is this unsuccessful *reductio ad absurdum* argument, coupled with the exasperating fabrication concerning the quasi-Newtonian universe discussed in §8.0, which constitutes the sagging foundation for Putnam's gratuitous claim "that the nature of the logical step from the special theory of relativity to the general theory of relativity is precisely what is obscured and not illuminated if one accepts Grünbaum's account" [pp. 235–236]. And it is on the same sagging foundation that Putnam rests his patently false belief that I am committed to *deny* his claim that ". . . if the world had been discovered to be quasi-Newtonian, Einstein would certainly have insisted on the requirement of invariance of the laws of nature under the Galilean group of transformations as the appropriate requirement, and would not have stressed this 'freedom' to adopt noncustomary definitions of simultaneity which so impresses Grünbaum" [p. 241].

Putnam distorts my claim concerning the community of assumptions between the STR and GTR into his straw man by alleging that I held Einstein "to have been asserting even in the special theory that we are free to define the to and fro velocities of light to be equal even when the reference system is not an inertial system" [p. 235]. Putnam goes on immediately to say "This is contrary to the way in which *everyone*, including Einstein himself, has always interpreted the special theory of relativity. As is well known, the demand of the special theory of relativity is that our reference system *must* be an inertial system." When he says this he runs afoul of the established broader construal of the scope of the STR discussed in §2.8 and documented there (see footnote 7) from the authoritative writings of Peter Bergmann.

2. The ratio of the to-and-fro velocities of light is given by

$$\frac{1-\epsilon}{\epsilon},$$

and it becomes unity for the choice $\epsilon = \frac{1}{2}$. Putnam tells us that in the STR, we are free to choose that ratio to be either 1 or some other "constant proportion," but he remarks obscurely that the latter choice "to define the to and fro velocities to be unequal . . . is equivalent to changing the reference system" [p. 235]. This obscure remark is either false or at best misleading: the reasoning which justifies our being free to render the to-and-fro velocities equal via $\epsilon = \frac{1}{2}$ in a given inertial system I_0 likewise countenances our rendering these velocities unequal via $\epsilon \neq \frac{1}{2}$ *in the very same reference system* I_0; these alternative choices of ϵ will merely issue in different time coordinatizations for the *same* system I_0.

Moreover, disregarding *descriptive* convenience, the reasoning which justifies our choosing the ratio

$$\frac{1-\epsilon}{\epsilon}$$

of the to-and-fro velocities to be either 1 or a constant other than 1 likewise justifies a choice of ϵ—and hence of the velocity ratio—which varies with *spatial direction* and/or *time*. Thus, Reichenbach gives an illuminating example apropos of a centro-symmetric process of light propagation in the STR in which "We . . . let the factor ϵ be different from $\frac{1}{2}$ and make it dependent on the direction" [40, p. 162]. Hence I claim that—aside from considerations of descriptive convenience which are not at issue here—Putnam is mistaken in asserting that in the STR "We are not free . . . to stipulate that the to and fro velocities of light are in a variable proportion to one another."

Putnam unwittingly documents his own desultoriness in dealing with my views on the status of metrical simultaneity when he quotes and comments on a statement from my 1962 essay which explicitly discusses the following *logical step:* Einstein's "abandonment" of the "customary classical [i.e., Newtonian] definition of metrical simultaneity, which is based on the transport of clocks," in favor of "Einstein's *particular synchronization scheme* within the *wider* framework of the *alternative consistent sets of rules for metrical simultaneity* any one of which is allowed by the [*absolute*] nonuniqueness of topological simultaneity" [1962, pp. 423–424; 1968, pp. 24–25]. It was concerning the elucidation of the nature of *this logical step* that I had said in the statement which Putnam [pp. 234–235] quotes from me: "Precisely this elucidation is given, as we have seen, by the thesis of the conventionality of metrical simultaneity" [1962, p. 424; 1968, p. 25]. In short, in the statement quoted from me by Putnam, I was arguing that his attempt to trivialize the conventionality of metrical simultaneity (in a manner analogous to EAN) as merely a general matter of correspondence rules fails, because this conventionality justifies and elucidates the nature of the logical step involved in supplanting Newtonian simultaneity by the particular metrical simultaneity used in the STR. In particular, it elucidates and justifies that different Galilean frames can legitimately disagree on simultaneity, whereas in the Newtonian theory they must agree: in the Newtonian theory, metrical simultaneity is absolute as a matter of temporal fact, whereas in the STR only topological simultaneity is thus absolute while also being nonunique.

But to say that the conventionality of metrical simultaneity elucidates how the STR can countenance the relativity of simultaneity as between different Galilean frames is surely *not* to say that it does so blind to the remaining theoretical context in which choices of ϵ are to function! It is *not* to say that the particular numerical way in which the option to choose ϵ will be exercised will be oblivious to the following optical fact for example: in *any* inertial system the *round*-trip time $\overline{ABA}$ of light for any path AB, whose one-way length is r, is $2r/c$, where c is the usual light velocity, a fact which assures that the choice of $\epsilon = \frac{1}{2}$ in each such system yields the universal one-way light velocity c. But it *is* to say that the conventionality of metrical simultaneity is *essential* to the justification of the particular way in which Einstein repaired the defective Newtonian notion of simultaneity.

The basis for Putnam's objection to my statement concerning the logical step from Newtonian to STR *simultaneity* is yet another invention of his: I am alleged to have offered the conventionality of metrical simultaneity as fully accounting for the *much larger* logical step involved in the transition from the Galilean transformations to the full-blown physical content of the Lorentz transformations. Thus, Putnam writes:

Let us consider precisely what was "the nature of the logical step" (a) on Grünbaum's view, and (b) on my view. . . .

I take it that Einstein was not pointing out the possibility of using metrics and coordinate systems which lead to unnecessarily complicated physical laws. He realized not only that simultaneity has to be defined before we can talk of empirically ascertaining simultaneity at a distance; but he realized furthermore that *as a matter of empirical fact* there exists a class of privileged space-time coordinate systems, i.e., coordinate systems relative to which the laws of nature assume not just a simplest form but the same form for all; and further that the class of transformations from one of these coordinate systems to another is not the Galilean group of transformations but the group of Lorentz transformations [pp. 234, 238–239].

Putnam makes this allegation in the face of the fact that my STR essay for the 1960 *Philosophy of Science* anthology edited by Danto and Morgenbesser [23, pp. 412–418] had explicitly emphasized the following: the conventionality of simultaneity legitimates, though does not entail, the particular choice $\epsilon = \frac{1}{2}$ in each inertial system which is required for equal to-and-fro light velocities, *but* this conventionality is *merely one* of several independent logical ingredients of the STR principle of the constancy of the speed of light, and a fortiori, merely one of the ingredients of the Lorentz transformations.

As we recall from §6.1 and the end of §7, Putnam inveterately misrepresents me as maintaining that the conventionality of a metrical attribute and the options exercised on the strength of it are matters of mere semantics or logic instead of depending on the full range of the known relevant facts of the world. He again engages in the same misrepresentation in the case of metrical simultaneity, ignoring that I rested its conventionality on the conjunction of two anti-Newtonian physical postulates: assumption (i) and the limiting assumption. I had done so not only in those of my publications that are cited in §8.0. I had also done so in Putnam's presence as part of a symposium paper which I read at the December 1962 meeting of the American Philosophical Association and published in a 1962 issue of the *Journal of Philosophy*. Specifically in that paper, I spoke of the need for "awareness that Einstein's philosophical characterization of distant simultaneity as *conventional* rests on *physical assumptions*" and of the importance of "the recognition of the logical role played by these particular physical assumptions in the very foundations of the STR" [italics in original; 28, p. 563]. And then I went on to enumerate the two relevant anti-Newtonian physical assumptions [28, p. 564]. Putnam therefore misrepresents me as denying what I have been at pains to emphasize before him, when he writes: ". . . the realization that we are free to define the to and fro velocities of light to be equal (assuming that we are in an inertial system) was simultaneously a logical and a physical insight—not *merely* an insight into the logical possibility of using space-time metrics which complicate the form of physical laws" [p. 239].

Putnam makes points which I regard to be fundamentally in error when he proceeds from this misrepresentation as follows:

If we interpret Einstein as emphasizing merely the logical possibility of using nonstandard definitions of simultaneity, then we miss the whole significance of the relation between temporal order and causal order. In Reichenbach's and Grünbaum's writings this relation is made quite mysterious. It is asserted that in some sense spatio-temporal order (topological properties) *really is* causal order, and this view has led to a series of attempts to *translate* all statements about spatio-temporal order into statements about causal order. In fact, the relation between the two is quite simple. According to the special theory of relativity, we are free to use any *admissible* reference system we like; however, the admissible reference systems are singled out by the two conditions that (a) the laws of nature must assume a simplest form relative to them, and (b) they must assume in fact the same form relative to all of them. It is, of course, assumed that our customary coordinate system is to be one of the admissible ones. Now, one of the *laws of nature* in our customary reference system is that *a cause always precedes its effects.* (Einstein was assuming that gravitation would turn out to have a finite velocity of propagation.) Since this law holds in one admissible reference system, it must hold in all. Hence the admissible reference systems are restricted to be such that a cause always does, indeed, precede its effect (according to the time order defined by that reference system) . . . We also see that the requirement imposed by Grünbaum, that all the admissible metrics should agree on the topology (the topology is "intrinsic") makes sense only within the framework of a desire to single out a class of permissible reference systems which will lead to the same laws of nature. If we are willing to change our definitions in a way which involves a change of our laws of nature, then no justification whatsoever can be given for insisting that the principle must be retained that causal signals should not be discontinuous [pp. 239–240, 241].

This statement prompts the following concluding comments of mine on Putnam's views on simultaneity:

1. Consider an empirical regularity *not* expressed by the vanishing of a tensor but known to hold in "our customary coordinate system." Putnam assumes here that such a regularity can be identified as qualifying to be a *law* in the theory of relativity, without already knowing that the regularity in question is invariant with respect to a specified set of reference frames linked by specified kinds of coordinate transformations. Putnam assumes this, for he purports to identify the statement "*a cause always precedes its effects*" (about which more below!) to be "one of the *laws of nature* in our customary reference system." And this identification is his basis for then *inferring* the invariant validity of the statement in all permissible reference systems. But his own criteria (a) and (b) for singling out the class of admissible reference systems show that his reasoning

here is unsound. Specifically, it is circular to *establish on the strength of being a "law of nature" in our customary system* that a presumed regularity holding in our customary reference system is *invariantly valid* with respect to the permissible systems specified by Putnam's (a) and (b). For the specified type of invariance on the part of the regularity is required *ab initio* in order to confer the kind of law status on it which alone can validate Putnam's *deduction* of the invariance of the cause-effect sequence.

The procedure in the STR is to see whether a regularity that is presumed to hold in one inertial system is in fact Lorentz-covariant, and only then does the regularity become eligible for elevation to the status of a law in virtue of its Lorentz covariance. And the empirical content of the special principle of relativity is that *there are specified regularities* such as Maxwell's equations which are in fact Lorentz-covariant. Thus the theory's empirical quest is for those specific regularities which do conform to the principle of relativity and only then do they qualify as "laws."

If Putnam's deduction of the invariance of cause-effect sequences is not circular, it is simply a *non sequitur*. For consider the following two regularities of our customary reference system, both of which present themselves as *candidates* for law status within the STR: (1) The velocity of sound in air (under standard conditions of temperature and pressure) is approximately ⅓ of a km./second *in our customary reference system;* and (2) The velocity of light is 3×10^{10} cm/sec *in that same system.* Suppose that the requirement of Lorentz covariance is not imposed *at the outset* to disqualify (1) as a "*law*" in our customary frame. Then if we accept (1) as a "law" and infer invariance in the manner employed by Putnam apropos of cause-effect sequences, we would reach the *false* conclusion that *our* sound waves have the same numerical velocity in all inertial systems. Without the tacit presupposition of invariance, the type of argument employed here by Putnam is a *non sequitur,* since it leads to a false conclusion. By contrast, candidate (2) is not disqualified because it satisfies the necessary covariance condition for being a "law." Note how Einstein goes to some lengths in the founding 1905 STR paper to emphasize that (2) is indeed a law. He writes:

> Let us take a system of co-ordinates in which the equations of Newtonian mechanics hold good [i.e., to the first approximation]. In order to render our presentation more precise and to distinguish this system of co-ordinates verbally from others which will be introduced hereafter, we call it the "stationary system." . . .
>
> The following reflexions are based on the principle of relativity and on the principle of the constancy of the velocity of light. These two principles we define as follows:—
>
> 1. The laws by which the states of physical systems undergo change are not affected, whether these changes of state be referred to the one or the other of two systems of coordinates in uniform translatory motion.
>
> 2. Any ray of light moves in the "stationary" system of co-ordinates with the determined velocity *c*, whether the ray be emitted by a stationary or by a moving body. . . .
>
> We now have to prove that any ray of light, measured in the moving system, is propagated with the velocity *c*, if, as we have assumed, this is the case in the stationary system; for we have not as yet furnished the proof that the principle of constancy of the velocity of light is compatible with the principle of relativity [12, pp. 38, 41, 46].

Note that, to begin with, Einstein enunciates "the principle of the constancy of the velocity of light" only with respect to the one "stationary" system. And the *law* status of this principle is then made to rest on its Lorentz covariance. By the same token, even before Newton's law of universal gravitation was supplanted by the new equations of the GTR, students of relativity rejected it as a *law* on the grounds that the distance and simultaneity ingredient in it are not Lorentz-covariant.

Hence Putnam cannot certify a regularity prevailing in our customary reference system to be a law without rendering his inference of cause-effect sequence invariance either circular or a *non sequitur*.

2. Can Putnam claim that "*a cause always precedes its effects*" is

an *empirical* regularity within the context of the theory of relativity in any frame? In particular, does he have a criterion for the time separation of two events which is independent of their being causally connectible? And moreover, can he identify one of two causally connected events asymmetrically as "*the*" (partial or total) cause and the other as "the" effect independently of the fact that the former is the earlier of the two? My answer to all of these questions is No!

In Newton's theory, clock transport guaranteed absolute simultaneity between certain events, and the resulting synchronism of clocks also furnished relations of earlier and later among other events. Hence it was an *empirical* fact in Newton's theory that *some* pairs of causally connectible events were absolutely time-separated. And in view of its two-way gravitational action-at-a-distance, Newton's theory was able to assert the existence of *other* pairs of causally connectible events which are absolutely simultaneous. But as the extensive discussion in the literature has shown [26, Chapter 7 and pp. 259–260], even in the Newtonian theory, the asymmetric relation of "being *the* cause" and its asymmetric converse of "being *the* effect" are respectively characterized by means of being the temporally earlier and later members of a pair of causally connected events. And if the world contains irreversible processes, the lower and higher time numbers assigned by absolutely synchronized Newtonian clocks will have added physical significance [26, Chapter 8].

But in the context of the STR, the presumed facts of clock behavior under transport destroy the physical foundation on which the *empirical* character of the question "are *all* pairs of causally connectible events time-separated?" depends. And thus causal connectibility becomes *constitutive* of *absolute time separation*, while causal nonconnectibility becomes constitutive of invariant *topological* simultaneity. This constitutivity of absolute time order by causal connectibility and nonconnectibility is *presupposed* by the light-signal method of synchronizing clocks and hence by the time relations which ensue from the time numbers thus assigned by clocks. Upon being mindful of the limiting assumption governing causal connectibility, this conception makes the state of affairs depicted by the Minkowski diagram *instantly perspicuous.* On the other hand, if Bridgman's proposed synchronism by infinitely slow clock transport, mentioned in §8.1 above, is a genuine *physical* (i.e., *non*conventional) basis for metrical simultaneity in the *special* theory of relativity, then it becomes an *empirical* instead of a definitional truth in the STR that *all* causally connectible events are absolutely time-separated. And, in that case, Einstein's conception would indeed require emendation in the context of the *special* theory. Since Bridgman's method of synchronism presumably does not work, however, in arbitrary reference systems of the *general* theory of relativity, Einstein's conventionalist conception of metrical simultaneity is not impugned by Bridgman's method in the context of the *general* theory. But, in any case, here the issue between Putnam and me concerns the contents rather than the merits of Einstein's own conception.

So much for Putnam's claim that "In Reichenbach's and Grünbaum's writings this relation [between temporal order and causal order] is made quite mysterious" and that he has shown it to be "quite simple" instead [p. 239]. But, alas, I can concur with Putnam when he says: "the disagreement between Grünbaum and myself can be revealingly brought out by contrasting our interpretations of the Special Theory of Relativity" [p. 233].

3. If the nonquantum assumptions of the theory of relativity are correct, then continuity is indeed intrinsic (built in) to the space-time intervals of events, and the theory asserts that continuity to hold *invariantly, prior to* the introduction of any particular coordinate systems. Contrary to Putnam, this state of affairs shows that the regularities of nature which the theory asserts to hold in various reference frames are *bound to agree on the topology*, even if they did not otherwise have the same descriptive form. Couple this with my account in §2.15 of the status of the topology. Then it appears that claims of purported regularities which affirm a different topology and assert the discontinuity of causal chains either would be false

or—if true—would merely be employing the nomenclature differently on the strength of TSC.

8.4. As an application of the quantitative relevance of the conventionality of metrical simultaneity, I shall now discuss the quantitative dependence of the length of a moving rod *on the definition of simultaneity.*

The Lorentz transformations of the STR incorporate the choice of $\epsilon = 1/2$ by each of the relatively moving inertial systems which they relate. And thence results the relativity of simultaneity as between any two such frames S and S' for any pair of events which are spatially separated along the line of their relative motion. In virtue of this relativity of simultaneity, the Lorentz transformations yield a length contraction of a rod moving with respect to either of these frames and at rest in the other, provided that the rod is not perpendicular to the line of their relative motion.

More generally, we can now exhibit the quantitative dependence of the length of a moving rod on the particular simultaneity convention used in the form of a specific value of ϵ, a dependence which Reichenbach [40, pp. 156–157] discussed only *qualitatively.*

Consider a rod of rest length l in an inertial system S whose velocity is v relatively to an inertial system K if we have chosen $\epsilon = 1/2$ in K. We know that the length of the rod in K is $l\sqrt{1 - \beta^2}$, if we have chosen $\epsilon = 1/2$ and the rod is parallel to the direction of the relative motion of S and K.

To determine the dependence of the length L of the moving rod relatively to the system K after we countenance choices of *any* ϵ between 0 and 1, we consider an auxiliary inertial system K' which is *at rest* with respect to K but in which ϵ may have any value between 0 and 1, while we fix ϵ as $1/2$ in K for the time being.

Let the simultaneity projection of the rod onto K be the points A and B at $t = 0$ in K. Let the points A' and B' be the points of K' which are adjacent to A and B respectively. Let the A' clock read zero when its adjacent A clock reads zero. If we chose $\epsilon \neq 1/2$ in K'

to synchronize the various clocks with the master clock at A', then the clock at B' will *not* read zero when the B clock reads zero, since the difference in the simultaneity criteria between K and K' requires that the pair of events at AA' and BB', which occur simultaneously in K at $t = 0$, *cannot* also occur simultaneously in K'. Hence, the length of the S rod as measured in K' cannot be given by the distance between A' and B' in our accompanying diagram. Instead, it will be

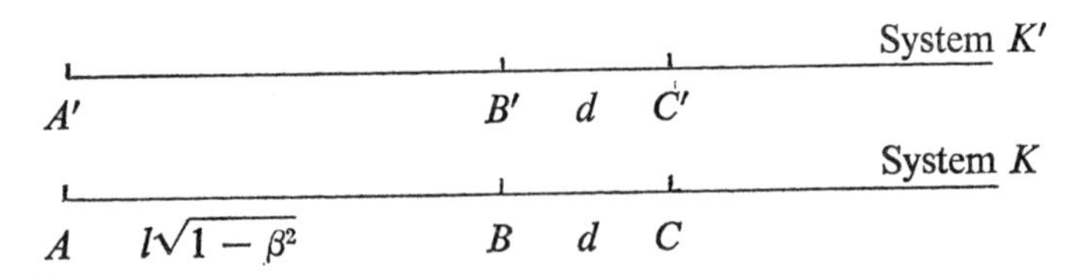

given by the projection $A'C'$ which is such that the K' clocks at A' and C' each read $t' = 0$. Let C' be at a distance d from B', which we chose to the left of C' in our accompanying diagram without loss of generality. The right end point of the rod will therefore move through the distance d to yield its K' simultaneity projection C', as contrasted with yielding its K simultaneity projection B. If C is the point in K which is adjacent to C', then we know the following from the §8.1 formula $t_2 = t_1 + \epsilon(t_3 - t_1)$: the event for which the C clock in K reads $t = 0$ is one for which the clock at C' reads $t' = -(\frac{1}{2} - \epsilon)T' = (\epsilon - \frac{1}{2})T'$, where T' is the round-trip time of light for the closed path $A'C'A'$. That time T' is the quotient of the round-trip distance and the round-trip velocity c:

$$\therefore T' = \frac{2[l\sqrt{1 - \beta^2} + d]}{c}.$$

And the event for which the B clock in K reads $t = 0$ is one for which the clock at B' reads $t' = -(\frac{1}{2} - \epsilon)\tau$, where τ is the round-trip time of light for the closed path $A'B'A'$, i.e.,

$$\tau = \frac{2l\sqrt{1 - \beta^2}}{c}.$$

These various clock readings are shown on the following diagram.

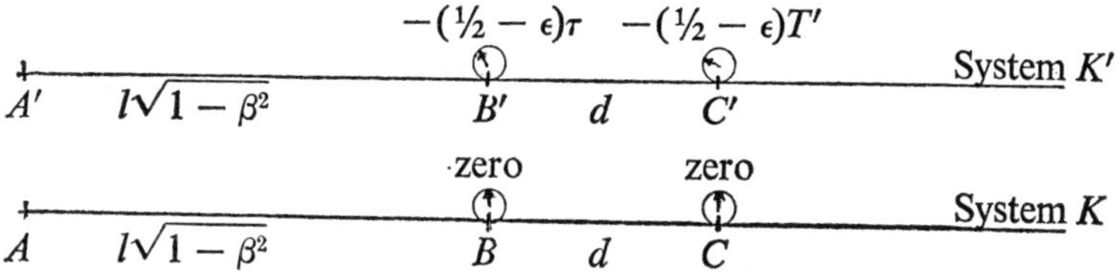

$-(\frac{1}{2}-\epsilon)\tau$ $-(\frac{1}{2}-\epsilon)T'$

System K'

A' $l\sqrt{1-\beta^2}$ B' d C'

zero zero

System K

A $l\sqrt{1-\beta^2}$ B d C

Our problem is to find the magnitude of the segment $A'C'$, which we'll call l_ϵ. We know that $l_\epsilon = l\sqrt{1-\beta^2} + d$, and hence we must determine the value of d. But d is the distance traversed by the right end of the rod when it travels through $B'C' = BC$ with the velocity v in K and arrives at C' when the C' clock reads $t' = 0$. Since the time *increments* on adjacent relatively stationary clocks are the same, the C clock will read a time $(\frac{1}{2}-\epsilon)T'$ when the contiguous C' clock reads 0 on the arrival of the right end of the rod. Hence the rod has traversed the distance $BC = d$ in K by traveling with the velocity v for a transit time $(\frac{1}{2}-\epsilon)T'$.

Therefore we can write

$$d = (\tfrac{1}{2}-\epsilon)\,v\,T' = 2\beta(\tfrac{1}{2}-\epsilon)[l\sqrt{1-\beta^2}+d].$$

Eliminating d from the right-hand side of the equation, we have

$$d = \frac{l\sqrt{1-\beta^2}\,(\frac{1}{2}-\epsilon)}{\dfrac{c}{2v} - (\frac{1}{2}-\epsilon)}.$$

Hence the total simultaneity projection l_ϵ onto K' is

$$l_\epsilon = l\sqrt{1-\beta^2}\left[1 + \frac{\frac{1}{2}-\epsilon}{\dfrac{c}{2v} - (\frac{1}{2}-\epsilon)}\right]$$

or

$$\boxed{l_\epsilon = l\sqrt{1-\beta^2}\cdot\frac{c}{c+v(2\epsilon-1)}.}$$

When $\epsilon = \frac{1}{2}$, this becomes $l\sqrt{1-\beta^2}$, as required by the Lorentz transformations. But when $\epsilon < \frac{1}{2}$, the fraction on the right-hand side becomes an *improper* fraction, thereby yielding a *larger* l_ϵ than for the case of $\epsilon = 1/2$, as shown in the diagram. And when $\epsilon > 1/2$, the right-hand fraction becomes a *proper* fraction, thereby yielding a *smaller* l_ϵ than in the case of $\epsilon = 1/2$, as we should expect for this case, since then the B' clock would be *ahead* of the B clock, so that the right end of the rod would have passed a K' clock reading $t' = 0$ which is located to the left of B'.

I wish to note, however, that in the context of Bridgman's method of synchronism by infinitely slow clock transport, the value $\epsilon = 1/2$ is uniquely singled out on *physical* grounds in each inertial system of the STR if the claims made by him are sound.

9. The Need for Universal Forces in the Case of Nonstandard Metrics: Refutation of Putnam's "Theorem"

9.0. As I indicated at the end of §6.0, Putnam assumes throughout the body of his paper the correctness of a very strong contention of his, which he claims to have proven as a theorem in the Appendix [pp. 247–255]. This contention of Putnam's denies a thesis of Reichenbach's that Einstein had espoused earlier in his 1921 lecture "Geometry and Experience" [11].

In the present section, I shall endeavor to show that Putnam's purported theorem is not only a *non sequitur* but false. And hence this demonstration will exhibit the unfoundedness of those of his claims which rest on his supposed theorem. A digest of the prior discussion of the issue in the literature will provide a useful basis for my subsequent critique of Putnam's theorem.

9.1. In his "Geometry and Experience" Einstein had written as follows:

Why is the equivalence of the practically-rigid body and the body of geometry—which suggests itself so readily—denied by Poincaré and other investigators? Simply because under closer inspection the real solid bodies in nature are not rigid, because their geometrical behaviour, that is, their possibilities of relative disposition, depend upon temperature, external forces, etc. Thus the original, immediate relation between geometry and physical reality appears destroyed, and we feel impelled toward the following more general view, which characterizes Poincaré's standpoint. Geometry (*G*) predicates nothing about the relations of real things, but only geometry together with the purport (*P*) of physical laws can do so. Using symbols, we may say that only the sum of (*G*) + (*P*) is subject to the control of experience. Thus (*G*) may be chosen arbitrarily, and also parts of (*P*); all these laws are conventions. All that is necessary to avoid contradictions is to choose the remainder of (*P*) so that (*G*) and the whole of (*P*) are together in accord with experience.

. . . The idea of the measuring-rod and the idea of the clock co-ordinated with it in the theory of relativity do not find their exact correspondence in the real world. It is also clear that the solid body and the clock do not in the conceptual edifice of physics play the part of irreducible elements, but that of composite structures, which may not play any independent part in theoretical physics. But it is my conviction that in the present stage of development of theoretical physics these ideas must still be employed as independent ideas; for we are still far from possessing such certain knowledge of theoretical principles as to be able to give exact theoretical constructions of solid bodies and clocks.

Further, as to the objection that there are no really rigid bodies in nature, and that therefore the properties predicated of rigid bodies do not apply to physical reality,—this objection is by no means so radical as might appear from a hasty examination. For it is not a difficult task to determine the physical state of a measuring-rod so accurately that its behaviour relatively to other measuring-bodies shall be sufficiently free from ambiguity to allow it to be substituted for the "rigid" body. It is to measuring-bodies of this kind that statements as to rigid bodies must be referred [11, pp. 191, 192].

After next giving the statement of Riemann's concordance assumption which I quoted in §6.0, Einstein continues:

It is true that this proposed physical interpretation of geometry breaks down when applied immediately to spaces of sub-molecular order of magnitude. But nevertheless, even in questions as to the constitution of elementary particles, it retains part of its importance. . . .

It appears less problematical to extend the ideas of practical geometry to spaces of cosmic order of magnitude. It might, of course, be objected that a construction composed of solid rods departs more and more from ideal rigidity in proportion as its spatial extent becomes greater. But it will hardly be possible, I think, to assign fundamental significance to this objection. Therefore the question whether the universe is spatially finite or not seems to me decidedly a pregnant question in the sense of practical geometry [11, p. 193].

As I have documented from the writings of Poincaré [1962, §6(i), pp. 493–505 and especially pp. 505 and 508; 1968, §6(i), pp. 106–120 and especially pp. 120 and 124], Einstein has here mistakenly associated with Poincaré a position which, it would appear, is more properly attributable to Duhem. Let us note the detailed criticism which Einstein offers here of that Duhemian position. In the light of his criticism of that position, Einstein can be presumed to have been merely an *advocatus diaboli* for it in 1949 when he effectively asked how Reichenbach would rebut it. Hence I was probably mistaken previously when I represented Einstein as himself *espousing* in 1949 the Duhemian position that he had erroneously ascribed to Poincaré as I did in §7(i) of the 1962 essay [1968, p. 125].

Let us note the if-clause in the following previously cited statement made by Einstein in 1916 (see my §2.11 above): "Thus Euclidean geometry does not hold even to a first approximation in the gravitational field, if we wish to take one and the same rod, independently of its place and orientation, as a realization of the same interval." And let us couple the point of Einstein's if-clause with those of his views which I cited from his 1921 lecture. We can then see the following: The substance of Reichenbach's account of the bearing of his "differential" and "universal" forces on the spatial geometry is an elaboration of Einstein's conception of physical geometry as furnished by "the practically rigid body."

By "differential" forces, Reichenbach understands thermal and other influences on solid rods which distort them in the sense of perturbing their otherwise concordant coincidence behavior as follows: the presence of differential forces makes the coincidence behavior of (transported) rods dependent on the chemical composition of these rods. The standard physics therefore corrects for these differential deformations on the basis of its correctional laws for thermal elongation, etc. For the sake of brevity, I shall speak of solid rods as "*DP*-corrected" if their lengths are adjusted for *differential* deformations (*D*) on the basis of the correctional laws of the physics *P*. The *DP*-corrected Paris rod is effectively free from differential forces. If *P* is the standard correctional physics, then the customary congruence convention can be given by stipulating the self-congruence of the *DP*-corrected Paris meter bar under transport. And, according to Reichenbach, it is the congruence standard furnished by such a differentially unperturbed rod (or by its equivalent) which constitutes the basis for the spatial geometry of the standard physics as follows: The *DP*-corrected coincidences of the rod (or their equivalent) yield a set of g_{ik} to within the usual inductive uncertainties inherent in the finitude of the data vis-à-vis the infinitude of space points.

Reichenbach goes on to point out, however, that the conventionality of congruence entitles us to specify *alternatively* that the *DP*-corrected Paris meter rod is *not* self-congruent but has a variable length in meters which is a specified function of the *independent* variables of spatial position, orientation, or time. Thus consider that a system of rectangular coordinates x, y has been assigned on a table top by means of a *DP*-corrected meter rod, but *without* commitment as to any *metrical* significance of the coordinate increments dx and dy. We are free to adopt the customary congruence convention according to which the length ds of the *DP*-corrected Paris bar is everywhere and always 1 meter. And if we adopt that customary congruence convention, then the objective coincidence behavior of the *DP*-corrected rod on the table top will issue in the standard metric $ds^2 = dx^2 + dy^2$. Suppose, however, that we adopt the alternative noncustomary stipulation that the length ds of the *DP*-corrected Paris rod is $1/y$ meters. In that case, the hyperbolic metric

$$ds^2 = \frac{dx^2 + dy^2}{y^2}$$

will result from the very coincidences of the *DP*-corrected rod with the table top which had yielded the Euclidean metric $ds^2 = dx^2 + dy^2$ in conjunction with the customary congruence stipulation. And the alternative stipulation that the *DP*-corrected Paris rod has the length $1/y$ meters will have specified the congruences among the space intervals in the following sense: Given the respective coordinate increments dx and dy for each of the intervals, this noncustomary convention stipulates the congruence of those intervals which have equal lengths ds according to the new metric

$$ds^2 = \frac{dx^2 + dy^2}{y^2}.$$

In this way, an alternative congruence convention issues in a suitably different metric tensor g'_{ik} and, in this case, in a new geometry as explained in §2.14.

Having defined universal forces as all-permeating forces *which affect all kinds of solids in the same way*, Reichenbach [40, p. 13] chose to invoke such forces metaphorically to characterize a *non*-customary congruence definition [26, Chapter 3, Section A]. Thus, in the case of the stipulation that the *DP*-corrected Paris rods are not self-congruent under transport but have the length $ds = 1/y$ in meters, the rods are said to have sustained the same "deformations" under the action of universal forces. On the other hand, in Reichenbach's metaphorical parlance, the use of the customary congruence convention is said to stipulate that the *DP*-corrected measuring rods are free from deforming universal forces. Thus, upon using Reichenbach's metaphorical terminology, we can render the point made by Einstein in his 1916 sentence above as follows: if we wish to stipulate that our *DP*-corrected rods are free from deforming universal forces (i.e., they are stipulated to "realize the same interval," which means that they are conventionally self-congruent), then they yield a non-Euclidean spatial metric tensor in his context of the GTR; by the same token, the *DP*-corrected rods could be claimed to yield a *Euclidean* metric tensor only if they were stipulated to be deformed by universal forces. And precisely this is indeed the position espoused by Reichenbach. That Reichenbach employs the term "universal forces" purely metaphorically *in the context of the congruence specification* is evident from the following statement by him of the role of these forces in that context: "The assumption of such forces means merely a change in the coordinative definition of congruence" [41, p. 133].

Having followed Einstein in mistakenly attributing the Duhemian thesis to Poincaré, Reichenbach addresses himself to the following claim made by that thesis: all actual and potential empirical findings provide sufficient latitude to replace the standard correctional physics P by a new one Φ which assures that (1) the coincidences of an *everywhere self-congruent* $D\Phi$*-corrected rod* will *always* yield an arbitrarily pre-selected metric tensor g'_{ik} and (2) all the Φ-physics deformations are *differential* in Reichenbach's sense. In response to Einstein's question on how he would deal with this claim [14, p. 677], Reichenbach comments on the special case of stellar parallax measurements which are required to be interpreted as yielding a pre-selected *Euclidean* g'_{ik}, saying: "But Poincaré overlooked the fact that such a requirement might compel the physicist to assume universal forces" [41, p. 135]. Thus Reichenbach denies the Duhemian thesis here. For he denies that there is an empirically tenable Φ such that it will *always* be possible to obtain the required kind of g'_{ik} by employing a congruence stipulation according to which the $D\Phi$-corrected rod is *self-congruent* under transport in space and time, while also maintaining that the Φ-physics deformations are always differential in his sense.

Reichenbach stated his denial without any proof. Believing this denial to be correct, I offered a proof of it in my 1962 essay [1962, pp. 514–517; 1968, pp. 132–135]. This proof was in the form of a thermal *counterexample* to the Duhemian thesis. As we shall see below, Putnam's theorem *asserts* the Duhemian thesis in a special way, and Putnam claims to have furnished an actual physico-geometric *proof* of it. It is therefore quite puzzling that Putnam did not deal at all with my thermal counterexample or even mention it; instead, he merely invoked incompatibility with his alleged theorem to dismiss [p. 209n] my suggested successive approximation procedure [1962, pp. 519–521; 1968, pp. 137–140] for determining the spatial metric tensor by means of a successively *DP*-corrected, everywhere self-congruent rod. But the validity of my thermal counterexample does *not* depend at all on the capabilities which I claimed for my successive approximation procedure. And since Putnam's theorem will turn out to be false, this theorem cannot serve to discredit my claim that the successive approximation procedure can yield only one geometry of constant curvature. But, as I explained in §2(ii) of Chapter II, this latter claim of mine does have to be weakened somewhat for reasons given in a valuable criticism by Arthur Fine [15].

It is unfortunate that Putnam felt that he could disregard my thermal counterexample. For I shall show that it can be *amended*, and then extended to the case of the totality of differential forces so as to exhibit the falsity of Putnam's theorem. Let me therefore first recapitulate the substance of my thermal counterexample while articulating further those of its features which bear on its subsequent extension. Suppose that the Duhemian has arbitrarily selected a specified strongly non-Euclidean metric tensor g'_{ik} and that he proposes to guarantee the existence of a correctional physics Φ, which must include both laws *and boundary conditions*, such that both of the following are the case in such local regions as our table top: (1) the conjunction of the arbitrarily selected metric tensor g'_{ik} (and hence of its associated strongly non-Euclidean geometry) with the required Φ furnishes a description of the coincidences of the rod which is equivalent to a description based on the conjunction of the Euclidean metric tensor g_{ik} with the correctional laws and boundary conditions P of the standard geometry and physics applicable to such local regions, (2) the nonstandard, non-Euclidean metric tensor g'_{ik} results from an everywhere *self-congruent* $D\Phi$-corrected rod, just as the standard Euclidean metric tensor g_{ik} results from an everywhere self-congruent DP-corrected rod, and the Φ-physics deformations are always differential in the sense of Reichenbach. To assess the feasibility of the required kind of Φ, I invite consideration of the simple idealized case in which our table top is subject only to thermal perturbations for some time and no other differential forces exist. Then we must inquire into the tenability of the conjunction Φ of a revised law of (linear) thermal expansion with thermal boundary conditions such that (i) the $D\Phi$-corrected *self-congruent* rod would yield the non-Euclidean metric tensor g'_{ik}, and (ii) the Φ-cum-g'_{ik} description of the rod coincidences would be equivalent to the P-cum-g_{ik} description.

More specifically, if the table top is covered by a network of rectangular coordinates, let g'_{ik} be the metric tensor of the hyperbolic metric

$$ds^2 = \frac{dx^2 + dy^2}{y^2}.$$

And assume that at some time or other the points on the line $y = 1$ as well as the points on the line $y = 2$ are each at the standard temperature T_0 of the usual physics at which the Paris meter bar is kept or at some other *identical* temperature T_1 within a range ΔT such that the following correctional law of the standard physics P is an excellent first approximation: $L = L_0(1 + \alpha \cdot \Delta T)$, where L_0 is the length of the body at the standard temperature, ΔT is the deviation from the standard temperature T_0, α is a coefficient depending on the *chemical composition* of the rod, and L the length at the temperature $T_0 + \Delta T$.[14] Let us use the Paris rod, so that L_0 is 1 meter in our example. Consider two positions of the rod: in position 1, it lies wholly on the line $y = 1$, and in position 2, it lies wholly on the line $y = 2$. In either position, the rectangular coordinates of the

[14] The argument which is about to be given by reference to this approximate form of the law can be readily generalized to forms of the law which allow for the temperature-dependence of the *rate* of thermal expansion and hence involve more than one coefficient of thermal expansion. This fact will be made evident later in this subsection.

intervals with which the rod will coincide differ only with respect to the x coordinate, dy being zero.

What then will be the coincidence behavior of the Paris rod along $y = 1$ and $y = 2$ respectively according to the assumedly true g_{ik}-cum-P description in which $ds^2 = dx^2 + dy^2$? And what must be the character of a nonstandard thermal physics Φ which is to function in an equivalent description in which the $D\Phi$-corrected self-congruent Paris rod yields the hyperbolic tensor g'_{ik}? According to the g_{ik}-cum-P description, on both $y = 1$ and $y = 2$ the rod would coincide with intervals for which $dx = 1$, if T were equal to T_0 in each of these places, all nonthermal differential forces being assumed to vanish. But if the temperature T of the rod has the same value T_1 other than T_0 in each of the two places, the rod will coincide instead with intervals for which $dx = 1 + \epsilon$ on both $y = 1$ and $y = 2$, where $\epsilon = \alpha(T_1 - T_0)$. It is here assumed, of course, that it is only the rod itself which is now at the higher temperature $T_1 > T_0$ at $y = 1$ and $y = 2$. The table top itself—which furnished the points to which the coordinate labels x, y are attached—is here assumed to be at the standard temperature, being presumed to be thermally insulated from the rod at the initial instant of their coincidence.

How must the contemplated thermal physics Φ differ from the standard one P above, if these very same coincidences on both $y = 1$ and $y = 2$ are to be consonant with the hyperbolic metric

$$ds'^2 = \frac{dx^2 + dy^2}{y^2}?$$

Φ-cum-g'_{ik} uses its own temperature scale T'. As equation (2) below will show, the new temperature T' is related by some transformation $T' = f(T, x^i)$ to the T scale of the P physics, where x^i represents the space coordinates. That the values of temperature assigned by the T' scale will depend not only on the values of the temperature T of the P physics but also on the space coordinates should not be surprising. For the T scale is associated with the Euclidean space metric g_{ik}, while the T' scale is associated with the hyperbolic space metric g'_{ik}: a thermometric mercury column which has the same length at $y = 1$ and $y = 2$ in the metric g_{ik} will *not* have the same length at these two places in the metric g'_{ik}. And while a Paris rod of given α which is at the *same* temperature T other than T_0 on $y = 1$ and then on $y = 2$ sustains the *same* g_{ik}-elongation at these two places, it cannot have the same g'_{ik} length ds' at the two places and hence will turn out not to be at the same temperature on the T' scale. If the coincidence behavior of the self-congruent unit Paris rod conforms to the *hyperbolic* metric in the case of $\Delta T' = 0$, then only intervals for which $\sqrt{dx^2 + dy^2} = y$ can have the unit length $ds' = 1$. Hence Φ-cum-g'_{ik} asserts that if the *unit Paris meter rod* is thermally (and otherwise differentially) undeformed, i.e., if the rod is at the standard temperature T'_0 of the T' scale, that rod coincides with $dx = 1$ on $y = 1$. And Φ-cum-g'_{ik} asserts further that under the condition $\Delta T' = 0$ of vanishing differential forces, the same unit rod would coincide with an interval $dx = y$ on any line $y =$ constant, where $dy = 0$. Thus, when $\Delta T' = 0$, the rod would coincide with $dx = 10$ on $y = 10$ and with $dx = 100$ on $y = 100$.

It is clear that any interval of coordinate increments dx and dy to which the Euclidean metric assigns the length $ds = \sqrt{dx^2 + dy^2}$ will be assigned the generally different length

$$ds' = \frac{\sqrt{dx^2 + dy^2}}{y}$$

by the hyperbolic metric. A Paris rod lying on a line $y =$ constant which is subject to a temperature difference ΔT and has the length $ds = L = 1 + \alpha\,\Delta T$ in P-cum-g_{ik} would therefore have the generally different length

$$ds' = L' = \frac{1 + \alpha\,\Delta T}{y}$$

in Φ-cum-g'_{ik}. But in the latter description, L' will also be given by $L' = 1 + \alpha'\,\Delta T'$, where α' and $\Delta T'$ will be the Φ-cum-g'_{ik} counterparts of α and ΔT. Equating the two expressions for L', we have the following expression for the deformation $L' - 1$ sustained by the Paris rod:

$$(1) \qquad L' - 1 = \alpha' \, \Delta T' = \alpha'(T' - T_0') = \frac{1}{y} + \frac{\alpha \, \Delta T}{y} - 1.$$

This equation yields the following *crucial results:*

1. Consider space points which happen to be at the standard temperature T_0 of the P physics, so that $\Delta T = 0$. In that case, the rod has length $ds = 1$ in P-cum-g_{ik} but length $L' = 1/y$ in Φ-cum-g'_{ik}. At *any* one such space point (x, y) other than those on $y = 1$ (or on $y = 0$), all Paris rods—*whatever their chemical composition!*—suffer the *same deformation*

$$\alpha' \, \Delta T' = \frac{1}{y} - 1$$

under the influence of the deviation $\Delta T'$ of the temperature T' at the given point from the standard T'-scale temperature T_0'. *But this shows that in the case of $\Delta T = 0$ the force on the rod arising from the Φ-physics "thermal" deviation $\Delta T'$ is not a differential force!*

Clearly, the same conclusion would follow, if one were to equate the deformation $L' - 1$ not to the single term $\alpha' \Delta T'$ but to a sum of a series of terms

$$\alpha' \Delta T' + \beta'(\Delta T')^2 + \gamma'(\Delta T')^3 + \ldots$$

Let us solve (1) for the Φ-physics temperature T' while recalling that $\Delta T = T - T_0$. We then obtain

$$(2) \qquad T' = T_0' + \frac{1}{\alpha' y} + \frac{\alpha}{\alpha' y} T - \frac{\alpha}{\alpha' y} T_0 - \frac{1}{\alpha'}.$$

And this equation shows that T' is so defined as to be a function of both the P-physics temperature T *and the space coordinate* y. Indeed equation (2) shows that the Φ-physics temperature T' will have the *standard* value T_0' at any space point y at which the P-physics temperature has the value

$$(3) \qquad T = \frac{y - 1}{\alpha} + T_0.$$

For the P physics tells us that it is precisely at space points y at which this temperature T prevails that the Paris rod will coincide on a line y = constant with an interval to which the metric

$$ds' = \sqrt{\frac{dx^2 + dy^2}{y^2}}$$

assigns the length unity. Since condition (3) reduces to $T = T_0$ only at $y = 1$, we see that only at $y = 1$ can the Φ-cum-g'_{ik} and P-cum-g_{ik} descriptions *agree* that the Paris rod is at the standard temperature, which is T'_0 and T_0 respectively. For it is only at $y = 1$ that *both* metrics can agree that the length of the Paris rod is unity.

Incidentally, (3) shows that the T-scale temperatures that would be needed, on lines where y is large, to have the rod coincide there with intervals that have g'_{ik}-length $ds' = 1$ may not be compatible with the survival of the rod as a *solid* body: the maintenance of the standard Φ-physics temperature T'_0 would melt (or even volatilize) the rod.

Furthermore, consider the points on *any one* given line y_k *other than* $y = 1$. Equation (3) tells us that two chemically different rods, whose respective P-physics coefficients of thermal elongation have the different values α_1 and α_2 would *both* be at the standard Φ-physics temperature T'_0 while being at the *different* T-scale temperatures

$$T_{\alpha_1} = \frac{y_k - 1}{\alpha_1} + T_0$$

and

$$T_{\alpha_2} = \frac{y_k - 1}{\alpha_2} + T_0.$$

If the Paris rod is at the standard P-physics temperature T_0 at any space point *other than* on $y = 1$, condition (3) tells us that it will *not* be at the standard Φ-physics temperature T'_0 there. And at all such points, the rod will have been "thermally" deformed so as to have the length

$$ds' = \frac{1}{y} \neq 1$$

independently of its chemical composition. Hence we see again that the Φ-cum-g'_{ik} description pays the following price for attempting to base its thermal physics Φ on a $D\Phi$-corrected self-congruent Paris rod which is held to yield the hyperbolic metric tensor g'_{ik} on the table top: *this description prevents its "thermal agencies" from qualifying as differential forces.* This conclusion is not, of course, gainsaid by the following fact: at a point at which $T \neq T_0$, so that $\Delta T \neq 0$, the deformation $\alpha' \, \Delta T'$ sustained by the rod does depend not only on its location y but also on its chemical composition. For in that case there is a nonvanishing dependence on α.

2. Consider a Paris rod *on* $y = 1$ which is at the *standard* temperature *there* according to *both* physical descriptions and hence qualifies there as being of length unity in *both* descriptions. Now consider *what happens to the rod as it is moved* to, say, $y = 2$ and brought into coincidence with that line there. The account of what happens given by Φ-cum-g'_{ik} is indeed *equivalent* to the one given by P-cum-g_{ik}.

For suppose that the Φ-cum-g'_{ik} description would say that $\Delta T' = 0$ as between $y = 1$ and $y = 2$. In that case, the standard temperature T'_0 prevails at both $y = 1$ and $y = 2$. Having been free from T'-scale thermal forces, the self-congruent unit rod coincided at $y = 2$ with an interval dx which satisfies the equation $ds' = 1 = dx/2$, i.e., with an interval $dx = 2$. But we know from equation (3) that the rod's still being at the standard Φ-physics temperature T'_0 at $y = 2$—which assures its coincidence there with an interval $dx = 2$—*is the same state* as its having acquired the higher nonstandard P-physics temperature

$$T = \frac{1}{\alpha} + T_0.$$

And what holds for the initial and final states at $y = 1$ and $y = 2$ is also true for the intervening states in transit. In other words, to be subjected to the P-physics temperature increase from the standard temperature T_0 at $y = 1$ to

$$T = \frac{1}{\alpha} + T_0$$

at $y = 2$ is *to be implicated in the same set of events* as to preserve the standard Φ-physics temperature T_0'.

On the other hand, suppose that Φ-cum-g_{ik}' were to assert that the rod's temperature dropped from T_0' at $y = 1$ to

$$T' = T_0' - \frac{1}{2\alpha'}$$

at $y = 2$. In that case

$$\Delta T' = -\frac{1}{2\alpha'},$$

and the rod was subjected to deformation instead of being self-congruent under displacement from $y = 1$ to $y = 2$. Since equation (1) tells us that for this case the rod will have a reduced length $L' = 1 + \alpha' \Delta T' = 1 - \frac{1}{2} = \frac{1}{2}$, it must coincide with an interval dx which satisfies the equation $ds' = \frac{1}{2} = dx/y$. At $y = 2$, the required interval is $dx = 1$. But, as confirmed by equation (1), coincidence with an interval $dx = 1$ at $y = 2$ at the specified *reduced* T' temperature $T' < T_0'$ is the same state as having preserved the standard T-scale temperature T_0 without any change in the g_{ik} length $ds = dx$.

However, this equivalence of the two descriptions with respect to what happens to the rod as it is moved in the region above $y = 0$ does not remedy in the least the failure of the "thermal" forces of the Φ physics to be differential.

3. It would be futile to try to assure the differential character of the thermal forces of the Φ-cum-g_{ik}' description by the stratagem of using the same temperature scale T as the P physics while introducing a space-dependent $\alpha' = \alpha \cdot F(x^i)$ in the equation

(4) $\quad L' = 1 + \alpha' \cdot \Delta T.$

For while this might work for $\Delta T \neq 0$, it would fail when $\Delta T = 0$. In the latter case, (4) would have the effect of requiring the rod to have the g_{ik}' length $ds' = 1$ at every point y under the same standard thermal conditions under which it has the g_{ik} length $ds = 1$. But it is logically impossible that the coincidence behavior of a unit rod, which is thermally at the standard temperature T_0, conform to both the Euclidean and hyperbolic metrics at points other than $y = 1$. To remove the contradiction and obtain the g'_{ik} behavior in the context of the temperature scale T, it would be necessary to modify (4) as follows:

$$(5) \qquad L' = F(x^i)[1 + \alpha \cdot \Delta T],$$

where $F(x^i)$ is a function of the space coordinates x^i, which is given by $F(x^i) = 1/y$ in the case of our particular hyperbolic metric.

But, according to (5), the length of the thermally *undeformed* rod is *not* everywhere unity; instead, the length of that differentially undeformed rod varies in the *same* way with the independent variable of spatial position *regardless of its chemical composition.* And such a revision of the thermal physics P thereby repudiates the Duhemian avowal that there is an equivalent description Φ-cum-g'_{ik} in which the *differentially* undeformed Paris rod everywhere has the same length unity, i.e., is everywhere self-congruent, while yielding the hyperbolic metric tensor.

As we shall soon see in §9.2 below, Putnam maintains that the Duhemian can in fact be rescued here by invoking differential gravitational and electromagnetic forces. More specifically, Putnam claims to have *proven* that for an *arbitrary* g'_{ik}, the deformations associated with $F(x^i)$ are *always* attributable to differential forces of a gravitational, electromagnetic, or interactional kind in a new physics Φ which has the following property: all of the Φ-physics deformations are *differential,* and if the Paris rod is $D\Phi$-corrected for these differential forces only, it will yield the arbitrary metric g'_{ik} while being assigned the length $ds' = 1$ everywhere, i.e., when free from Reichenbach's metaphorical deforming *universal* forces. And I shall refute that claim by extending the thermal argument of this subsection.

4. We must guard against the attempt to save the differential character of the thermal forces needed for Φ-cum-g'_{ik} by a device of the following kind: the introduction of what is a new temperature scale T' *in name only* which has the effect of *legislating away* the

thermal *boundary condition* $\Delta T = 0$ of the P physics in favor of a suitably different one $\Delta T \neq 0$ of the T scale as required to yield the metric tensor g'_{ik} from the deformational law (4) above. Clearly this is an inadmissible *deus ex machina*, since temperature fluctuations such that the standard temperature T_0 prevails at various space points at some time or other cannot be ruled out a priori by fiat. The Duhemian cannot simply manufacture the thermal *sources* so as to obtain the thermal boundary conditions relevant to the rod's coincidence behavior to suit the requirements of his thesis.

It would seem that at least under the oversimplified condition in which thermal forces are the only differential forces, Reichenbach's denial of the Duhemian thesis is correct.

9.2. Putnam writes: "It is difficult to show the full nature and extent of Reichenbach's error without going into details on the question of the transformations between metrics" [p. 250]. It is indeed unfortunate that Putnam felt satisfied to assess Reichenbach's claim without a calculation of the kind carried out in our preceding subsection.

Let me now quote the relevant portions of the fuller text in which this statement of Putnam's occurred. Putnam writes:

> Let g_{ik} be the customary metric tensor and let P be the customary physics. The combination of the two I shall refer to as a world system. Suppose we introduce an alternative metric tensor g'_{ik} leading to a quite different geometry and modify the physics from P to P', so that the new world system $g'_{ik} + P'$ is equivalent to the world system $g_{ik} + P$, i.e., the two world systems are thoroughly intertranslatable. Reichenbach points out that if the standard of congruence in the system $g_{ik} + P$ is the solid rod corrected for differential forces, then going over to $g'_{ik} + P'$ it will be necessary to postulate that the rod undergoes certain additional deformations. These "additional" deformations will be the same for all solid rods independently of their chemical composition *relative to the original description* $g_{ik} + P$. Overlooking the fact that it is only *relative to the original description* that these additional forces affect all bodies in the same way, Reichenbach supposes that the new physics P' must be obtained by simply taking over the physics P and postulating

an additional force U which affects all bodies the same way. Thus if the total force acting on a body according to the world system $g_{ik} + P$ is $F = E + G + I$, where E is the total electromagnetic force acting on the body, G is the total gravitational force acting on the body, and I is the sum of the interactional forces acting on the body, i.e., the forces due to the interaction of the electromagnetic and the gravitational field, if such exist, then he supposes that according to the physics P', the total force F' acting on the body will be given by $F' = E + G + I + U$, where U is the "universal force." There are a number of things wrong with this account. For example, "universal forces" *must depend on the chemical composition of the body they act on*, since no force which is independent of the chemical composition of the body will have the effect of producing the same *deformations* in all solid bodies *independently of their resistance to deformation*. For example, gravitation is not a "universal" force in this sense, since the surface gravity of Jupiter will produce quite *different* deformations in (a) a man and (b) an iron bar.

However, the main point overlooked by Reichenbach is this. When we construct P', we cannot in general simply take over the physics, P, and introduce an additional force, U. This is not, in general, the simplest way of obtaining an equivalent world system based on the metric g'_{ik}. What will in general happen is that *P' changes the laws obeyed by the differential forces*. In other words, if the total electromagnetic force acting on a body according to P is E, then the total electromagnetic force acting on the same body according to P' will not be E but $E' = E + \Delta E$. Similarly, the gravitational force acting on a body according to P' will not be G but $G' = G + \Delta G$, and the interactional force will not be I but $I' = I + \Delta I$. It is perfectly possible that the additional force $\Delta E + \Delta G + \Delta I$ introduced by the change in the laws for the differential forces will be perfectly sufficient to account for the additional deformations required to take us from the world system $g_{ik} + P$ to the world system $g'_{ik} + P'$. According to $g_{ik} + P$, the total force acting on a body at any given time is $F = E + G + I$; according to $g'_{ik} + P'$, the total force acting on a body at any time is $F' = E' + G' + I'$. According to both systems, there are no universal forces in Reichenbach's sense and bodies change their lengths as they are moved about only slightly and only due to the differential forces including differential forces accounted for by their own atomic constitutions.

If we change from the world system $g_{ik} + P$ to the world system

$g'_{ik} + P'$, then we have to postulate universal deformations in the sizes and shapes of measuring rods as they are moved about which are the same for all measuring rods independently of their chemical composition relative to our original description, i.e., the description provided by $g_{ik} + P$; but symmetrically, if $g'_{ik} + P'$ was our original system and we change to $g_{ik} + P$, then we'll have to postulate additional deformations in the sizes and shapes of measuring rods which will be the same for all measuring rods independently of their chemical composition relative to what we described their size and shape as being in $g'_{ik} + P$. Thus the so-called *congruence definition*—"*the standard of congruence is the solid body corrected for differential forces*"—may be compatible with every admissible metric tensor. Moreover, even if we specify a metric tensor g_{ik} and a world system $g_{ik} + P$ to which all admissible world systems are to be equivalent, the physics P that "goes with" the metric tensor g_{ik} is not uniquely determined. We can use, by hypothesis, the physics P to get a correct world system, $g_{ik} + P$; but there will be many other systems of physics, including systems P^* which postulate universal forces, such that $g_{ik} + P^*$ is equivalent to $g_{ik} + P$. In other words, even if g_{ik} is the normal metric, there will be systems of physical theory P^* according to which universal forces are present and according to which we have to obtain the corrected length of the measuring rod by correcting not only for differential forces but also for some universal force. Thus, once again we see that in fact it may be the case that for every metric g_{ik} there is both a correct physics P such that universal forces are absent and an equally correct physics P^* (i.e., a physics P^* such that $g_{ik} + P$ and $g_{ik} + P^*$ are equivalent descriptions) such that universal forces are present according to $g_{ik} + P^*$.

It is difficult to show the full nature and extent of Reichenbach's error without going into details on the question of transformations between metrics. But the essence of the mistake has already been indicated. If Reichenbach's universal force, U, can be expressed as a sum $U = \Delta E + \Delta G + \Delta I$ of increments in the differential forces introduced by the new physics P', then the new physics P' will not have to postulate any such universal force U at all. Indeed, the simplest way of going over from the physics P to the new physics P' required by the change in the metric tensor from g_{ik} to g'_{ik} is in general *not* to introduce a universal force, since it is very difficult to construct a force which will produce the same deformations in bodies independently of their chemical constitution, but rather to suitably modify the equations for the differential forces. Conversely, if, according to the normal physics P, $F = E + G + I$, then it is always possible to introduce new quantities E', G', I' such that $E = E' + \Delta E'$, $G = G' + \Delta G'$, $I = I' + \Delta I'$ and such that the sum $U = \Delta E' + \Delta G' + \Delta I'$ is a universal force in Reichenbach's sense, and thus to obtain a physics P^* according to which the forces acting on any body are not the forces E, G, and I but rather the differential forces E', G', and I' plus the universal force U. This will lead to the same total resultant force $U + E' + G' + I' = E + G + I$. In sum, the forces acting on a body can always be broken up so that the total resultant force includes a *component* universal force; and a component universal force acting on a body can always be broken up into components which can be combined with the differential forces. That this is indeed the case is the content of the following.

Theorem. Let P be a system of physics (based on a suitable system of coordinates) and E be a system of geometry. Then the world described by E *plus* P can be redescribed in terms of an arbitrarily chosen metric g_{ik} (compatible with the given topology) *without postulating "universal forces,"* i.e., forces permanently associated with a spatial region and producing the same deformations (over and above the deformations produced by the usual forces) independently of the composition of the body acted upon. In fact, according to the new description g_{ik} *plus* P' (which has exactly the same factual content as E *plus* P):

(1) All deformations are ascribed to three sources: the electromagnetic forces, the gravitational forces, and gravitational-electromagnetic interactions.

(2) All three types of forces are dependent upon the composition of the body acted upon.

(3) If there are small deformations constantly taking place in solid bodies according to E *plus* P (as there are, owing to the atomic constitution of matter), then no matter what geometry may be selected, the new g_{ik} can be so chosen that the deformations according to g_{ik} *plus* P' will be of the same order of magnitude. Moreover, it will be impossible to transform them away by going back to E *plus* P.

(4) If it is possible to construct rods held together by only gravitational forces or only electromagnetic forces, then (in the absence of the other type of field) the interactional forces of the third type (postulated by P') will vanish.

(5) If there are already "third type forces" according to *E plus P*, then the situation will be thoroughly symmetrical, in the sense that (i) going from the old metric to g_{ik} involves postulating additional deformations (relative to the description given in *E plus P*) which are the same for all bodies, and (ii) going from g_{ik} back to *E plus P* involves postulating additional deformations which are also the same for all bodies, *relative to the description given in* g_{ik} *plus* P'; and the same number and kind of fundamental forces are postulated by both P and P'.

Proof: It suffices to retain the original coordinate system, and replace the original notion of distance, wherever it occurs in physical laws, by the appropriate function of the coordinates. The second law of motion is now destroyed, since it now reads

$$\mathbf{F} = m\ddot{\mathbf{x}}$$

and $\ddot{\mathbf{x}}$ (the second derivative of the position vector) is no longer "acceleration," owing to the "arbitrary" character of the coordinate system as viewed from the new g_{ik} tensor. (We assume the new g_{ik} is compatible with the original topology.) But we can restore the second law by construing "force" in the old laws as not force at all, but some other quantity—say "phorce" (P). Then the above law is rewritten as

$$\mathbf{P} = m\ddot{\mathbf{x}}$$

and the law $F = ma$ is reintroduced as a definition of "force." (Here a must be defined in terms of the new g_{ik} tensor.) The difficulty is that so far we have only defined *total resultant force*. To obtain a resolution into component forces, we proceed as follows: Obtain the total resultant force F on the body B by determining its mass and acceleration (the latter can be found from the law $P = m\ddot{\mathbf{x}}$, once we know the total "phorce" acting on B and the g_{ik} tensor). Now set the gravitational field equal to zero, determine the total "phorce" that would now be acting on B, and determine from this the total force that would be acting on B. Call this E (electromagnetic force). Similarly, set the electromagnetic field equal to zero, and obtain the total force that would be acting on B. Call this G (gravitational force). Finally define I (interactional force) from the equation $F = E + G + I$. . . .

Comment. If *E plus P* is the "normal" system, one may attempt to rescue the claim that "every other system g_{ik} *plus* P' contains universal forces" in various ways. For example, Reichenbach considers letting the external *and interior* forces approach zero "in the limit." But then our forces I go to zero as well.* Alternatively, one might eliminate the external forces, and then let the interior forces become *constant* (instead of zero). Universal forces are present if deformations *still* take place. But the difficulty is that whether the interior forces are constant or not depends upon whether P or P' is used. (An experimental criterion is ruled out, beyond a certain point, by the atomic constitution of matter.) One might say that the coordinating definition should be changed to: "The differential forces obey laws which do not depend upon the coordinates but only upon coordinate differences." But this requirement can be formally complied with as well. For example, if the curvature of space varies from place to place (as in general relativity), then any reference to a particular place can be made coordinate independent by *describing* the place in questions as e.g., "the place where the metrical field has such-and-such values." (There is a tendency in Grünbaum to emphasize the constant curvature case, which is not physically realistic, since general relativity is well confirmed.) [Pp. 247–253, 254.]

I claim that Putnam's theorem is not only a *non sequitur* but false. And my reasons for the respective charges of *non sequitur* and falsity consist of the following series of considerations.

1. What is necessary to show that a force is differential in Reichenbach's sense is to show that the magnitude of the *deformation* effected by the force depends on the chemical composition of the body; it will *not* do to show that the magnitude of the *force* has that kind of dependence. The reason is that the magnitude of the *force* normally also has that kind of dependence in the case of Reichenbach's *universal* forces, so that this dependence cannot distinguish his differential forces. Thus, suppose that two Paris rods of different chemical composition and hence, normally, different densities are kept at the standard T-scale temperature at both $y = 1$ and $y = 2$, and are both said to have undergone the *same deformation* $1/y - 1$ at $y = 2$. Then the *different masses* associated with their differing

* This depends upon the identity of gravitational and inertial mass. The point is that in order to make the interior *gravitational* forces approach zero, we must make the mass of B approach zero and hence $F = ma \to 0$.

densities will be subject to the same "accelerations." Hence if the "agencies" which are said to have produced this like deformation are to be deserving of the metaphorical label "universal forces," their magnitudes can depend on the chemical composition of the rod on which they act.

Reichenbach had rested the distinction between universal and differential forces on the irrelevance versus relevance of chemical composition to the magnitude of the *deformation*, not to the magnitude of the force. For he allowed that some kinds of Paris rods may require more "persuasion" or "force" than others to undergo the universal (i.e., same) deformation $1/y - 1$. Speaking of universal forces, he said: "They affect all materials in the same way" [40, p. 13]. As I pointed out in my original essay, however, Reichenbach's classification of forces into universal and differential is not mutually exclusive, unless it is *relativized* to specifiable kinds of conditions involving the absence or presence of certain kinds of constraints. For I wrote there:

It is entirely correct, of course, that a uniform gravitational field (which has not been transformed away in a given space-time coordinate system) is a universal force in the *literal* sense *with respect to a large class of effects* such as the free fall of bodies. But there are other effects, such as the bending of elastic beams, with respect to which gravity is clearly a *differential* force in Reichenbach's sense: a wooden book shelf will sag more under gravity than a steel one. And this shows, incidentally, that Reichenbach's classification of forces into universal and differential is not mutually exclusive. Of course, just as in the case of any other force having differential effects on measuring rods, allowance is made for differential effects of gravitational origin in laying down the congruence definition [1962, p. 441; 1968, pp. 45–46].

Hence I consider it unjust to Reichenbach on Putnam's part to say that one of "a number of things wrong with this [i.e., Reichenbach's] account" is that " 'universal forces' *must depend on the chemical composition of the body they act on*" [p. 248].

We see that Putnam notes himself that the magnitudes of universal forces (no less than those of differential ones) do depend on the composition of the body on which they act. And this fact can now be seen to make part (2) of his theorem unavailing as a basis for his claim that his Φ physics P' (which is my "Φ physics") dispenses with universal forces in favor of particular kinds of differential "electromagnetic," "gravitational" and "interactional" forces. For his part (2) says of these forces: "All three types of forces are dependent upon the composition of the body acted upon" [p. 251]. And since the latter characteristic applies to universal forces as well, part (2) cannot establish the differential character of the three kinds of forces which allegedly make universal forces dispensable.

Hence part (2) would be unavailing for Putnam's purposes, even if he had proven that it is true. It is clear that the electromagnetic, gravitational, and interactional forces *of the P physics* do produce *differential* effects and hence are differential forces in respect of these effects. But we shall soon see that nothing in Putnam's argument shows that the *homonymous* three new Φ-physics forces *as defined by him* individually depend on the composition of the body acted upon, let alone that the magnitudes of the respective deformations produced by them have that kind of dependence.

2. In the interest of notational continuity with the discussion of the thermal example in §9.1, let me retain the name "g_{ik}-cum-P" for the *original* description of the customary physics which Putnam calls "E plus P," while using the designation "g'_{ik}-cum-Φ" for the *new* description, which Putnam calls "g_{ik} plus P'." Thus, Putnam's P' is my Φ, but I trust that it will not be confusing that he uses the *un*-primed "g_{ik}" to denote the arbitrary metric tensor of the *new* description while I use instead the *primed* g'_{ik}. In particular, as in my thermal example, the new metric tensor will be the one of Poincaré's hyperbolic metric on the table top.

Consider such differential forces of the P physics as electrostriction, magnetostriction, and gravitational sagging of bookshelves. Let us assume with Putnam that all of the differential forces of the P physics (e.g., thermal ones) are at bottom either electromagnetic, gravitational, or otherwise interactional. It is then clear that in the P physics, the Paris rod will be differentially undeformed at space points where the electromagnetic and gravitational forces are zero. Thus, just as the rod is thermally undeformed at the *standard* temperature, so it is electromagnetically and gravitationally (and interactionally) undeformed with respect to differential effects in Reichenbach's sense at points where the electromagnetic and gravitational fields vanish. In saying this, I am assuming without any detriment to my impending counterexample to Putnam's theorem that the magnitude of the differential deformation sustained by a rod at a place at a given time depends only on the field intensities then prevailing at that point and *not* also on the field intensities to which the rod was subjected earlier at other places and times. That is to say, I am not considering those special kinds of rods which the P physics describes as exhibiting, for example, the hysteresis behavior of ferromagnetic materials. In this respect, I am on common ground with Putnam. And it is clearly a fair test of Putnam's theorem whether it can handle the large class of solid rods which do not exhibit hysteresis in the P physics.

Now consider two places, say on $y = 1$ and $y = 2$ at both of which the *external* electromagnetic and gravitational forces of the P physics *vanish* at *some* times when a transported Paris rod coincides with these lines $y =$ constant. Let us denote the electromagnetic and gravitational forces of the P physics by "E_P" and "G_P" respectively. Being differentially undeformed in both places, the rod will have the length $ds = 1$ in *both* places. According to g_{ik}-cum-P, the rod will therefore coincide with an interval $dx = 1$ on both $y = 1$ and $y = 2$. But according to the hyperbolic metric, the rod will have the lengths $ds' = dx/y$ and will therefore have the respective unequal lengths 1 (at $y = 1$) and ½ (at $y = 2$). More generally, according to g'_{ik}-cum-Φ, at any place on a line $y =$ k, where k is a constant different from 1, a Paris rod free from the differential forces E_P and G_P of the P physics will undergo a deformation of magnitude $1/y - 1$ *independently of its chemical composition*. And hence the *crucial question* is whether Putnam has shown that, though independent of chemical composition, this deformation can be *the sum of three Φ-physics deformations whose respective magnitudes each do depend on the chemical composition of the rod.* We saw in our oversimplified thermal example that if this deformation could be attributed to only one kind of force, i.e., if $\alpha' \Delta T' = 1/y - 1$, then clearly that one kind of deforming force cannot be differential.

We must see therefore whether *as Putnam has defined them,* the Φ-physics forces E_Φ, G_Φ, and I_Φ, which he calls respectively "electromagnetic," "gravitational," and "interactional," can be held to produce deformations which have the following two properties: (i) their sum is equal to the required $1/y - 1$, and (ii) the respective individual magnitudes of these *deformations* demonstrably depend on the composition of the rod, so that E_Φ, G_Φ, and I_Φ respectively can each be claimed to be Reichenbachian differential forces. Let us investigate whether Putnam's E_Φ, G_Φ, and I_Φ have these two properties.

The question is therefore whether the E_Φ, G_Φ, and I_Φ defined by Putnam are such functions of E_P, G_P, and I_P that the following equality holds:

$$(1) \qquad \frac{1}{y} - 1 = D_{E_\Phi} + D_{G_\Phi} + D_{I_\Phi},$$

where D_{E_Φ} is the deformation of the rod due to the electromagnetic field E_Φ at $y = k$, and similarly for the others. (These three deformations now jointly play the role of the deformation $\alpha' \cdot \Delta T'$ of our thermal example.) To answer our question, we recall the relevant part of Putnam's statement above: "Now set the gravitational field equal to zero . . . and determine from this the total force that would be acting on [the body] B. Call this E (electromagnetic force). Similarly, set the electromagnetic field equal to zero, and obtain the total force that would be acting on B. Call this G (gravitational force)" [p. 253]. This statement is susceptible of two interpretations: (a) The gravitational and electromagnetic fields which Putnam wants to turn off (i.e., set equal to zero) one at a time are the fields G_P

and E_P of the P physics; or (b) The fields which are to vanish one at a time are the G_Φ and E_Φ of the new physics. Let us examine the consequences of these two possibilities in turn.

a. Since E_Φ is defined as equal to the *total* force when $G_P = 0$, then G_Φ must clearly be zero under the condition $G_P = 0$. And, by the same token, since G_Φ is defined as equal to the total force when $E_P = 0$, then E_Φ must clearly be zero under the condition $E_P = 0$. But this means that when both G_P and E_P are *zero* on $y = k$, then so also are G_Φ, E_Φ, and hence I_Φ. And in that case the three deformations in our equation (1) above each vanish, thereby failing to yield the required nonvanishing total deformation $1/y - 1$ at loci $y \neq 1$. But then the deformations for which the description Φ-cum-g'_{ik} must correct the rod are not attributable solely to differential forces! Indeed, as is evident from our thermal example, the deformation $1/y - 1$ would not be differential even if only *one* of the fields G_Φ or E_Φ were to vanish. For the vanishing of only one of them would make I_Φ vanish, thereby requiring that only *one* of the fields E_Φ or G_Φ account for a deformation $1/y - 1$ *which is independent of the composition of the body*. But the force produced by such a field could then *not* be differential.

There are a number of other difficulties. Putnam tells us that the assumption of conditions contrary to the GTR "is not physically realistic, since general relativity is well confirmed" [p. 254]. In §2.12 above, I have already commented on this surprising and unfounded statement. But let us take Putnam at his word and assume that our P physics is the GTR. In that case, electromagnetic energy has equivalent mass, and the field equations show that no E_P field is possible without a G_P field, though there can be a G_P field without an E_P field. Now suppose that—*per impossibile*—Putnam could make the G_P field vanish at will *throughout the given coordinate system as required by his program*. In that hypothetical event, the E_P field would thus have to vanish everywhere as well, and thereby all differential forces would have to vanish throughout the coordinate system in the P physics and —as we saw—therefore also in the Φ physics. But if this could materialize, a *unique* determination of the spatial metric tensor could be made by the differentially undeformed rod. And this contradicts Putnam's claim that an arbitrary metric tensor is allowed by the requirement that the differentially undeformed rod be self-congruent. In any case, however, it is totally impossible to turn off the E field, let alone the G field, at will *throughout* the coordinate system, as required by Putnam's program for determining the forces on the rod at the various places. And hence this boundary condition constitutes a *deus ex machina* in his program.

Note that Putnam needs to be able to *turn off* the G_P and E_P fields one at a time at *every* point of his given coordinate system at will. By contrast, my counterexample merely requires the following physically reasonable boundary condition: at some time or other, the E_P and G_P fields happen to vanish *within an interval* on each of two lines $y = k$ with which the Paris rod can be brought into coincidence.

My counterexample here is not rebutted at all, it seems to me, by the following comment of Putnam's:

> . . . one might eliminate the external forces and then let the interior forces become *constant* (instead of zero). Universal forces are present if deformations *still* take place. But the difficulty is that whether the interior forces are constant or not depends upon whether P or P' is used. (An experimental criterion is ruled out, beyond a certain point, by the atomic constitution of matter.) [P. 254.]

The foundation of the entire discussion has been that the P-cum-g_{ik} description does correctly describe the coincidence phenomena, and the question before us is whether the alternative *equivalent* descriptions Φ-cum-g'_{ik} (required by the various arbitrary metric tensors g'_{ik}) need to correct only for *differential* deformations D_{E_Φ} etc., of the rod. Now the P-cum-g_{ik} physics is therefore presumed to have asserted correctly that the metric g_{ik} results from the rod behavior when the rod is a unit rod after being P-corrected for only those *external* forces which *act differentially on the rod as a whole*. For precisely the reasons pointed out in Einstein's 1921 lecture [11, p. 192] and cited in §9.0 above, in the present state of the P physics the measuring rod plays the role of an independent single whole, i.e., nothing needs to be assumed about the distribution of the *internal* forces other than that they are compatible with the solid state of the rod. In particular, in my counterexample we need *not* assume, as Putnam has it, the *constancy* of those kinds of internal forces whose external counterparts act differentially *on the rod as a whole.* But let us grant merely for argument's sake that in my counterexample we did need to assume their internal constancy during rod transport from $y = 1$ to $y = 2$ in the presumably true P physics. Then this could not impugn in the least the force of my counterexample as a proof that if Putnam's alternative description Φ-cum-g'_{ik} is to be equivalent to P-cum-g_{ik}, the Φ physics cannot confine its corrections to deformations which are differential. Hence I see no merit at all in Putnam's unclear pronouncement that "the difficulty is that whether the interior forces are constant or not depends upon whether P or P' (i.e., Φ) is used" [p. 254]. Incidentally, it is a plain travesty on his part to say that "Reichenbach considers letting the external *and interior* forces approach zero 'in the limit' " [p. 254]. Reichenbach had made the following altogether different statement: "*Change of shape is called small if the exterior forces are small relative to the interior forces.* The more nearly this condition is realized, the more rigid is the body; but only at the unattainable limit where the exterior forces disappear relative to the interior forces would the rigid body be realized in the strict sense" [40, p. 23]. And note that my counterexample is not predicated on the attainment of this limit in the course of *transport throughout a region* but only on the physically reasonable vanishing of the external forces in two spatially separated locations at different times.

b. Let us now turn to the alternative interpretation of what Putnam had in mind in proposing to turn off the gravitational and electromagnetic fields one at a time, viz., that the fields which are to vanish one at a time are the G_Φ and E_Φ of the *new* physics.

As A. Janis has pointed out, the fundamental weakness here is that we must already know whether the magnitude of the G_Φ field is zero or not in order to know whether it is turned off ("set the gravitational field equal to zero"), and yet the magnitude of G_Φ is first going to be determined after a series of steps which include (1) turning it off in order to obtain the total force that will be *called* "E_Φ," (2) turning off E_Φ in order to first obtain the total force that will be *called* "G_Φ"! Clearly, this is a vicious circle. Moreover, as Putnam notes, an essential step in the determination of the acceleration (not to speak of the mass) is the determination of the second time derivative $\ddot{\mathbf{x}}$ of the g'_{ik} position vector $\mathbf{x}$. Putnam tells us that this latter quantity, in turn, "can be found from the law $\mathbf{P} = m\ddot{\mathbf{x}}$, once we know the total 'phorce' acting on B" [p. 253]. But this is again circular, since there is no independent way of determining the left-hand side of this equation, as is evident from the way in which Putnam has defined "phorce." And yet knowledge of this enters essentially at several stages of the determination of the total force as outlined by Putnam. Indeed this latter circularity exists on interpretation (a) as well.

Even ignoring these insurmountable circularities, the argument is a *non sequitur* in the following decisive respect: it does not follow at all that, *as defined here by Putnam*, the forces E_Φ and G_Φ depend on the composition of the body, let alone that these forces are *differential.* And yet, as we saw, it is the *latter* property which is to be proven. If the E_Φ and G_Φ defined here by Putnam are to establish the complete dispensability of universal forces, they must be the respective g'_{ik} counterparts of E_P and G_P *and* be demonstrably differential forces. What does Putnam offer to this end? He claims to have shown that "(2) All three types of forces are dependent upon the composition of the body acted upon" [p. 251]. And his sole basis for this conclusion is the mere gratuitous assertion: "(2) is clear since E and G depend upon the composition of the body acted upon" [p. 253]. But this is a mere equivocation on the meaning of "E" and "G": what is clear is that the rod deformations produced by E_P

and G_P are differential. But it hardly follows from this that, *as defined by Putnam,* the homonymous quantities E_Φ and G_Φ are dependent on the rod's composition, let alone that they are differential forces.

It appears therefore that on interpretation (a) of Putnam's definitions of the new E and G forces, his theorem is false, whereas on interpretation (b), it is a *non sequitur* with respect to *unspecified* quantities E_Φ and G_Φ. Since I cannot believe that Putnam intended interpretation (b), I maintain that his theorem is false.

3. I conclude that the arbitrary metric tensors g'_{ik} cannot be obtained by Putnam's method from a rod which is corrected solely for differential deformations: as between the description P-cum-g_{ik} and the class of alternate ones Φ-cum-g'_{ik} proposed by Putnam, only the former description is such that all deformations are due to external forces which are differential. Hence Putnam has not furnished viable reasons for doubting the following claim: the exclusion of universal deformations, i.e., the requirement that all deformations of the rod be *differential,* does indeed restrict the metric. Specifically, Putnam has not succeeded in discrediting the following assumption: in the context of the rod's actual coincidences, the customary congruence convention, viz., the solid body is everywhere self-congruent under transport *after being corrected for differential deformations alone*, does single out a unique congruence class among those intervals in physical space which are large enough to coincide with a solid body however small. (This formulation recognizes, as did my §2.13, that the solid body must have the "assistance" of, say, standing micro-waves in submolecular intervals.) According to this assumption, the customary congruence convention allows only those systems of correctional physics P which are compatible with these unique congruences. The proponent of this assumption recognizes, of course, no less than Putnam, that the g_{ik} tensor is already ingredient in at least some of the correctional laws and that solid rods generally yield concordant metrical findings only when suitably corrected. But Putnam has not supplied a foundation for the following criticisms of his.

i. He says:

> By a "correspondence rule" I shall mean such a statement as . . . "congruence as established by transporting a solid rod is usually correct to within such-and-such accuracy" . . . the term "correspondence rule" in this sense should be distinguished from Grünbaum's "coordinative definitions," e.g., "the length of a solid rod corrected for differential forces does not change." This last statement is an intratheoretic statement, and does not, Grünbaum to the contrary, restrict the choice of a metric at all unless considerations of *descriptive* as well as *inductive* simplicity are allowed to operate [p. 206n].

ii. He says further that the customary congruence convention "is compatible with the adoption of any metric whatsoever, although one has to change the physical laws appropriately (in a manner explained in the Appendix) if one changes the metric" [p. 225]. And again: ". . . Reichenbach was right in maintaining that the standard of congruence in physical theory is the corrected solid body. What it is necessary to add is that this is an *intratheoretic* criterion of congruence, and that the 'empirical' findings vastly underdetermine the metric" [p. 247]. Moreover, my account in §9.1 of the thermal example shows that the defense of Reichenbach's thesis against Putnam's theorem does *not* depend on the neglect of the following fact, correctly noted by Putnam: "When we construct P', we cannot in general simply take over the physics, P, and introduce an additional force, U. This is not, in general, the simplest way of obtaining an equivalent world system based on the metric g'_{ik}. What will in general happen is that P' *changes the laws obeyed by the differential forces*" [p. 249].

4. Let us suppose merely for argument's sake that Putnam's theorem were true. Could this impugn the validity of the Riemann-Clifford view that the congruences entailed by $ds^2 = g_{ik}\, dx^i\, dx^k$ importantly involve a conventional or "definitional" ingredient? It is clear from my defense of the Riemann-Clifford thesis in §2 that it could not: even if Putnam's theorem *were true*, it would have estab-

lished only that: ". . . the metric is implicitly specified by the whole system of physical and geometrical laws and 'correspondence rules.' No very small subset by itself fully determines the metric . . ." [p. 206]. But it would not have discredited in the least the Riemann-Clifford conception of the status of the congruences asserted by the metric.

This point is of substantial importance, since Putnam is concerned to emphasize that "nothing that one could call a 'definition' " determines the metric [p. 206; see also pp. 210–211]. But when I speak of giving a "definition" of congruence, I do so first and foremost in order to emphasize the ontological and epistemic status which Riemann and Clifford assigned to the congruence relations asserted by the metric. And I use the term "definition" only secondarily because I reject Putnam's theorem. That is to say, my *additional* reason for speaking of a "definition" of congruence is my belief that (1) the differentially undeformed Paris rod does determine congruences among macro-intervals of physical space and (2) the congruences can therefore be stipulated in a relatively cryptic manner. Hence even if Putnam had convinced me of the validity of his theorem, I would have wished to say the following: intervals which are assigned the same *ds* numbers by the metric sustain the equality relation of congruence on the strength of self-congruences which hold as a matter of "definition" or "convention." In short, my use of the term "definition" or "convention" is prompted not so much by the mode of specifying the congruences as by the character of the self-congruences ingredient in the specification.

This brings me directly to the misunderstanding inherent in Putnam's account of the sense in which I think that physical geometry is "about" differentially undeformed solid bodies. He writes:

> . . . correspondence rules are necessary in physical geometry, or at any rate in connection with physical theory as a whole. And it would be a gross error to conclude from the fact that these correspondence rules mention, say, solid bodies, that physical geometry is *about* solid

bodies. It is misleading to say that physical theories are about the objects mentioned in the correspondence postulates rather than about the theoretical entities and magnitudes mentioned in the theoretical postulates. As we explained above, in connection with the discussion of explanation *versus* articulation, physical theories are said to explain the behavior of the ordinary middlesized objects mentioned in the correspondence rules by reference to the theoretical entities and magnitudes that they postulate [p. 217].

And speaking of my reference to physical geometry as "articulating" relations between differentially undeformed solid rods and other physical entities, Putnam says: ". . . this error is analogous to the error we would be making if we described electromagnetic theory as 'the articulation of the relations holding among voltmeters quite apart from perturbing influences'" [pp. 206–207].

But Putnam overlooks completely that I had *not inferred* what I consider physical geometry to be *about* from what is mentioned in the correspondence rules in the fashion of his voltmeter example. Instead, my Riemannian account of the intrinsic metrical amorphousness of physical space led me to conclude in §2 that (1) pre-GTR physical geometry is about the relations of one or more concordant extrinsic congruence standards to intervals and other parts of physical space, and (2) GTR space-time physical geometry is about these now varying relations as functionally connected with the other physical states of affairs which *explain* them. Thus, I do not say in the manner of the voltmeter example that the relations sustained by solid rods (and by other extrinsic metric standards like travel times of light) are constitutive of the relations expressed by $ds^2 = g_{ik}\, dx^i\, dx^k$ *because* solid rods (and light rays) are mentioned in the correspondence rules for spatial distance ds; instead, the intrinsic metrical amorphousness of physical space is my reason for my claim regarding what physical geometry is about.

5. Is Putnam entitled to claim that "to emphasize the constant curvature case" if one accepts the GTR "is not physically realistic" [p. 254]? He alludes here to the fact that, as I first conceived of it, the method of successive approximation for determining the spatial

metric tensor, if successful, would work only for the case of constant curvature. To this I have two things to say. First, as I have noted in Chapter II, §2(ii), the method of successive approximation which I outlined, if successful, could be extended to the case of a spatial g_{ik} which makes for a geometry of variable curvature. For I set forth there how there can be convergence to a particular set of functions g_{ik} (in a particular coordinate system) in the case of variable curvature. Second, as was explained in a recent treatise on the GTR [1, pp. 338–349], the three-dimensional physical spaces yielded by the GTR cosmological models associated with the so-called Robertson-Walker metric [1, p. 345] are all Riemannian spaces of *constant* curvature. But, in any case, my suggested procedure of successive approximation was *not* predicated on the GTR, and I have indicated in §2.12 why that cannot be held to render it physically unrealistic.

6. Finally, I should add that since Putnam's theorem is at the very best a *non sequitur*, my suggestion of a *possible* discovery procedure for the correct metric tensor of physical space is not subject to the following indictment by him: Speaking of high-level theory construction in which choices are made between two fully developed alternative explanations of the phenomena, Putnam says: ". . . the determination of the g_{ik} tensor requires theory construction in this 'high-level' sense, and no cookbook full of recipes for the discovery of these high-level theories can be or ought to be given" [p. 210].

Bibliography

1. Adler, R., M. Bazin, M. Schiffer, eds. *Introduction to General Relativity.* New York: McGraw-Hill, 1965.
2. Bergmann, P. "The General Theory of Relativity," in *Handbuch der Physik*, Vol. IV, S. Flügge, ed. Berlin: Springer, 1962. Pp. 203–272.
3. Bergmann, P. *Introduction to the Theory of Relativity.* New York: Prentice-Hall, 1946.
4. Clifford, W. K. *The Common Sense of the Exact Sciences.* New York: Dover Publications, 1955.
5. D'Abro, A. *The Evolution of Scientific Thought from Newton to Einstein.* New York: Dover Publications, 1950.
6. Dicke, R. H. *The Theoretical Significance of Experimental Relativity.* New York: Gordon and Breach, 1964.

7. Eddington, A. S. "The Cosmological Controversy," *Science Progress*, 34:225–236 (1939).
8. Eddington, A. S. *The Mathematical Theory of Relativity*. Cambridge: Cambridge University Press, 1952.
9. Eddington, A. S. *Space, Time and Gravitation*. Cambridge: Cambridge University Press, 1953.
10. Einstein, A. "The Foundations of the General Theory of Relativity," in *The Principle of Relativity, a Collection of Original Memoirs*. New York: Dover Publications, 1952. Pp. 111–164.
11. Einstein, A. "Geometry and Experience," in *Readings in the Philosophy of Science*, H. Feigl and M. Brodbeck, eds. New York: Appleton-Century-Crofts, 1953. Pp. 189–194.
12. Einstein, A. "On the Electrodynamics of Moving Bodies," in *The Principle of Relativity, a Collection of Original Memoirs*. New York: Dover Publications, 1952. Pp. 37–65.
13. Einstein, A. "Prinzipielles zur allgemeinen Relativitätstheorie," *Annalen der Physik*, 55:241 (1918).
14. Einstein, A. "Reply to Criticisms," in *Albert Einstein: Philosopher-Scientist*, P. A. Schilpp, ed. Evanston: The Library of Living Philosophers, 1949. Pp. 665–688.
15. Fine, A. "Physical Geometry and Physical Laws," *Philosophy of Science*, 31:156–162 (1964).
16. Fock, V. *The Theory of Space, Time and Gravitation*. 2nd rev. ed.; New York: Macmillan, 1964.
17. Grünbaum, A. "A Consistent Conception of the Extended Linear Continuum as an Aggregate of Unextended Elements," *Philosophy of Science*, 19:288–306 (1952).
18. Grünbaum, A. "The Denial of Absolute Space and the Hypothesis of a Universal Nocturnal Expansion," *Australasian Journal of Philosophy*, 45:61–91 (1967).
19. Grünbaum, A. "Epilogue" for P. W. Bridgman's *A Sophisticate's Primer of Relativity*. Middletown: Wesleyan University Press, 1962. Pp. 165–191.
20. Grünbaum, A. "The Falsifiability of a Component of a Theoretical System," in *Mind, Matter, and Method: Essays in Philosophy and Science in Honor of Herbert Feigl*, P. K. Feyerabend and G. Maxwell, eds. Minneapolis: University of Minnesota Press, 1966.
21. Grünbaum, A. "The Falsifiability of Theories: Total or Partial? A Contemporary Evaluation of the Duhem-Quine Thesis," *Synthèse*, 14:17–34 (1962).
22. Grünbaum, A. "The Genesis of the Special Theory of Relativity," in *Current Issues in the Philosophy of Science*, H. Feigl and G. Maxwell, eds. New York: Holt, Rinehart and Winston, 1961. Pp. 43–53.
23. Grünbaum, A. "Logical and Philosophical Foundations of the Special Theory of Relativity," in *Philosophy of Science: Readings*, A. Danto and S. Morgenbesser, eds. New York: Meridian, 1960. Pp. 399–434.
24. Grünbaum, A. "Modern Science and Refutation of the Paradoxes of Zeno," *Scientific Monthly*, 81:234–239 (1955).
25. Grünbaum, A. *Modern Science and Zeno's Paradoxes*. Middletown: Wesleyan University Press, 1967. 2nd ed., London: Allen and Unwin, 1968.

26. Grünbaum, A. *Philosophical Problems of Space and Time*. New York: Knopf, 1963.
27. Grünbaum, A. "The Philosophical Retention of Absolute Space in Einstein's General Theory of Relativity," *Philosophical Review*, 66:525–534 (1957).
28. Grünbaum, A. "The Relevance of Philosophy to the History of the Special Theory of Relativity," *Journal of Philosophy*, 59:561–574 (1962).
29. Grünbaum, A. "The Special Theory of Relativity as a Case Study of the Importance of the Philosophy of Science for the History of Science," in *Philosophy of Science* (*The Delaware Seminar*, Vol. 2), B. Baumrin, ed. New York: Interscience Publishers, 1963. Pp. 171–204.
30. Hobson, E. W. *The Theory of Functions of a Real Variable*. Vol. 1. New York: Dover Publications, 1957.
31. Hume, D. *Treatise of Human Nature*, Part II, Section IV. Oxford: Oxford University Press, 1941.
32. Landau, L., and E. Lifschitz. *The Classical Theory of Fields*. 2nd rev. ed. (tr. from the Russian by M. Hamermesh). Reading: Addison-Wesley, 1962.
33. Møller, C. *The Theory of Relativity*. Oxford: Oxford University Press, 1952.
34. Newton, I. *The Mathematical Principles of Natural Philosophy*. Introduction by Alfred del Vecchio. New York: Citadel Press, 1964.
35. North, J. *The Measure of the Universe*. Oxford: Oxford University Press, 1965.
36. Page, L. *Introduction to Theoretical Physics*. New York: Van Nostrand, 1935.
37. Putnam, H. "An Examination of Grünbaum's Philosophy of Geometry," in *Philosophy of Science* (*The Delaware Seminar*, Vol. 2), B. Baumrin, ed. New York: Interscience Publishers, 1963. Pp. 205–255.
38. Putnam, H. "Three-Valued Logic," *Philosophical Studies*, 8:73–80 (1957).
39. Reichenbach, H. *Axiomatik der relativistischen Raum-Zeit-Lehre*. Braunschweig: F. Vieweg & Sons, 1924.
40. Reichenbach, H. *The Philosophy of Space and Time*. New York: Dover Publications, 1958.
41. Reichenbach, H. *The Rise of Scientific Philosophy*. Berkeley: University of California Press, 1951.
42. Riemann, B. "On the Hypotheses Which Lie at the Foundations of Geometry," in *A Source Book in Mathematics*, Vol. II, David E. Smith, ed. New York: Dover Publications, 1959.
43. Schlick, M. "Are Natural Laws Conventions?" in *Readings in the Philosophy of Science*, H. Feigl and M. Brodbeck, eds. New York: Appleton-Century-Crofts, 1953. Pp. 181–188.
44. Schlick, M. *Grundzüge der Naturphilosophie*. Vienna: Gerold & Company, 1948.
45. Synge, J. *Relativity: The General Theory*. Amsterdam: North Holland Publishing Company, 1960.
46. Tolman, R. *Relativity, Thermodynamics and Cosmology*. Oxford: Oxford University Press, 1934.
47. Weyl, H. *Space-Time-Matter*. New York: Dover Publications, 1950.
48. Whitrow, G. *The Natural Philosophy of Time*. London: Thomas Nelson & Sons, 1961.

INDEX

INDEX